Advanced Information and Knowledge Processing

Series editors

Lakhmi C. Jain
Bournemouth University, Poole, UK and
University of South Australia, Adelaide, Australia

Xindong Wu
University of Vermont

Information systems and intelligent knowledge processing are playing an increasing role in business, science and technology. Recently, advanced information systems have evolved to facilitate the co-evolution of human and information networks within communities. These advanced information systems use various paradigms including artificial intelligence, knowledge management, and neural science as well as conventional information processing paradigms. The aim of this series is to publish books on new designs and applications of advanced information and knowledge processing paradigms in areas including but not limited to aviation, business, security, education, engineering, health, management, and science. Books in the series should have a strong focus on information processing—preferably combined with, or extended by, new results from adjacent sciences. Proposals for research monographs, reference books, coherently integrated multi-author edited books, and handbooks will be considered for the series and each proposal will be reviewed by the Series Editors, with additional reviews from the editorial board and independent reviewers where appropriate. Titles published within the Advanced Information and Knowledge Processing series are included in Thomson Reuters' Book Citation Index.

More information about this series at http://www.springer.com/series/4738

Sergei V. Chekanov

Numeric Computation and Statistical Data Analysis on the Java Platform

 Springer

Sergei V. Chekanov
HEP Division
Argonne National Laboratory
Lemont, IL
USA

Additional material to this book can be downloaded from http://extras.springer.com/.

ISSN 1610-3947   ISSN 2197-8441   (electronic)
Advanced Information and Knowledge Processing
ISBN 978-3-319-28529-0   ISBN 978-3-319-28531-3   (eBook)
DOI 10.1007/978-3-319-28531-3

Library of Congress Control Number: 2016932001

Printed on acid-free paper

This Springer imprint is published by SpringerNature
The registered company is Springer International Publishing AG Switzerland

This book is dedicated to my family

Preface

Numerical and statistical algorithms are typically confined within a specific programming language. For example, the R open-source data-analysis software uses a specialized scripting language, which is an implementation of the "S" programming language. Many commercial mathematical programs follow this trend. This book is about a platform for statistical calculations using algorithms that are not confined by a chosen language. For example, this platform allows mixing Python and Java numerical libraries, or using them on their own. Or, one can use this book to program statistical code using other languages, such as Groovy, Ruby, and BeanShell. This book is about an approach to scientific programming and visualization that does not set strict requirements on specific programming languages, nor on operating systems where such calculations are performed.

There are many books written about Java—one of the most popular programming languages. There are many books written about Python, which is another very popular programming language. This book explains how to mix them, bringing incredible algorithmic power and cutting-edge numeric libraries to scientific computations and data visualization.

In this book I did not go deep inside particular scientific research area, since the aim was to give concrete examples which illustrate which Java libraries should be used to perform computations. In the cases when I could not cover the subject in detail, a sufficient number of relevant references was given, so the reader can easily find necessary information for each chapter using external sources.

Thus this book presents practical approaches to numerical computations, data analysis, and knowledge discovery, focusing on programming techniques. Each chapter describes the conceptual underpinning for numerical and statistical calculations using Java libraries, covering many aspects from simple multidimensional arrays and histograms to clustering analysis, curve fitting, neural networks, and symbolic calculations. To make the examples as simple as possible from the computational point of view, I fully embrace the scripting approach in the course of this book. This leads to short and clear analysis codes, so you could concentrate on the logic of analysis flow rather than on language-specific details.

This book uses Python as the main programming language, since it is elegant and easy to learn. It is a great language for teaching scientific computation. For developers, this is an ideal language for fast prototyping and debugging. The book discusses how to design code snippets for numeric computation and statistics on the Java platform. To be more exact, we will use Jython (Python implemented in Java), a language that uses not only native Python modules, but can also access very comprehensive Java classes. The reader will learn how to write analysis codes, while numerous code snippets will give you some ideas on numeric algorithms which can easily be incorporated into realistic research application. The book includes more than 300 code snippets to produce data-visualization plots in 2D and 3D.

I am almost convinced myself that this book is self-contained and does not depend on detailed knowledge of computing language, although knowledge of Python and Java is desirable. However, the reader may still need some programming background in order to use this book with other languages, such as Groovy, BeanShell, and Ruby, since I did not give very detailed coverage of these languages.

Who Is This Book for

This book is intended for general audiences, for those who use computing to make sense of data surrounding us. It can be used as a source of knowledge on data analysis and statistical calculations for students and professionals of all disciplines. This book was written for undergraduate and graduate students, academics, professors, and professionals of any field and any age. The book could be used as a textbook for students.

We also hope that this book will be useful for those who study financial markets, since the numeric algorithms discussed in this book are undoubtedly common to any knowledge discovery research. This book equips readers with the description of a computational platform for statistical calculations which can be viewed as an inexpensive alternative to costly commercial products used by financial-market analysts.

I assume the readers are not familiar with Python/Jython, the main programming language used for code snippets in this book. But some basic understanding of statistics and mathematics would be very helpful to understand the material of this book.

All example codes of this book can easily be transformed to Java, Groovy, Ruby/JRuby, or BeanShell codes. You are presumed to have knowledge of programming in Java, if you will choose the path of moving the examples to Java, or if you will decide to create Java libraries to be deployed as jar files for a new project. The book will discuss how to do this, and a few Java examples will be provided.

Transformations of the example snippets to scripting languages, such as Groovy, Ruby/JRuby, or BeanShell, may require some knowledge of these scripting languages. The good thing is that the analysis algorithms and numerical libraries will be exactly the same, so a little effort is required to move to other languages. Again, we will show you how to convert Jython codes to these languages. In most cases, our examples should be sufficient to get started with a new language. The more knowledge about Groovy and Ruby/JRuby you can bring, the more you will get out of this book.

Books You May Read Before

The material of this book is self-contained. However, to understand the material deeper, you may need to look at other sources. First of all, there are plenty of good books [1–5] on Python and Jython, which are more complete for language-specific topics than the information given in this book. If you program in Java that forms the backbone of numerical and graphical libraries discussed in this book, a great deal of supplementary information can be found in Java books [6–10].

Secondly, there are several books on Groovy, a popular scripting language that can be used to work with the Java numerical libraries discussed in this book [11, 12]. If your choice is JRuby, the Ruby programming language on the Java platform, look at these books [13, 14] to get started.

Thirdly, as you read, you may need to look at external sources to understand the material better, especially when we come to statistical interpretations of data. We will supply the reader with the necessary references, so he or she can choose the most appropriate (and affordable) books to discover the world of data analysis and data mining.

References

1. Pilgrim M (2004), Dive into Python, Apress
2. Guzdial M (2005) Introduction to computing and programming in Python, a multimedia approach, Prentice Hall
3. Martelli A (2006) Python in a nutshell (in a nutshell (O'Reilly)), O'Reilly Media, Inc.
4. Lutz M (2007) Learning Python, 3rd edn. O'Reilly Media, Inc.
5. Langtangen H (2008) Python scripting for computational science. Springer-Verlag, Berlin, Heidelberg
6. Richardson C, Avondolio D, Vitale J et al (2005) Professional Java, JDK 5 Edition, Wrox
7. Arnold K, Gosling J, Holmes D (2005) Java(TM) programming language. In: The Java Series, 4th edn. Addison-Wesley Professional
8. Flanagan D (2005) Java in a nutshell, 5th edn. O'Reilly Media, Inc.
9. Eckel B (2006) Thinking in Java, 4th edn. Prentice Hall PTR
10. Bloch J (2008) Effective Java. In: The Java Series, 2nd edn. Prentice Hall PTR
11. Subramaniam V (2013) Programming Groovy 2: dynamic productivity for the Java Developer, Pragmatic Programmers, LLC
12. King D, Glover A (2013) Groovy in action, Manning Publications, 2007, https://www.manning.com/books/groovy-in-action
13. Edelson J, Liu H (2008) JRuby cookbook, O'Reilly Media
14. Bini O (2007) Practical JRuby on Rails Web 2.0 Projects: Bringing Ruby on Rails to Java, Apress

Acknowledgements

This book describes a software which is a collective work of many developers who have dedicated themselves to scientific computing. The author is grateful to all people who contributed to scientific software, and for their inspiration and dedication to science and knowledge-discovery software.

Many numeric and graphic libraries discussed in this book were released as open-source projects. I am grateful to the authors of such open-source programs for their enthusiasm to share their work, and for making their software publicly available.

You can find a list of contributions to the software packages described in this book on the jWork.ORG web page (http://jwork.org/dmelt/). A special note of thanks to those of you who reported bugs in a constructive way, helped with solutions, and shared your knowledge and experience with others.

Much of this project grew out of fruitful collaboration with many of my colleagues who devoted themselves to high energy physics. Over the course of the past twenty-five years I have learned a lot about programming aspects of scientific research. I would like to thank my colleagues for checking and debugging the examples shown in this book, and here the list will be endless.

I would like to thank everyone at Springer for their help with the production process. In particular, managing editors H. Desmond and J. Robinson, who helped start this book in its present form.

Not least, personal thanks go to my dear wife, Tania, and my sons, Alexey (Alosha) and Roman, for their love and patience to a husband and father who was only half (mentally) present after coming from his work. Without their patience and understanding, this book would not have been possible. Finally, I also thank my parents and sister for their support of my interests in all aspects of science.

Chicago   Sergei V. Chekanov
January 2016

Contents

Conventions and Acronyms

This book uses the following typographical convention: A box with a code inside usually means interactive Python/Jython commands typed in the "Jython Shell." All such commands start with the symbol $>>>$ which is the usual invitation in Python to type a command. This is shown in the example below:

```
>>> print "Hello, Jython"
```

Working interactively with the Jython prompt has the drawback that it is impossible to save typed commands. In most cases, the code snippets are not so short, although they are still much shorter than in any other programming language. Therefore, it is desirable to save the typed code in a file for further modification and execution. In this case, we use Jython macro files, i.e., we write a code using the DMelt (or any other) editor [15], save it in a file with the extension ".py", and run it using the keyboard shortcut [F8] or the button "run" from the DMelt tool bar menu. Such code examples are also shown inside the box, but code lines do not start with the Python invitation symbol $>>>$. In such situations, the example codes will be shown as:

```
print "Run this code from a file" # this is a comment
```

For examples written in the Python language, double quotes and apostrophe are interchangeable. For Java and other languages, this is not the case. So, to make our code to be easily convertible to Java or Groovy, we will use double quotes around strings. As in the above example, we will try to comment code lines as much as we can. For Python, comments are preceded by the hash character.

If a code snippet is used as a Python/Jython module by other programs, then we should write our code inside a file. A Python code always imports an external module using its file name. Since the file names are important, we will indicate exactly which file name should be used under the box with a code. For example, if a program code is considered a module that has to be imported by another code example, we will show it as:

```
print "Hello, please put this code to a file"
```

Listing 1 File "hello.py"

with the description indicating the file name. For instance,

```
>>> import hello
Hello, please put this code to a file
```

imports the file "hello.py" and executes it, printing the string. In other cases, we will use arbitrary file names for the code snippets.

We use typewriter font for Jython and Java classes and methods. For file names and directories, we also use the same font style with additional parentheses.

We remind that the directory name separators are backward slashes for Windows, and slashes for Linux and Mac computers. For example, the directory with examples will be shown as:

macro/examples/

For Windows computers, the same directory should be shown as:

macro\examples\

The dots in this example are used to indicate the upper-level directory.

We will try to avoid using abbreviations. When we use abbreviations, we will explain their meaning directly in the text. When space allows, we will use meaningful names for variables. This is all.

Chapter 1
Java Computational Platform

1.1 Introduction

Java is both a programming language and a computing platform which runs Java code. This book uses both. But the Java programming language is not necessary for the approach adopted in this book, since the Java platform allows the usage of scripting languages, such as Jython/Python, Groovy, Ruby/JRuby, BeanShell, and others.

The heart of the Java platform is the Java Virtual Machine (JVM) that runs programs converted to Java bytecode programs. The conversion to bytecode is done by Java compiler. Bytecode is the optimized and effective machine language of JVM. The JVM reads this bytecode, interprets it, and executes the program.

In fact, even if you write your code using other programming languages, such as Python and Groovy, which are simpler than the Java language, your code still will be converted to Java bytecode programs.

The JVM is ported to different platforms and insulates the program from the underlying hardware and operating system. Thus it provides hardware- and operating-system independence. The Java application programming interface (API) is also a part of the Java platform. Java API classes are used for building software applications.

1.1.1 Programming in Java

First, let us discuss the Java programming language, one of the most popular object-oriented programming languages in use. The statistics of SourceForge reports that the number of open-source applications written in Java is close to those written in C++. According to the TIOBE software index (http://www.tiobe.com/), a

© Springer International Publishing Switzerland 2016   1
S.V. Chekanov, *Numeric Computation and Statistical Data Analysis*
on the Java Platform, Advanced Information and Knowledge Processing,
DOI 10.1007/978-3-319-28531-3_1

programming-language popularity website, Java is among the most popular object-oriented languages at the time when this book is written.

Let us briefly discuss the main features of Java that make this language number one in the industry:

- Java is multi-platform with the philosophy of "write once, run anywhere";
- Well structured, clean, efficient, simpler (no pointers);
- Stable, robust, and well supported: Java programs written (or compiled) many years from now can be compiled (or executed) without modifications even today. This is true even for Java source code with graphic widgets. In contrast, C/C++ programs usually require continuous time-consuming maintenance in order to follow up the development of C++ compilers and graphic desktop environment;
- Java has the reflection technology, which is not present in C++. The reflection allows an application to discover information about created objects, thus a program can design itself at runtime. In particular, this is considered to be an essential feature for building integrated development environments (IDEs);
- Java has several "intelligent" IDEs, which are indispensable tools for large software projects. Some of them, such as Eclipse or Netbeans, are free. We probably should note that one can also use free IDEs for C/C++, but they are not as intelligent as those for Java and usually miss many important features;
- Automatic garbage collection: Having in hand this feature, a programmer does not need to perform a low-level memory management;
- Extensive compile-time and runtime checking;
- Java always passes references to objects instead of objects themselves, therefore, independent of how you program in Java, your code will be rather efficient;
- Java is truly multithreaded. This significantly simplifies the development of applications that should run in parallel on multi-core machines;
- Advanced serialization: Almost any Java object can be written/retrieved to/from a file;
- Programs written in Java can be embedded to the Web. This is important for distributed analysis environment (Java WebStart, plugins, applets), especially when data-analysis tools are not localized in one single place but scattered over the Web (nowadays, this is the most common situation).

Sometimes, one can hear that Java is slower than C++. The subject itself is controversial, since the answer totally depends on the nature and the goal of an application. Nowadays, most people agree that after introduction of Just-in-Time compiler (JIT), Java is as fast as C++. Probably, in some areas, Java is still slower than C++, but the nature of such controversy is already a sign that the performance gap is now quite small and there is no alarming difference in speed between Java and C++ programs. And anyway, the proper comparisons with C++ is usually unfair: Java does a tremendous amount of runtime checks, such as array bound checking, thread synchronization, runtime checking, garbage collection, etc., to make sure that a Java code runs without problems, and without putting extra stress on a programmer.

Execution of a code on the JVM has a feel of "slowness" since it takes a few milliseconds for the JVM to startup. But when JVM runs a bytecode code, the execution is as fast as for any other program. The JIT compilation converts Java bytecode into native machine code at runtime. The conversion step can be slow; however, this does not matter as much for numerical calculations involving large loops due to JIT compilation.

One should however mention that Java uses more memory than C or FORTRAN. The main reason—JVM does a lot of internal bookkeeping for garbage collection, program optimization at runtime, and providing a safeguard for the Java program. Well, it is better to assign such tasks to the JVM—people who need to use Java will have more time to think about numeric algorithms and how to advance their respective applied disciplines.

1.1.2   The DMelt Software Platform

Numerical and statistical calculations explained in this book use the DataMelt (shorter, DMelt) software platform [1] that runs on the Java platform. It is a collection of libraries integrated with different programming languages. Unlike other statistical programs, it is not limited to a single programming language: DMelt can be used with several scripting languages such as Python/Jython, Groovy, Ruby/JRuby, as well as with Java. Generally, the DMelt computational platform extends the standard Java software platform in several areas:

- Adds a support for Jython, Groovy, JRuby, BeanShell, and GNU Octave high-level scripting languages.
- Adds an IDE and interactive shells to work with these scripting languages and with Java. It also adds a support to process programs in the command line (i.e., in a batch mode).
- Adds comprehensive Java libraries for numeric computation and visualization, incorporating free scientific packages from more than a hundred Java developers around the world. At the moment when this book is written, DMelt includes more than 30,000 Java classes from more than a hundred open-source Java libraries.
- DMelt includes online resources for library updates, class documentation, and for example databases. The Web-based package descriptions are directly accessible from the DMelt IDE. We will discuss this topic later.

Figure 1.1 illustrates the DMelt program structure. DMelt includes a support for several scripting languages that can be run on JVM, third-party numerical libraries integrated with IDE, and online services for update and documentation.

DMelt was designed to enable researches to spend their time thinking about problems and their solutions, rather than diving into low-level coding using programming languages. DMelt analysis macros for data manipulations are based on Jython, an implementation of the high-level language Python. Thus, one can fully benefit from a variety of programming possibilities offered by Python, including its syntax clarity and high-level libraries. But Jython is not a prerequisite for this framework: Java and other languages supported by DMelt can also be used to access the mathematical and graphical libraries of DMelt.

Fig. 1.1 The structure of the DMelt program. It includes Java-based DMelt API, third-party numerical libraries, DMelt IDE, and Web services for documentation and updates. This plot was created using a DMelt Jython script shown in Listing 8.11 of Sect. 8.6.1

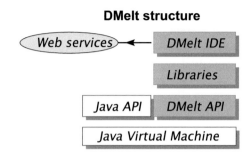

With time, any computational framework based on a simple-to-learn programming language naturally gets large and difficult to handle; this is quite an inevitable feature of modern life. Properly chosen computation language is essential to maintain simplicity of user communication with exponentially growing programs. This is where Java comes to its power: Java virtual machine and various integrated development environments (Eclipse, NetBeans, IntelliJ, and other IDE) can help develop programing codes, tell about errors or mistyped classes, and in general, provide a layer of intelligent activity between a human, who writes a code or interprets its algorithmic logic, and a machine designed for program execution. This is rather different from low-level languages like C/C++ or FORTRAN which are often used for numerical calculations. For such languages, a researcher is usually on his own with a text editor and a programming language itself which typically requires good programming skills and several manuals on a bookshelf.

DMelt is by no means a simple framework, although it is based on Java and high-level Python language. As mentioned before, it has tens of thousands Java classes and methods designed for scientific computation, data analysis, and data visualization. The DMelt library core for statistical and graphical analysis is based on the `jhplot` package, which contains more than a thousand Java classes and methods. However, you will be surprised to find how easy it is to work with this program. Partially, this is because of Python language implemented in Java (Jython) and, partially, because of Java itself.

DMelt [1] is a successor of the jHepWork program that has its origin in high-energy physics in the middle of 2000. The "Hep" part of jHepWork abbreviates "High-Energy Physics." The jHepWork program was described in the book [2]. DMelt substantially extends jHepWork in many areas, but remains backward compatible with jHepWork.

1.1.3 Some Warnings

We should immediately warn you: the DMelt numerical and graphical libraries can be considered neither as most efficient nor error-free. The code of DMelt does not always follow the coding recommendations for Java developers including naming conventions and code layout. We even admit that some parts were not designed with the highest possible performance for code execution in mind. The reason is simple: it was not written by professional programmers. The numerical libraries were written by many people at different times, most of them were students and scientists who had to develop numerical and data visualization algorithms for their own research programs, since commercial software companies either could not offer similar programs or their products were too expensive. Many contributed packages have been discontinued many years ago, but have been brought to life after their inclusion into DMelt. In addition, some packages were written using Java 1.1, and this had also some impact on the coding style of certain libraries.

Thus, a professional programmer may immediately find some parts of the code that look unprofessionally written. This is true even for some examples shown in this book. The reason for this was not because we were not aware of such coding issues. In some cases, we did not find appealing reasons to keep very strict coding standard at the expense of simplicity. For example, in most cases, we import all classes inside a package using the statement:

```
>>>    from PackageName import *
```

instead of importing only certain classes as

```
>>>    from PackageName import class1,class2
```

We did not enforce the latter case to keep the examples of this book short and concise, so we could fit the code snippets into the pages of this book. Also, it is possible that you may not like to type long lists of imported classes during a code prototyping (personally, I do not like this style), since this can be done later during code deployment.

A professorial programmer might find some other odds, like why some object containers are designed to store only double values (like the P1D class to be discussed below), while it is more practical to store integer values when necessary. Again, the motivation was not because of omissions. The reason was that the reader may not want to dive into extra complexity of dealing with different types, since integers are only a subset of float values. There are plenty of other classes which are well suited for storing integer values (we will discuss them in this book).

The main motivation for the DMelt project was to develop an accessible and friendly tool to be used in scientific search, with a syntax oriented toward scientists rather than programmers. The design of this project was mainly motivated by simplicity: there are many programming languages which are required to learn for many years before starting to write useful scientific and engineering projects. The approach discussed in this book is very different: generally, the reader does not need to know any programming language to start writing analysis codes using DMelt libraries. However, if it happens that the reader knows either Java or Python (or both) already, he or she will find this book to be also interesting, since DMelt is not just a simplified entry to the world of the Java and Python computer programming. It shows how to use programming for practical purposes such as numeric calculations, statistics, and data analysis.

The reader may also notice that a little attention has been paid to how to write and use Java or Jython classes. Of course, classes are necessary for any object-oriented language. The reason for this is the following: for the majority of scientific data analysis programs, the logic of scripting programs is linear, i.e., an analysis code typically consists of a well-defined sequence of statements to be evaluated one by one, from the top to the bottom of the code. It is very unlikely that data analysis logic will contain highly parallel algorithmic branches as those for the usual graphical user interface (GUI) development.[1] Certainly, the classes are necessary when one develops Java libraries to be used by a scripting language. But, in this book, we mainly concentrate on the scripting examples based on the existing Java libraries of DMelt, rather than discussing how to write classes for numerical computation to be deployed as external libraries.

1.1.4 Errors

This book may contain typos, omissions, or even errors. DMelt can also contain bugs. If you notice any errors or if you have suggestions regarding the book and code examples, I would be happy to hear from you. You can send your comments to:

```
dmelt@jworks.org
```

One can also post bug reports to the DMelt forum accessible from the main Web page:

```
http://jwork.org/dmelt/
```

DMelt is not a software that stands still. Therefore, this book represents a snapshot of the time when the DMelt version 1.4 was in use, therefore, some examples may fall out of date. Therefore, the reader is encouraged to look at the Web page given above to find corrected examples.

---

[1]We should probably say that this may not be totally true in future when multi-core machines will be rather common and one will face with the question of how to parallelize analysis codes to gain high performance. We briefly discuss this topic in this book.

1.2 Scripting with DMelt

Let us say a few words about numeric calculations and data analysis. Scripting in a scientific research is essential. There are plenty of programs heavily based on graphical user interfaces (GUI), where a researcher should go over many mouse clicks before reaching a designed result. Typical examples are Microsoft Excel, Origin, and many other commercial products. The scripting approach is somewhat different: it requires from a developer to type only short commands and store them in files so one can easily reproduce the results by executing such macro files later. During the program development, an analysis framework should help find a proper method for a particular class instance and to supply with a comprehensive description of its methods, which can fit to the program logic. It should control your code while you are typing and correct it when needed!

In this respect, DMelt is similar to Wolfram Research Mathematica or Maple. However, unlike these commercial products, DMelt is more data oriented. In addition, unlike software based on specialized languages, DMelt is more oriented toward general-purpose programming languages, such as Java, Python, or Groovy. Being Java-based, DMelt is also more GUI oriented since all the power of Java graphical widgets for user interfaces is in your hands. In addition, DMelt uses Python, which is a very popular programming language for science and engineering. Finally, DMelt is free.

One should bear in mind that DMelt was designed for data manipulations and visualizations in which speed of scripting language is not essential, since the bulk of numeric computations are moved to dedicated libraries. Scripts are designed for calling numeric libraries. The main bottleneck in this approach is human interaction with graphical objects using the mouse, keyboard, or the network latency in the case of remote data or programs.

In practice, results obtained with Jython programs can be obtained faster than those designed in C++/Java, because code development is so much easier that a user often winds up with a much better algorithm based on Jython syntax and DMelt high-level objects than he/she would in case of C++ or Java. For CPU intensive tasks, like large loops over primitive data types, reading files, etc., one should use high-level structures of Jython and DMelt or user-specific libraries that can be developed using the DMelt IDE. This is the basic idea. The rest of this book will spell it out more carefully.

1.2.1 Learning by Example

You will be surprised to know that even the most realistic numerical and statistical examples given in this book have rather short source-code snippets. I promise that all our example codes typically fit to 2/3 of the printed page of this book at most. This came to be possible using Python syntax and its high-level built-in data structures.

This was also possible due to known Python capabilities to "glue" different programming languages. In the case of its Java implementation, one can seamlessly integrate Python, Java, and Java-based numerical libraries.

As you walk through the examples, you may decide to type all the listed codes in by hand. This is the best way to get familiar with the coding techniques. But, even though Jython examples of this book are short, you may still avoid typing them when following the book pages. The example source codes of this book (for each section separately) can be downloaded without charge from:

http://jwork.org/dmelt/

or from the mirror:

http://sourceforge.net/projects/dmelt/

Look at the section called "Documentation" which gives you the necessary location of the tar file with the all code examples.

1.2.2   Using Jython for Code Examples

DMelt can be used with a number of programming languages but, as mentioned earlier, many scripts given in this book are written in Python [3], which is an increasingly popular programming language in science and engineering [4]. It has simple and easy to learn syntax which reduces the cost of program maintenance. Jython [5] is an implementation of Python in Java. In contrast to the standard Python (or CPython) written in C, Jython is fully integrated with the Java platform, thus Jython programs can make full use of extensive built-in and third-party Java libraries. Therefore, Jython programs have even more power than the standard Python implemented in C. Finally, the Jython interpreter is freely available for both commercial and noncommercial use.

You may ask the question: What is the point of using Jython and Java if CPython is also portable and can be installed on Linux and Windows platforms? This answer is this: CPython calls libraries implemented in C/C++ or FORTRAN, but these libraries, by definition, should be compiled separately for each platform (in fact, as CPython itself). Thus, CPython cannot provide a genuine multi-platform framework. In case of Jython, libraries developed using Java are truly multi-platform and do not require separate deployment for each computer platform.

Programs written using DMelt are usually rather short due to the simple Python syntax and high-level constructs implemented in Python and in the core DMelt libraries. As a front-end data-analysis environment, DMelt helps to concentrate on interactive experimentation, debugging, rapid script development, and finally on work-flow of scientific tasks, rather than on low-level programming.

1.2.3 Differences with Other Math Software

How does DMelt compare with other commercial products? Throughout the years there have been many commercial products for data analysis, but it is important to realize that they are typically platform specific. On top of this, commercial products are either costly and/or do not provide a user with the source code, or both.

We would not be too wrong in saying that it is difficult to find a commercial product with the same functionality in certain analysis areas, and with such a variety of methods existing in DMelt. For example, only one single class used to build, manipulate, and display one-dimensional histogram has about eighty methods (plus dozens of methods inherited from other classes). Usually, commercial software is not competitive enough for such specialized tasks as histogramming and processing of large data samples. Together with a high price, this was one of the reasons why commercial products did not significantly penetrate into the software environment of high-energy physics and other natural sciences where data reduction and data mining are the most important tasks. Here, we should also add that Linux and Unix are the most common platforms in universities and laboratories and this also had a certain impact on the number of data analysis packages to be used in scientific research.

As mentioned before, DMelt is a computational platform, not just a program that uses specific language. This means you can use different programming languages and operating systems to perform numeric and statistical analysis. This is quite different from other free programs, such as the R-project [6], a free software environment for statistical computing and graphics. The R-project is a programming language and, at the same time, a software environment based on this language. DMelt is not limited by a certain programming language. While most of the examples in this book are based on the Python language, Chap. 16 will discuss how to use DMelt with Java, BeanShell, Groovy, and Ruby.

There are a number of data analysis and statistics programs written in C++/C, such as ROOT [7, 8], GNU Octave [9], and the above-mentioned R-Project [6], to name a few. Compared to C/C++ based programs, DMelt:

- inherits the well-known Java robustness. For example, the source code of this project developed more than ten years ago can easily be compiled without any changes even today. Even jar libraries compiled many years from now can run on modern Java Virtual Machine. In contrast, C++ programs with user graphical interfaces require a constant support to meet hardware and operating system requirements. A typical lifetime of unsupported C++ code with graphic widgets is several years for Linux-based platforms;
- being Java-based, provides a cross-platform capabilities. DMelt does not require compilation and installation. This is especially useful for plugins distributed via the Internet in the form of jar files;
- can be integrated with the Web in the form of applets or Java Web-start applications, thus it is better suited for distributed analysis environment via the Internet.

This is an essential feature for large scientific communities or collaborations working on a single project;

- has an automatic garbage collection. This is a significant advantage over C++/C as the user does not need to perform low-level memory management;
- is designed from the ground up to support programming with multiple threads. It is a truly multithreaded language. This makes parallel programming easier and leads to a more efficient use of modern computers with multi-core processors. Unlike C++, the Java virtual machine takes care of low-level threads according to the host multi-core computers;
- comes with the Java reflection technology, i.e., the ability to examine or modify the behavior of applications at runtime. This feature is missing in C/C++;
- includes an advanced help system with a code completion based on the Java reflection technology. With increasingly large numbers of classes and methods in C++ programs, it is difficult to access information on classes and methods without advances IDE. Using the DMelt IDE, it is possible to access the full description of classes and methods during editing Jython scripts;
- includes powerful Java serialization mechanism that allows saving complex objects in files, such as complex arrays with data and together with graphical attributes.

You can find a more detailed discussion on differences between Java and C++ in this Website [10].

1.3 Installation

The good news is that you do not need to install anything to make all the examples discussed in this book to work. But to run DMelt, you need to have the Java development kit (JDK) installed. You can also use Java Runtime Environment (JRE), which is very likely already installed on your computer but, in this case, there will be some limitations (for example, you will not be able to compile Java code).

The Java software is available at:

https://java.com/en/

Installing the JDK or JRE is rather simple for any platform (Windows, Solaris, Mac, and Linux). Once installed, check Java by typing java -version on the prompt. If Java is installed, you should see

```
java version "1.8.X"
Java(TM) 2 Runtime Environment, Standard Edition
Java HotSpot(TM) Client VM (build 1.8.X, mixed mode)
```

or a similar message ("X" indicates a subversion number). You will need at least Java 1.7 or above.

DMelt does not require installation. Download the package from the following location:

http://jwork.org/dmelt/

or

http://sourceforge.net/projects/dmelt/

The package for the download has the name "dmelt-VERSION.zip", where "VERSION" is a version number.[2] Unzip this file to a folder. You will see several files and the directories "lib", "macros" and "doc". For Windows, just click on the file "dmelt.bat" which brings up the DMelt IDE windows. For Linux, Unix, and Mac, run the script "dmelt.sh".

For Mac, Linux, and UNIX, you can put the file "dmelt.sh" to the "$HOME/bin" directory, so one can start DMelt from any directory. In this case, one should set the variable JEHEP_HOME (defined inside the script "dmelt.sh") to the directory path where the file "dmelt.jar" is located.

The first time you execute the "dmelt.sh" (for Linux/Mac OS) or "dmelt.bat" (Windows) launch programs, DMelt takes more time to start than usual. This is normal since Jython cashes the Java jar libraries, i.e., it creates a new directory "cachedir" inside the directory "lib/jython" with the description of all classes located in jar files defined in the CLASSPATH variable. Next time when you execute the startup script, the DMelt IDE will start fast as the package cache is ready (of course, if you did not modify the Java CLASSPATH before starting the DMelt IDE).

If you want to run a Jython script in a batch mode, without launching the IDE, you can do it as:

```
./dmelt_batch.sh macro.py arg1 arg2 ..
```

where "macro.py" is a Jython/Python macro to be processed and "arg1 arg2 .." is an optional list of command-line arguments passed to this file. On Windows, use "dmelt_batch.bat".

As mentioned before, DMelt is a computational platform—it can be used with different programming languages on different operating systems. One can also run data analysis programs in Groovy, JRuby, BeanShell, Octave, and Java programming languages. To run scripts written in different languages, make sure that the files have the correct extensions:

```
./dmelt_batch.sh macro.groovy  # runs  Groovy script
./dmelt_batch.sh macro.rb      # runs  JRuby / Ruby  script
./dmelt_batch.sh macro.bsh     # runs  BeanShell script
./dmelt_batch.sh macro.java    # runs the standard Java code
./dmelt_batch.sh macro.m       # runs Octave script
```

---

[2]The current book uses the DMelt version 1.4.

Chapter 16 discusses how to program scripts in these programming languages using the common Java API of DMelt.

1.4 DMelt Workbench

Feel free to skip this section and jump to Chap. 2, since for readers with some programming experience, this section could be too obvious. For those who are just entering the computational world, I will try to explain here several tricks that could be useful for source code editing and execution. DMelt comes with a lightweight integrated development environment (IDE) which includes a source code editor with code completion and a code analyzer, a JythonShell (the "JythonShell" tab), a BeanShell (the "BeanShell" tab), and a panel with the file manager. The script "dmelt.sh" (for Linux/UNIX/Mac) or "dmelt.bat" (Windows) starts the DMelt IDE. After initial initialization, you will see the DMelt IDE with a source code editor as shown in Fig. 1.2.

It should be noted that the source code development using the DMelt libraries can be done using any text editor, while the execution of Jython scripts or compiled

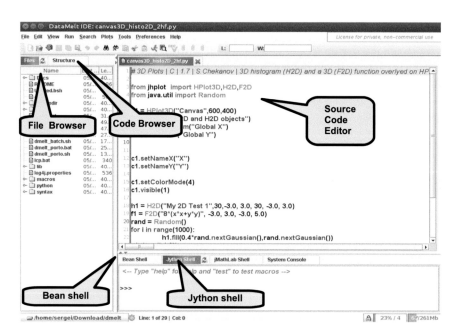

Fig. 1.2 DMelt workbench ("DMelt IDE"). It includes a source code editor and interactive prompts

codes can be done using a shell prompt after specifying the CLASSPATH environmental variable. This part should be considered for advanced users and will be discussed later.

The DMelt workbench has three main windows:

- The Source Code Editor (central window);
- The Toolbar menu (above the text area);
- File and code browser window (left window);
- BeanShell and JythonShell windows (at the bottom of the main editor).

When DMelt starts for the first time, it creates files with preferences located in the directory:

```
$HOME/.dmelt
```

for Linux/Mac, or

```
$HOME\dmelt.ini
```

on Windows. There are several preference files inside this directory: the main file, "jehep.pref" with all source code editor preferences, a user dictionary file, a JabRef preference file and other files. If you need to reset all settings to their default values, just remove the directory with the preference files (or just the file "jehep.pref").

1.4.1   Source Code Editor

The source code editor can be used to edit files, and it has all features necessary for effective programming: syntax highlighting, syntax checker, and a basic code completion. For bookmarks, one should click on the right margin of the source code editor. A blue mark should appear that tells that the bookmark is set at a given line. One can click on it with the mouse in order to jump to the bookmarked text location.

The file browser is used to display files and directories. By clicking on the selected file name one can open a new tab for the text editor. For most types of files (LaTeX, C++, Java, Python, Groovy, etc.), the code browser shows the structure of the currently opened document.

1.4.2   DMelt Libraries

Although this topic is probably for advances users, we feel that it is necessary to describe the DMelt library structure here. Generally, the program contains Java jar

libraries and Python-based libraries. After the installation, the main DMelt directory
contains the following subdirectories:

- "lib"—contains Java libraries. When you start DMelt, this directory is scanned
  by DMelt and all libraries inside this directory are put to the Java CLASSPATH
  environmental variable. The directory lib contains several subdirectories with
  jar files: freehep—contains only FreeHEP Java libraries; the directory
  "jython" contains Jython libraries; system contains libraries necessary to run
  DMelt, including third-party libraries. User-specific libraries can be put under the
  user directory (which will also be scanned by the DMelt startup script).
- "python"—contains Python libraries. By default, they are not imported by
  Python modules, so it is up to you to import them into your programs. There
  are several ways to do this: (1) put the directory name in the file "registry"
  located in the directory "lib/jython". You should define the variable
  python.path as:

```
python.path=[dir]/python/packages
```

where [dir] the installation directory where DMelt is installed (it should
contain the file "dmelt.jar"). Or, alternatively, one can set the location of
Python libraries at the beginning of your Jython code as:

```
import sys
sys.path.append("[dir]/python/packages")
```

If you run a Jython module inside the DMelt IDE, one can specify the current
installation directory using the variable SystemDir which always points to the
installation directory. Thus the line above will look as:

```
import sys
sys.path.append(SystemDir+"/python/packages")
```

I will come back to this point later in the text where several Python-based scientific libraries will be discussed.

- "macros"—contains Python modules necessary to run DMelt. It also contains
  examples in the subdirectory "examples". When you start DMelt, the directory
  "macros/system" is put inside the Jython class path automatically by the
  script "sysjehep.py" located inside the subdirectory "system".

A user can put new macros in the "macros/user" directory that has several
macros: For example, one can replace a string with another string in the current text
just by calling the method replace(str1,str2), where str1 and str2 are
input strings; In fact, all user macros are rather similar to those used by the jEdit[3]
editor, as long as the textArea class is used.

---

[3]jEdit is one of the most popular Java-based text editors.

The Jython class path is defined via the module "sysjehep.py" located in the "macros/system" directory. This variable specifies the location of Jython modules which can be visible for the "import" statement. The module "sysjehep.py" is loaded automatically every time you start DMelt (or reload the Jython console). One can specify the location of the user modules in this file. By default, every file which is put into the installation directory or to "macros/system" or "macros/user", should be visible for DMelt macros and normally you do not need to import them.

1.4.3 Jython and BeanShell Consoles

The BeanShell and JythonShell can help quickly prototype small pieces of codes without using text editors. By default, the BeanShell window is active. To activate the JythonShell window, click on the JythonShell tab.

BeanShell and JythonShell can be used to run external programs. Just type the exclamation mark "!" in front of the program name you want to execute. For example:

```
>>> !latex <file>   #   latex for the file <file>
>>> !make           #   compile C++/Fortran
>>> !ls             #   list files (linux/unix/max)
>>> !dir            #   list files (windows)
```

One can use the command history (keyboard key: [Up] and [Down]) and Java and Python code completion using the key combination [Ctrl] + [Space]. Use the help options to learn more about additional features for interactions with the native OS. Both shells contain their own help system.

There are several predefined variables available for the both consoles. They have been exported by the scripts located in the "macros/system" directory: "sysjehep.py" (for Jython) "sysjehep.bsh" (for BeanShell).

Using the JythonShell, one can access properties of this file using the Editor class and its methods. This class holds currently opened document and allows a manipulation with its text. For example, if one types print Editor.DocDir() in the JythonShell, one can display the full path to the currently opened document. One can access all public methods of the Editor static class using the DMelt code assist for the Editor class:

```
>>> Editor. [Ctrl]+[Shift] # show all methods
```

or

```
>>> dir(Editor) # show all methods
"DocDir", "DocMasterName", "DocMasterNameShort",
"DocName", "DocStyle" ..
```

(below we print only a few first methods). As you can see, having in hand this class, one can access a broad variety of methods for text manipulation. More information about this class can found using the code assist to be discussed in Sect. 1.4.10.

Below we give several examples of how to access the information about an opened file using several predefined variables. Use the JythonShell to type the following commands:

```
>>> print "Java classpath",ClassPath
>>> print "File name=",DocName -
>>> print "Directory=",DocDir
>>> print "File separator=",fSep
>>> print "Name with complete path",DocMasterName
>>> print "Name without extension",DocMasterNameShort
```

All these variables are automatically imported by the IDE.

But how one can find the predefined variables available in DMelt while working with Jython macros files? This should be easy. Click on JythonShell and type the following command:

```
>>> dir()
[ClassPath", "DicDir", "DocDir", "DocMasterName",
 "DocMasterNameShort", "DocName", "DocStyle",
 "Editor", "ObjectBrowser", "ProjDir", "SetEnv",
 "SystemDir", "SystemMacrosDir", "UserMacrosDir"
 ...
```

(again I show only a few first variables). The best approach to learn about them is just to print them out. Analogously, one can print all such variables using the BeanShell commands (but using the BeanShell syntax).

It is useful in many cases to clean up messages from the interactive consoles. To reload either the `File Browser` or Jython (BeanShell), one should use the `Reload` buttons located directly on the small blue tabs of the console windows.

DMelt uses the following aliases for the BeanShell macros:

[CLASSPATH]	java class path;
[FILE_SHORT]	returns the filename without the extension;
[FILE_SHORT_NODIR]	returns the file name without the extension and the path;
[FILE]	returns the full name of the file including the path;
[FILE_NODIR]	returns the full name of the file excluding the path;
[DIR_FILE]	returns file directory;
[DIR_SYS]	returns system directory.

For example, typing `print [FILE]` prints the name of the currently opened file. Such substitutions can be used in macros. For example, if a macro contains [FILE], it will be automatically replaced by the current file name. We will discuss how to write programs using the BeanShell language in Sect. 16.3.

1.4.4 Accessing Methods of Instances

A user can view the available methods by typing `obrowser` in the BeanShell. This will open an object browser window with all objects. If one needs to add some object, one should type `obrowser.add(obj)`, where "`obj`" is an instance of a Java class.

```
>>> obrowser
>>> obj=new JLabel("OK")
>>> obrowser.add(obj)
```

For a similar task in the JythonShell prompt, a user should use the code assist, see Sect. 1.4.10, or the Python `dir("obj")` method. The DMelt code assist cannot be used in BeanShell. In this case, one should use the `obrowser` class.

One can manually execute the `obrowser` class by running the script "`obrowser.bsh`" or "`obrowser.py`" from the directory `macros/system`.

1.4.5 Editing Jython Scripts

First, a new file with the extension "`.py`" should be created. There are several ways to do this: (1) Select [File]-[New]-[jHPlot script] in the main menu. A new template Jython script should appear; (2) Use the menu [File]-[New]-[Text document]. Then, save it as a Jython file with the extension "`.py`"; (3) Click on any Jython file with the extension "`.py`" in the File Browser.

If a script with the user analysis program is ready, one should save it. But before this, you may check syntax errors without actual execution of the macro file. Look at the menu [Run] and [Check Jython syntax]. In case of errors, the IDE will point to a line with an error. For the syntax checker, it is impossible to identify runtime errors, i.e. errors which may happen during the execution of the script. Once you know that there are no errors, save the file again using the menu `[File]`.

1.4.6 Running Jython Scripts

To run a Jython script, use the `[run]` button from the main toolbar of DMelt. One can also use the keyboard key `[F8]` for fast script execution. In case of an error, the DMelt main editor will move the cursor to the appropriate line with the detected error. Press any key to remove the line highlighting (red color). One can also execute a Jython file line-by-line using the `[run]` menu of the main toolbar.

During the execution, all program outputs will be redirected to the JythonShell window. It is appropriate mention here one feature: JythonShell is not designed for very heavy output, as it is mainly a debugging tool. If you want to print a lot of messages, then you should be careful: you may wipe out significant resources by doing this and your code execution will be rather slow. For example, consider a simple Jython program: printing integer values:

```
for i in range(1000000):
  print "Test=",i
```

Save these lines in the file "test.py". Then open it in the DMelt editor and run it by clicking on the button [run]. You will immediately see that the memory monitor in the right corner of the IDE will show an increase in the memory usage. So, try to avoid such situations. If you need to debug loops like this, use a few iterations. If you still want to run the loop over many iterations, you should not print many debugging messages for every single iteration inside a loop. For example, one can print a status line every 100 iteration as:

```
for i in range(1000000):
  if i%100 == 0: print "Test=",i
```

As you may guess, % means a remainder of $i/100$, which should be 0 for the string to be printed.

1.4.7 Macro Files of the DMelt IDE

In your Jython programs, you can a full access to the text that you see in the DMelt editor. And, one can even load files to the editor using interactive shells.

To show a text file in the DMelt editor, one can use JythonShell of the IDE. Simply type the line:

```
>>> view.open("news.html",0)
```

which opens the file "news.html" (downloaded using the previous example) in a new tab of the editor. The object view represents an "editor" text component of the DMelt IDE. This instance is imported by default during the IDE startup. One can find the methods of this editor instance using the dir(view) method.

In order to find what can be done with the object view, run the program "test" using JythonShell. It illustrates the major operations with the text editor and shows how to manipulate with the example text loaded to the editor using various macros.

Let us show an example of how to work with text files using the object view. Assuming that a file is loaded into the editor, use these commands from JythonShell:

```
>>> view.selectAll()      # select all text in the editor
>>> mydoc=view.getText() # get text to mydoc string
>>> view.setText(" ")     # clear the editor
```

```
>>> view.setText("text") # set new text
```

In this example, `mydoc` is a string with the text from the editor. To find and replace a string is rather easy as shown in this code snippet:

```
>>> mydoc=mydoc.replace("old","new");
>>> view.setText(mydoc)
```

One can also find the current caret position and insert a text at a specified position as:

```
>>> cc=view.getCaretPosition()
>>> view.insertString("inserted text", 4)
```

where "4" is the position where the text will be inserted. One can move the caret to a specific line using the method `view.goToLine(line)`, where `line` is the line number.

As usual, you can learn about the `view` class using the standard approach:

```
>>> type(view)
>>> dir(view)
```

1.4.8   Running BeanShell Scripts

One can run BeanShell scripts and Java source codes in the same way as Jython scripts discussed above. Remember to use the file extension `.bsh` for BeanShell programs. However, DMelt has much less advanced error handling compared to the JythonShell. As in the Jython case, the [F8] key can be used for fast execution of BeanShell scripts. During the execution, the output from the scripts will be redirected to the BeanShell window.

1.4.9   Compiling and Running Java Code

Java files can be edited in the same way as Jython macros. Java source code can be compiled into bytecodes using the menu from the [Run] toolbar menu as [Run]-[Javac current file]. Similarly, one can also compile all Java source codes located in the same directory or build a jar library.

We should be more specific about how to run Jython scripts from DMelt without the main text editor. For this, one can set the "CLASSPATH" variable to the libraries located in the "lib" To run a Jython file, say file.py, run the command:

```
bash#: dmelt_batch.sh file.py
```

from the Linux/UNIX or Mac prompt.

1.4.10 DMelt Code Assist

DMelt has several methods to access the documentation of Java classes. Here we
will discuss a few approaches that are the most popular while working with Jython
code.

1.4.10.1 Using the Dot Method

If a user is working with the JythonShell, all methods associated with a particular
object are shown in a table after typing a dot after the name of an object and pressing
[Space] after holding down the [Ctrl] key. For example, if a Jython or Java
object obj was instantiated, type:

```
>>> obj.   # press [Ctrl]+[Space] for the help
```

to bring up a table with all methods associated with a class instance obj.

However, the help system works differently for files loaded in the code editor. In
this case, one can get the information on available public methods using the built-
in DMelt Code Assist. When a user types a name of some object followed by a
dot, [F4] key can be used to check the methods associated with this object. For
example:

```
>>> from jhplot import POD
>>> obj=POD()      # create a Java object
>>> obj.           # press [F4] for  POD methods
>>> obj="string"  # create a Jython object
>>> obj.           # press [F4] for "str" methods
```

Pressing the [F4] key after each object name, followed by a dot, brings up a table
with all methods that belong to a particular Jython or Java class. In the above example, the table shows the methods of the class POD.

One can also search for a particular method using the pop-up table of the Code
Assist. For this, the standard Java regular expressions [11, 12] can be used. To sort
rows with methods, one should click on the column headers. Then one can push
a selected method to the text editor using a double-click or a mouse menu. The
selected method will be inserted into the code editor at the line after the dot.

The Code Assist also allows you to look at the full API documentation of the jHPlot classes or methods. Click on a selected line in the Code Assist table and use the right mouse button to get the associated JavaDoc information. One can also push a selected method into the editor window by clicking on a pop-up mouse menu.

1.4.10.2 Using the Help Class

If you work with Java objects, one can call a convenient static class that brings up a Web browser with complete javadoc documentation for a give class. This class is called `jhplot.help` (note it uses lower case letters). When a Java object is instantiated, use the method help.doc(obj), where `obj` is created Java object. Let us show an example which illustrates how to look at the complete Javadoc API of the class `HPlot` used to create a plot frame:

```
>>> from jhplot import *
>>> c=HPlot()       # initialize a canvas
>>> c.visible()     # show it
>>> help().doc(c) # starts a web browser with Javadoc
```

Of course, one can also look at the Javadoc of the `help` class itself:

```
>>> from jhplot import *
>>> help().doc(help())
```

or look at any class supported by the standard Java platform. For example:

```
>>> from jhplot import *
>>> from java.util import ArrayList
>>> a=ArrayList()
>>> help().doc(a) # Look at Javadoc of ArrayList()
```

The same approach works for Groovy, JRuby, and Java codes. However, this approach does not work for native Python classes.

1.4.10.3 Using the Import Statements

The user can also find the complete Java API for a given class using alternative approach. This works for the standard Java API and for most of third-party Java libraries.

Almost every Jython code has the statement such as

```
>>> from package import class
```

where "package" is a given Java package, and "class" is a class inside this package. Navigate the mouse cursor to this line (which is typically at the very beginning of your code), and with the mouse right button select "Lookup API" from the pop-up menu. This brings up a Web browser with the displayed Java API for the given class. This method works for the DMelt classes, as well as for the standard Java classes.

Similarly, for Java, Groovy, and BeanShell source codes, navigate the cursor to the lines such as

```
>>> import package.class;
```

and, using the mouse, select "Lookup API" from the pop-up menu. In the cases when the regular expression ("*") is used to import all classes, this method will attempt to open the package summary page of Java API, if it is available.

1.4.10.4 Using the doc() Method

Many (not all!) classes of DMelt have their own help, usually based on the original work of the developers. To access the API help documentation, use the method doc(). For example, after creation of HPlot() canvas as c1= HPlot(), one can look at the API documentation of its methods by executing the line c1.doc().

The code assist of the BeanShell window is somewhat different. Please read the corresponding help when using this shell.

1.4.11 Other Features

The DMelt GUI has several other features that should be useful while working with a text or data:

- The menu [Search] can be used for searching particular strings or substrings in an opened document;
- The menu [Run] is useful for compilation of Java source codes, building Java jar libraries, execution of Jython scripts, and checking Jython syntax;
- The menu [Tool] is designed for working with LaTeX files. It provides a bibliographic manager (based on the JabRef program), LaTeX tools for common LaTeX symbols. It can be used also to start a graphical canvas for plotting and a basic image editor based on the popular ImageJ program.

It is worth exploring the menu options of the IDE. Especially, if you would like to see DMelt examples, look at the menu [Tool] and select [Examples]. This brings up a dialog with the list of available Jython examples. One can run the example macros to get a clue on how to use the main package core DMelt with numerical and graphical libraries.

DMelt also supports an online example database with more than 400 code snippets written in Jython, Java, Groovy, JRuby, BeanShell, and GNU Octave languages.

Before jumping to the data analysis sections, I would like to give one advice. The DMelt IDE is only an experimental tool and, probably, it was not tested as carefully as any other professional IDE. It was designed as a lightweight source code editor with IDE-like features to help new users to get started. Therefore, if you are not too satisfied with the DMelt IDE, you are likely to be a professional programmer. In this case, use ether Eclipse or NetBeans (well, if you are a professional programmer, you must know about them!). Both Eclipse and Netbean can be used for editing DMelt scripts if you set the $CLASSPATH to the DMelt jar libraries inside the directory "lib" (again, I do not need to explain this to advanced users).

Finally, you may even use popular code editors such as jEdit, Emacs, or VI (Linux and Windows) or Notepad (Windows). How to run DMelt scripts in the console without using the IDE has been described before.

1.4.12 Working with Images

We should remind that in order to export a canvas into an image file, one can use the export(file) method after calling draw(obj) or update() methods. This was discussed in Sect. 8.2.6. Below we will continue with this example.

If an image is saved in JPG, GIF, or PNG format, one can view it using the IView class. Below we will show how to use it.

Assume we have created a script which plots a histogram or any other object on the HPlot canvas (or any other canvas). We assume that the instance of this canvas is c1. If you want to create an image file and open it immediately after creation in the IView frame, put these lines at the end of your Jython script:

```
>>> file=Editor.DocMasterName()+".png"
>>> c1.export(file)    # export to a PNG file
>>> IView(file)        # view the image
```

where the method Editor.DocMasterName() accesses the file name of the currently opened script. Execution of this file brings up a frame with the PNG image inside.

If an image is saved in one of the raster formats, such as JPG, GIF, PNG, one can edit it (crop, resize, add a text, etc.) using the IEditor class. This class is built on the top of the ImageJ program [13], which is the most advanced image editor implemented in Java.

Let us show a typical example when a canvas, such as HPlot or HPlot3D, is first saved in an image file and then the IEditor class is used to edit and analyze the created image:

```
>>> file=Editor.DocMasterName()+".png"
>>> c1.export(file)          # export to a PNG file
>>> IEditor(file)            # edit the created image
```

where c1 is a canvas object and `Editor.DocMasterName()` is the name of the
currently opened document in the IDE. The example shows how the image created
by the c1 object is redirected to the `IEditor` program for further editing.

1.4.13 DMelt License

DMelt includes many third-party libraries and many reused classes that have been
rewritten and adopted for the use together in DMelt. The reader can find a list of
contributed packages on the DMelt website. Many libraries are not supported any
longer by the original authors, but they were very useful at the time when DMelt was
under heavy development. The author apologizes in advance if some references to
the used libraries are missing; such omissions were wholly unintentional. It is hard
to keep track of all contributions over the last 10 years of DMelt development. We
will provide the necessary references when we discuss specific topics on numerical
and statistical computations.

All packages listed below are subject to their licenses (as well as the packages
cited in the following sections). The vast majority of the included packages are
GNU-licensed or have very permissive open-source licenses. Some documentation
libraries, examples, installer, code assist databases and language files integrated
into the DMelt software package are free for noncommercial purposes (academic
research, science, and education).

In order to use DMelt Java libraries in commercial programs with possibly
closed-source environment, you will need to receive a redesigned version of this
program suitable for commercial purpose. You should become a DMelt member and
request for redesigned DMelt without GPL libraries. The business-friendly DMelt
will include libraries under "permissive" licenses, such as LGPL, MIT, Apache, and
BSD. In addition, the membership allows the use of the documentation and examples for commercial purpose.

1.4.13.1 Disclaimer of Warranty

DMelt is *not a commercial* product, although it is professionally written and many
libraries have been tested by a large scientific community. I cannot guarantee that it
is fault-free in all possible foreseeable situations. Therefore, you use this package at
your own risk.

The author and publisher make no warranties, express or implied, that the
programs contained in this book are free of errors, or are consistent with any

particular standard of merchantability. They should not be relied on for solving a problem whose incorrect solution could result in injury to a person or loss of property. If you do use the program in such a manner, it is at your own risk. The authors and publisher disclaim all liability for direct or consequential damages resulting from your use of the programs.

References

1. Computation and visualization environment. DataMelt. http://jwork.org/dmelt/
2. Chekanov S (2010) Scientific data analysis using Jython scripting and Java (advanced information and knowledge processing). Springer, London
3. Python programming language. http://www.python.org/
4. Langtangen H (2008) Python scripting for computational science. Springer, Heidelberg
5. The Jython project. http://www.jython.org/
6. The R project for statistical computing. http://www.r-project.org
7. Brun R, Rademakers F, Canal P, Goto M (2003) Root status and future developments, ECONF C0303241, MOJT001
8. Brun R, Rademakers F (1997) ROOT: an object oriented data analysis framework. Nucl Instrum Method A389:81. http://root.cern.ch/
9. GNU octave high-level interpreted language. http://www.gnu.org/software/octave/
10. Wikipedia, comparison of Java and C++. http://em.wikipedia.org/wiki/Comparison_of_Java_and_C
11. Java regular expressions, ReGex package. http://java.sun.com/j2se/1.6.0/docs/api/
12. Stubblebine T (2007) Regular expression pocket reference: regular expressions for Perl, Ruby, PHP, Python, C, Java and .NET (Pocket reference (O'Reilly)). O'Reilly Media Inc, Sebastopol
13. ImageJ Java library. http://rsb.info.nih.gov/ij/

Chapter 2
Introduction to Jython

We have already pointed out that Jython is an implementation of the Python programming language, unlike CPython which is implemented in C/C++. While these
implementations provide almost identical Python language programming environment, there are several differences. Since Jython is fully implemented in Java, it
is completely integrated into the Java platform, so one can call any Java class and
method using the Python language syntax. This has some consequences for the way
you would program in Jython. During the execution of Jython programs, the Jython
source code is translated to Java bytecode that can run on any computer that supports
the Java virtual machine.

This chapter gives a short introduction to the Jython programming language.
We cannot give a comprehensive overview of Jython or Python in this chapter: this
chapter aims to describe a bare minimum which is necessary to understand the Jython
language, and to provide the reader with sufficient information for the following
chapters describing numerical methods and statistical techniques using the DMelt
libraries.

2.1 Code Structure and Jython Objects

As for CPython, Jython programs can be put into usual text files with the extension
".py". A Jython code is a sequence of statements that can be executed line by line,
from the top to the bottom. Jython statements can also be executed interactively using
the Jython shell (the tab "JythonShell" of the DMelt IDE).

Comments inside Jython programs can be included using two methods: (1) To
make a single-line comment, put the sharp "#" at the beginning of the line; (2) To
comment out a multiline block of a code, use a triple-quoted string.

It is good idea to document each piece of the code you are writing. Documentation
comments are strings positioned immediately after the start of a module, class, or

© Springer International Publishing Switzerland 2016   27
S.V. Chekanov, *Numeric Computation and Statistical Data Analysis*
on the Java Platform, Advanced Information and Knowledge Processing,
DOI 10.1007/978-3-319-28531-3_2

function. Such a comment can be accessed via the special attribute __doc__. This
attribute will be considered later when we will discuss functions and classes.

Jython statements can span multiple lines. In such cases, you should add a back-
slash at the end of the previous line to indicate that you are continuing on the next
line.

As for any dynamically typed high-level computational language, one can use
Jython as a simple calculator. Let us use the DMelt IDE to illustrate this. Start the
DMelt and click the "JythonShell" tab panel. You will see the Jython invitation $>>>$
to type a command. Let us type the following expression:

```
>>>    100*3/15
```

Press [Enter]. The prompt returns "20" as you would expect for the expression
(100*3)/15. We did not use any assignment in this mathematical expression,
therefore Jython assumes that you just want to see the output. Now, try to assign this
expression to some variable, say W:

```
>>> W=100*3/15
```

This time, no any output will be printed, since the output of the expression from the
right side is assigned to the variable W. Jython supports multiple assignments, which
can be rather handy to keep the code short. Below we define three variables W1, W2
and W3, assigning them to 0:

```
>>> W1=W2=W3=0
```

One can also use the parallel assignments using a sequence of values. In the example
below we make the assignment W1=1, W2=2 and W3=3 as:

```
>>> W1,W2,W3=1,2,3
```

At any step of your code, you can check which names in your program are defined.
Use the built-in function dir() that returns a sorted list of strings

```
>>> dir()
["W","W1","W2","W3","__doc__",  "__name__"  ...
```

(we have truncated the output in this example since the actual output is rather long).
So, Jython knows about all the defined variables starting from the character W. You
will see more variables in this printed list which are predefined by DMelt.

One can print out variables with the print() method as:

```
>>>    print W1
1
```

One can also append a comment in front:

```
>>> print "The output =",W1
The output = 1
```

You may notice that there is a comma in front of the variable W1. This is because "The output" is a string, while "W1" is an integer value, so their types are distinct and must be separated in the print statement.

How do we know the variable types? Use the type() method which determines the object type:

```
>>> type(W1)
<type "int">
```

The output tells that this variable holds an integer value (the type "int").

Let us continue with this example by introducing another variable, "S" and by assigning a text message. The type of this variable is "str" (string).

```
>>> S="The output ="
>>> type (S)
<type "str">
```

So, the types of the variables W1 and S are different. This illustrates the fact that Jython, as any high-level language, determines types based on assigned values during execution, i.e., a variable may hold a binding to any kind of object. This feature is the most useful for scripting: now we do not need to worry about defining variable types before making assignments to a variable. Clearly, this significantly simplifies program development.

Yet, the mechanics behind such useful feature is not so simple: the price to pay for such dynamical features is that all variables, even such simple as "W1" and "S" defined before, are objects. Thus, they are more complicated than simple types in other programming languages, like C/C++ or FORTRAN. The price to pay is slower execution and larger memory consumption. From the other hand, this also means that you can do a lot using such feature! It should also be noted that some people use the word "value" when they talk about simple types, such as numbers and strings. This is because these objects cannot be changed after creation, i.e., they are immutable.

First, let us find out what can we do with the object "W1". We know that it holds the value 10 (and can hold any value). To find out what can be done with any objects in the JythonShell window is rather simple: type "W1" followed by a dot and press the [Space] key by holding down [Ctrl]:

```
>>> W1. [Ctrl]-[Space]
```

You will see a list of methods attributed to this object. They usually start as __method__, where "method" is some attribute. For example, you will see the method like __str__, which transforms an object of type integer to a string. So, try this

```
>>> SW=str(W1); type(SW)
<type "str">
```

Here we put two Jython statements on one line separated by a semi-column. This, probably, is not very popular way for programming in Jython, but we show this example to illustrate that one can use this syntax as well. In some cases, however, a program readability can significantly benefit from this stile if a code contains many similar and short statements, such as a1=1; a2=2. In this case, the statements have certain similarity and it is better to keep them in one single logical unit. In addition, we will use this style in several examples to make our code snippets short.

The last expression in the above line does not have "=", so Jython assumes that what you want is to redirect the output to the interactive prompt. The method type() tells that SW is a string. As before, you may again look at the methods of this object as:

```
>>> SW. [Ctrl]-[Space]
```

This example shows the list of methods attributed to this object. One can select a necessary method and insert it right after the dot.

In addition to the JythonShell help system, one can discover the attributes of each Jython object using the native Jython method dir():

```
>>> dir(SW)
```

In the following sections we will discuss other methods useful to discover attributes of Java objects.

2.1.1 Numbers as Objects

Numbers in Jython are immutable objects (i.e., cannot be changed after they are created), rather than simple types as in other programming languages (C/C++, Fortran or Java). There are two main types: integers (no fractional part) and floats (with fractional part). Integers can be represented by long values if they are followed by the symbol "L". Try this:

```
>>> Long=20L
<type "long">
```

The only limit in the representation of the long numbers is the memory of Java virtual machine.

Table 2.1 Methods of floating point objects in Python/Jython

Summary of most popular methods for values

$abs(x)$	__abs__	Absolute value
$pow(x, y)$ or $y**x$	__pow__	Raise x to the power y
$-x, +x$	__neg__, __pos__	Negative or positive
$+, -$	__radd__, __rsub__	Add or subtract
$*, /$	__rmul__, __rdiv__	Add or subtract
$x < y, x > y$	__com__	Less or larger. Returns 0 (false) or 1 (true)
$cmp(x, y)$	__com__	Compare numbers. Returns 0 (false) or 1 (true)
$x <= y, x >= y$	–	Comparison: less (greater) or equal.
$x == y, x! = y$	–	Comparison: equal or not equal
$str(x)$	__str__	Convert to a string
$float(x)$	__float__	Convert to float
$int(x)$	__int__	Convert to integer
$long(x)$	__long__	Convert to long

Let us take a look again at the methods of a real number, say 20.2 (without any assignment to a variable).

```
>> 20.2. [Ctrl]-[Space]
```

or, better, you can print them using the method "dir()":

```
>>>   dir(20.2)
```

Again, since there is no assignment, Jython just prints the output of the dir() method directly to the same prompt. Why are they needed and what can you do with them? Some of them are rather obvious and are shown in Table 2.1.

There are more methods designed for this object, but we will not go into further discussion. Just to recall: any number in Jython is an object and you can manipulate with it as with any object to be discussed below. For example, Jython integers are objects holding integer values. This is unlike C++ and Java where integers are primitive types.

Is it good if simple entities, such as numbers are, have properties of objects? For interactive manipulation with a code and fast prototyping, probably we do not care so much, or even can take advantage of this property. But, for numerical libraries, this feature is unnecessary and, certainly, too heavy for high-performance calculations. We will address this issue later in the text.

2.1.2   Formatted Output

In the above examples we have used the `print` command without setting control over the way in which printed values are displayed. For example, in the case of the expression `"1.0/3.0"`, Jython prints the answer with 17 digits after the decimal place!

Obviously, as for any programming language, one can control the way the values are displayed: for this, one can use the `%` command to produce a nicely formatted output. This is especially important if one needs to control the number of decimal places the number is printed to. For example, one can print `1.0/3.0` to three decimal places using the operator `%.3f` inside the string:

```
>>>   print "The answer is %.3f"%(1.0/3)
The answer is 0.333
```

As you can see, Jython replaces the character `"f"` with the variable value that follows the string. One can print more than one variable as shown in the example below:

```
>>>   print "The answer is %.3f and %.1f"% (1.0/3, 2.0/3)
The answer is 0.333 and 0.7
```

One can also use the operator `%` to control the width of the displayed number, so one can make neatly aligned tables. For example, the string `"10.1f"` forces the number to be printed such that it takes up to ten characters. The example below shows how to do this using the new-line character to print the second number on a new line. As one can see, we align this second number with that printed on the first line:

```
>>> print "The answer: %.3f \n   %13.1f"% (1.0/3, 2.0/3)
The answer: 0.333
            0.7
```

2.1.3   Mathematical Functions

To perform mathematical calculations with values, one should use the Jython `math` module which comes together with the standard specification Python programming language. Let us take a look at what is inside this module. First, we have to import this module using the `"import math"` statement:

```
>>> import math
```

Use the usual approach to find the methods of this module:

```
>>> dir(math)
["acos", "asin", "atan", "atan2", "ceil",
 "classDictInit", "cos", "cosh", "e", "exp",
 "fabs", "floor", "fmod", "frexp", "hypot",
 "ldexp", "log", "log10", "modf", "pi", "pow",
 "sin", "sinh", "sqrt", "tan", "tanh"]
```

Most of us are familiar with all these mathematical functions that have the same names in any programming language. To use these functions, type the module name `math` followed by the function name. A dot must be inserted to separate the module and the function name:

```
>>> math.sqrt(20)
4.47213595499958
```

As before for JythonShell, one can pick up a necessary function as:

```
>>> math.    [Ctrl]-[Space]
```

It should be noted that, besides functions, the `math` module includes a few well-known constants: π and e:

```
>>> print "PI=", math.pi
PI= 3.141592653589793
>>> print "e=", math.e
e= 2.718281828459045
```

If you have many mathematical operations and want to make a code shorter by skipping the "`math`" attribute in front of each function declaration, one can explicitly import all mathematical functions using the symbol "`*`":

```
>>> from math import *
>>> sqrt(20)
4.47213595499958
```

2.2 Complex Numbers

Jython has support for complex numbers which come from the Python language specification. One can define a complex number in Python as $x+y\mathbf{j}$, where $\mathbf{j}$ indicates an imaginary part.

```
>>> C=2+3j
>>> type(C)
<type "complex">
```

Once a complex number is defined, one can perform mathematical manipulations as with the usual numbers. For example,

```
>>> 1j*1j
(-1+0j)
```

Mathematical operations with complex numbers can be performed using the "cmath" module, which is analogous to the "math" module discussed above. The example below demonstrates how to calculate hyperbolic cosine of a complex value:

```
>>> import cmath
>>> print cmath.cosh( 2+3j )
(-3.724+0.511j)
```

The output values for the real and imaginary parts in the above example were truncated to fit the page width.

In addition, DMelt libraries provide a comprehensive support for the standard complex number arithmetic operators and the necessary mathematical functions. You can import such libraries to your Jython code using the import statement. Because complex numbers and associated functions are implemented in Java, one can use these libraries in pure Java applications, or other scripting languages on the Java platform.

Complex numbers in DMelt are based on the class Complex from the Java package jhplot.math. Unlike Python, the imaginary part is denoted by the conventional character "i". A complex number is defined by the constructor Complex(real,img), where real is a real part and img is an imaginary part of a complex number. Either part can be zero. Below we show a simple example of how to deal with complex numbers using the jhplot.math package:

```
from jhplot.math import *

a=Complex(10,2)    #  10 + 2i
print a.toString()
b=Complex(20,1)    #  20 + 1i
cc=a.minus(b)      #  a - b
print cc.toString()
cc=a.times(b)      #  a * b
print cc.toString()
```

Listing 2.1 Working with complex numbers

There are many more methods associated with the class `Complex`. Look at the API description of the class `Complex`, or use the DMelt code assist to learn about its methods while working with this class.

Complex numbers are also accessible from third-party Java libraries. For example, Apache Common Math [1] provides complex numbers, as well as complex versions of transcendental functions. Let us consider a simple example:

```
from org.apache.commons.math3.complex import *
a = Complex(2.0, 4.0)    # 2 + 4i
b = Complex(1.0, 2.0)    # 1 + 2i
c=a.add(b)               # add
print c
print c.log()            # natural logarithm
print c.cos()            # cosine
```

Listing 2.2 Complex transcendental functions

The output of this code is shown below:

```
(3.0, 6.0)
(1.90333, 1.1071)
(-199.69, -28.465)
```

Matrices of complex numbers are also supported by the Apache Commons Math and by the Jama package discussed in Sect. 5.1.

2.3 Strings as Objects

Strings are also treated as values since they are immutable. To define a string, one should enclose it in double (") or single (') quotes. Our preference is to use the double quotes since the example code snippets can easily be converted to Java code where the difference between double and single quotes are important. The escape character is a backslash, so one can put a quote character after it. The newline is given by *n* directly after the backslash character. Two strings can be added together using the usual "+" operator.

As mentioned above, an arbitrary value, `val`, can be converted into a string using the method `str(val)`. To convert a string into `int` or `float` value, use the methods `int(str)` or `float(str)`. Below we illustrate several such conversions:

```
>>> i=int("20")
>>> type(i)
<type "int">
>>> f=float("20.5")
>>> type(f)
<type "float">
```

As before, all the methods associated with a string can be found using [Ctrl]-[Space] or the `dir()` method:

```
>>> dir("s")
...
"capitalize", "center", "count","decode",
"encode", "endswith",  "expandtabs", "find",
"index", "isalnum", "isalpha", "isdecimal",
"isdigit", "islower", "isnumeric", "isspace"...
```

(we display only a few first methods). Some methods are rather obvious and do not require explanation. All methods that start from the string `"is"` check for a particular string feature.

Below we list more methods:

`len(str)`	gives the number of characters in the string `str`.
`string.count(str)`	counts the number of times a given word appears in a string.
`string.found(str)`	numeric position of the first occurrence of word in the string.
`str.lower()`	returns a string with all lower case letters.
`str.upper()`	returns a string with all upper case letters.

Strings can be compared using the standard operators: $==, !=, <, >, <=,$ and $>=$.

2.4 Import Statements

There are several ways that can be used to import a Java or Python package. One can use the `"import"` statement followed by the package name. In the case of Python, this corresponds to a file name without the extension `".py"`. The import statement executes the imported file, unlike lower level languages, like C/C++, where the `import` statement is a preprocessor statement. The consequence of this is that the `"import"` statement can be located in any place of your code, as for the usual executable statement. We have seen already how to import the Python package `"math"`. Here is an example illustrating how to import the Java Swing package (usually used to build a GUI):

```
>>> from javax.swing import *
```

In the above example we use the wildcard character "*" to import all packages from Java Swing. In this book, you will see that we use "*" wildcard almost for all examples, since we would like to keep our examples short. This is often considered as a bad style since it "pollutes" the global namespace. However, if you know that the code is not going to be very long and complicated, we should not worry too much about this style.

Let us give another example. Now we would like to import a Java class from some Java package into our program. This is where the major difference with CPython becomes apparent. Java classes can be called in the same style as any Python module: for example, we can import the Java class `javax.swing.JFrame` as simple as this:

```
>>> from javax.swing import JFrame
```

Of course, one can also import all classes of the `javax.swing` package using the "*" wildcard.

Another way to import classes is to use the `"import"` statement without the string `"from"`. For example:

```
>>> import javax.swing
```

In this case, we should use the qualified names inside a program, i.e.,

```
>>> f=javax.swing.JFrame("Hello")
```

Although it takes more typing, we have avoided in this example polluting the global namespace of our code.

2.4.1 Executing Native Applications

Section 1.4.3 has shown that native applications can be run using JythonShell by appending "!" in front of an external command. In addition, one can also use Jython `"os.system"` package to run an external program.

The code below shows how to run an external command. In this example, we bring up the Acroread PDF file viewer (it should be found on the current PATH if this program installed on your system):

```
>>> import os
>>> rc=os.system("acroread")
>>> if rc == 0:
>>> ... print "acroread started successfully"
```

The statement `"if"` checks whether the execution has succeeded or not. We will discuss the comparison tests in the next section. The same operation is equivalent to the command `!acroread` when using the JythonShell.

2.5 Comparison Tests and Loops

2.5.1 The "if-else" Statement

Obviously, as in any programming language, one can use the `"if-else"` statement for decision capability of your code. The general structure of comparison tests is

```
if [condition1]:
     [statements to execute if condition1 is true]
elif [condition2]:
     [statements to execute if condition2 is true]
....

else:
     [rest of the program]
```

The text enclosed in square brackets represents some Jython code. After the line with the statement `"if"`, the code is placed farther to the right using white spaces in order to define the program block. Either space characters or tab characters (or even both) are accepted as forms of indentation. In this book, we prefer two spaces for indentation. It should also be noted that the exact amount of indentation does not matter, only the relative indentation of nested blocks (relative to each other) is important.

The Python language uses indentation instead of braces to structure the code. The indentation is a good Python feature: the language syntax forces to use the indentation that you would have used anyway to make your program readable. Thus even a lousy programmer is forced to write understandable code!

Now let us come back to the comparison tests. The `[condition]` statement has several possibilities for values `"a"` and `"b"` values as shown in Table 2.2.

Let us illustrate this in the example below:

```
>>> a=1; b=2;
>>> if a*b>1:
>>> .. print "a*b>1"
>>> else:
>>> .. print "a*b<=1"
```

In the case if you will need more complex comparisons, use the Boolean operators such as `"and"` and `"or"`:

```
>>> a=1; b=0
>>> if a>0 and b=0:
>>> ..print "it works!"
```

One can also use the string comparisons = (equal) or ! = (not equal). The comparison statements are case sensitive, i.e., `"a" == "A"` is false.

Table 2.2 Most popular Python/Jython comparison tests

Comparison tests	
$a == b$	a is equal to b
$a! = b$	a is not equal to b
$a > b$	a is greater than b
$a >= b$	a is greater than or equal to b
$a < b$	a is less than b
$a <= b$	a is less than or equal to b
$a == b$	a is equal to b
$a! = b$	a is not equal to b

2.5.2   Loops. The "for" Statement

The need to repeat a statement or a code block is an essential feature of any numerical calculation. There is, however, one feature you should be aware of: Python should be viewed as an "interface" language to access libraries, rather than that used for heavy repeated operations like long loops. According to the author's experience, if the number of iterations involving looping over values is larger than several thousands, such part of the code should be moved to an external library to achieve a higher performance and a lower memory usage compared to the Python code operating with loops over objects. In the case of Jython programs, such libraries should be written in Java.

In this section we will be rather brief. One can find more detailed discussion on this topic in any Python textbook.

The simplest loop which prints, say, 10 numbers is shown below:

```
>>> for i in range(10):
>>> ... print i
```

This "for" loop iterates from 0 to 9. Generally, you can increment the counter by any number. For example, to print numbers from 4 to 10 with the step 2, use this example:

```
>>> for i in range(4,10,2):
>>> ... print i
```

2.5.3   The "continue" and "break" Statements

The loops can be terminated using the "break" statement, or some iterations can be skipped using the "continue" statement. All such control statements are rather convenient, since they simplify programming by avoiding various "if" statements which make the Python code difficult to understand. This is illustrated in the example below:

```
>>> for i in range(10):
>>> ... if (i == 4): continue
>>> ... if (i == 8): break
>>> ... print i
```

In this loop, we skip the number 6 and break the loop after the number 8.

2.5.4 Loops. The "while" Statement

One can also construct a loop using the "while" statement, which is more flexible since its iteration condition could be more general. A generic form of such a loop statement is shown below:

```
while CONDITION:
... <Code Block as long as CONDITION is true>
```

Let us give a short example which illustrates the while loop:

```
>>> a=0
>>> while a<10:
>>> ... a=a+1
```

The while loop is terminated when a=10, i.e., when the statement after the "while" statement is false. As before, one can use the control statements discussed above to avoid overloading the execution block with various "if" statements.

One can also create an infinite loop and then terminate it using the "break" statement:

```
>>> a=0
>>> while 1:
>>> ... print "infinite loop!"
>>> ... a=a+1;
>>> ... if a>10:
>>>         break
>>> ... print i
```

In this example, the "break" statement together with the "if" condition controls the number of iterations.

2.6 Collections

Numeric and statistical computations are usually based on object collections, since they have been designed for various repetitious operations on sequential data structures—exactly what we normally do when analyzing data from multiple measurements. In addition, a typical measurement consists of a set of observations which can be conveniently stored in a data container as a single unit.

Unlike other languages, we consider Python collections to be useful mainly for storing and manipulation with other high-level objects, such as collections with a better optimized performance for numerical calculations. In this book, we will use the Jython collections to store sets of DMelt histograms, mathematical functions, Java-based data containers, and so on.

Of course, one can use Jython collections to keep numerical values, but this approach is not going to be very efficient: an analysis of such values will require Python loops that are known to be slow. Second, there are no too many pre-built Jython libraries for object manipulation.

Nevertheless, in many parts of this book, we will use collections that contain numeric values: this is mainly for pedagogical reasons. Besides, we do not care too much about the speed of our example programs when analyzing tens of thousands of events.

2.6.1 Lists

As you may guess, a list is an object which holds other objects, including values. The list belongs to a *sequence*, i.e., an ordered collection of items.

2.6.1.1 List Creation

An empty list can be created using squared brackets. Let us create a list and check its methods:

```
>>> list=[]
>>> dir(list) #  or list. + [Ctrl]+[Space]
```

One can also create a list which contains integer or float values during the initialization:

```
>>> list=[1,2,3,4]
>>> print list
[1, 2, 3, 4]
```

The size of this list is accessed using the len(list) method. The minimum and maximum values are given by the min(list) and max(list) methods, respectively. Finally, for a list which keeps numerical values, one can sum up all the list elements as sum(list).

One can even create a mixed list with numbers, strings, or even other lists:

```
>>> list=[1.0,"test",int(3),long(2),[20,21,23]]
>>> print list
[1.0, "test", 3, 2L, [20, 21, 23]]
```

One can access each element of the list as list[i], where "i" is the element index, $0 < i < \text{len(list)}$. One can select a slice as list[i1:i2], or even select the entire list as list[:]. A slice which selects index 0 through "i" can be written as list[:i]. Several lists can be concatenated using the plus operator "+", or one can repeat the sequence inside a list using the multiplication "*".

As before, one can find the major methods of the list using by pressing [Ctrl] + [Space] keys. Some methods are rather obvious.

To add a new value, use the method append():

```
>>> list.append("new string")
```

A typical approach to fill a list in a loop would be:

```
>>> list=[]
>>> for i in range(4,10,2):
>>> ...     list.append(i)
```

(here, we use Step 2 from 4 to 10). The same code in a more elegant form looks like:

```
>>> list=range(4, 10, 2)
>>> print list
[4, 6, 8]
```

If one needs a simple sequence, say from 0 to 9 with Step 1, this code can be simplified:

```
>>> list=range(10)
>>> print "List from 0 to 9:",list
List from 0 to 9: [0,1,2,3,4,5,6,7,8,9]
```

One can create a list by adding some condition to the range statement. For example, one create lists with odd and even numbers:

```
>>> odd =range(1,10)[0::2]
>>> even=range(1,10)[1::2]
```

Another effective "one-line" approach is to fill a list with values is demonstrated below:

```
>>> import math
>>> list = [math.sqrt(i) for i in range(10)]
```

Here we created a sequence of `sqrt(i)` numbers with $i = 0...9$.

Finally, one can use the `"while"` statement for adding values in a loop. Below we make a list which contains ten zero values:

```
>>> list=[]
>>> while len(list)<10:
>>> ... list.append(0)
```

2.6.1.2 Iteration Over Elements

Looping over a list can be done with the `"for"` statement as:

```
>>> for i in list:
>>> ...print i
```

or calling its elements by their index `"i"`:

```
>>> for i in range(len(list)):
>>> ...print list[i]
```

2.6.1.3 Sorting, Searches, Removing Duplicates

The list can be sorted with the `sort()` method:

```
>>> list.sort()
>>> print list
[1.0, 2L, 3, [20, 21, 23], "new string", "test"]
```

To reverse the list, use the method `reverse()`.

To insert a value, use the `insert(val)` method, while to remove an element, use the `remove(val)` method. One can delete either one element of a list or a slice of elements. For example, to remove one element with the index `i1` of a list use this line of the code: `"del list[i1]"`. To remove a slice of elements in the index range i1–i2, use `"del list[i1:i]"`. To empty a list, use `"del list[:]"`. Finally, use `"del list"` statement to remove the list object from the computer memory.

In should be noted that the list size in the computer memory depends on the number
of objects in the list, not on the size of objects, since the list contains pointers to the
objects, not objects themselves.

Advanced statistical analysis will be considered in Sect. 10.2, where we will show
how to access the mean values, median, standard deviations, moments, etc., of distributions represented by Jython lists.

Jython lists are directly mapped to the Java ordered collection `List`. For example,
if a Java function returns `ArrayList < Double >`, this will be seen by Jython
as a list with double values.

To search for a particular value `"val"`, use

```
>>> if val in list:
>>> ...print "list contains", val
```

For searching values, use the method `index(val)`, which returns the index
of the first matching value. To count the number of matched elements, the method
`count(val)` can be used (it also returns an integer value).

2.6.1.4 Removal of Duplicates

Often, you may need to remove a duplicate element from a list. To perform this task,
use the so-called dictionary collection (will be discussed below). The example to be
given below assumes that a `list` object has been created before, and now we create
a new list (with the same name) but without duplicates:

```
>>> tmp={}
>>> for x in list:
>>> ...tmp[x] = x
>>> list=tmp.values()
```

This is usually considered to be the fastest algorithm (and the shortest). However, this
method works for the so-called hashable objects, i.e., class instances with a "hash"
value which does not change during their lifetime. All Jython immutable built-in
objects are hashable, while all mutable containers (such as lists or dictionaries to be
discussed below) are not. Objects which are instances of user-defined Jython or Java
classes are hashable.

For unhashable objects, one can first sort objects and then scan and compare them.
In this case, a single pass is enough for duplicate removal:

```
>>> list.sort()
>>> last = list[-1]
>>> for i in range(len(list)-2, -1, -1):
>>> ...if last==list[i]:
>>>        del list[i]
>>> ...else:
>>>        last=list[i]
```

The code above is considered to be the second fastest method after that based on the dictionaries. The method above works for any type of elements inside lists.

2.6.1.5 Examples

Lists are very handy for many data analysis applications. For example, one can keep names of input data files which can be processed by your program in a sequential order. Or, one can create a matrix of numbers for linear algebra. Below we will give two small examples relevant for data analysis:

A matrix. Let us create a simple matrix with integer or float numbers:

```
>>> mx=[
...     [1,  2],
...     [3,  4],
...     [5,  6],
...     ]
```

One can access a row of this matrix as mx[i], where "i" is a row index. One can swap rows with columns and then access a particular column as:

```
>>> col=[[x[0] for x in mx], [x[1] for x in mx]]
>>> print col
[[1, 3, 5], [2, 4, 6]]
```

Advanced linear algebra matrix operations using pure Jython approach will be considered in Sect. 5.1.8.

Records with measurements. Now we will show that the lists are very flexible for storing records of data. In the example below we create three records that keep information about measurements characterized by some identification string, a time stamp indicating when the measurement is done and a list with actual numeric data:

```
>>> meas=[]
>>> meas.append(["test1","06-08-2009",[1,2,3,4]])
>>> meas.append(["test2","06-09-2009",[8,1,4,4,2]])
>>> meas.append(["test3","06-10-2009",[9,3]])
```

This time we append lists with records to the list holding all event records. We may note that the actual numbers stored in a separate list can have an arbitrary length (and could also contain other lists). To access a particular record inside the list meas use its indexes:

```
>>> print meas[0]
>>> ["test1", "06-08-2009", [1, 2, 3, 4]]
>>> print meas[0][2]
[1, 2, 3, 4]
```

2.6.2   Tuples

Unlike lists, tuples cannot be changed after their creation, thus they cannot grow or
shrink as the lists. Therefore, they are *immutable*, similar to the values. As the Jython
lists, they can contain objects of any type. Tuples are very similar to the lists and can
be initiated in a similar way:

```
>>> tup=()                      # empty tuple
>>> tup=(1,2,"test",20.0)  # with 4 elements
```

The usual operations that can change the object (such as `append()`), cannot be
applied, since we cannot change the size of this container.

In the case if you need to convert a list to a tuple, use this method:

```
>>> tup=tuple([1,2,3,4,4])
```

Below we will discuss more advanced methods which add more features to manipulations with the lists and tuples.

2.6.3   Dictionaries

Another very useful container for numeric computation and data analysis is the so-
called dictionary. If one needs to store some objects (which, in turn, could contain
other objects, such as more efficiently organized collections of numbers), it would
be rather a good idea to annotate such elements. Or, at least, to have some human-
readable description for each stored element, rather than using an index for accessing
elements inside the container as used for lists or tuples. Such a description, or the
so-called "key", can be used for fast element retrieval from a container.

Dictionaries in Jython (as in Python) are designed for one-to-one relationships
between keys and values. The keys and the corresponding values can be any objects.
In particular, the dictionary value can be a string, numeric value, or even other
collections, such as a list, a tuple, or other dictionary.

Let us give an example with two keys in the form of strings, `"one"` and `"two"`,
which map to the integer values "1" and "2", respectively:

```
>>> dic={"one":1,  "two":2}
>>> print dic["one"]
1
```

In this example, we have used the key `"one"` to access the integer value "1". One
can easily modify the value using the key:

```
>>> dic["one"]=10
```

It should be noted that the keys cannot have duplicate values. Assigning a value to the existing key erases the old value. This feature was used when we removed duplicates from the list in Sect. 2.6.1.3. In addition, dictionaries have no concept of order among elements.

One can print the available keys as:

```
>>> print dic.keys()
```

The easiest way to iterate over values would be to loop over the keys:

```
>>> for key in dic:
>>> ... print key, "corresponds to", dic[key]
```

Before going further, let us rewrite the example given in the previous section when we discussed the lists. This time we will use record identifications as keys for fast retrieval:

```
>>> meas={}
>>> meas["test1"]=["06-08-2009",[1,2,3,4]]
>>> meas["test2"]=["06-09-2009",[8,1,4,4,2]]
>>> meas["test3"]=["06-10-2009",[9,3]]
>>> print meas["test2"]
["06-09-2009", [8, 1, 4, 4, 2]]
```

In this case, one can quickly access the actual data records using the keys. In our example, a single data record is represented by a list with the date and additional list with numerical values.

Let us come back to the description of the dictionaries. Here are a few important methods we should know about:

`dic.clear()`	clean a dictionary;
`dic.copy()`	make a copy;
`has_key(key)`	test, is a key present?;
`keys()`	returns a list of keys;
`values()`	returns a list of values in the dictionary.

One can delete entries from a dictionary in the same way as for the list:

```
>>> del dic["one"]
```

One can sort the dictionary keys using the following approach: Convert them into a list and use the `sort()` method for sorting:

```
>>> people = {"Eve":10, "Tom": 20, "Arnold": 50}
>>> list = people.keys()
>>> list.sort()
>>> for p in list:
>>> ...   print p," is ",people[p]
```

```
Arnold  is    50
Eve  is    10
Tom  is    20
```

2.6.4 Functional Programming

Functional programming in Python/Jython performs various operations on data structures, like lists or tuples. For example, let us create a new list by applying the formula:

$$\frac{b[i] - a[i]}{b[i] + a[i]} \tag{2.1}$$

For each element of two lists, a and b, you would write a code such as:

```
>>> a=[1,2,3]
>>> b=[3,4,5]
>>> c=[]
>>> for i in range(len(a)):
>>> ... c.append( b[i]-a[i] / (a[i]+b[i]) )
```

To circumvent such unnecessary complexity, one can reduce this code to a single line using functional programming:

```
>>> a=[1.,2.,3.]
>>> b=[3.,4.,5.]
>>> c= map(lambda x,y: (y-x)/(y+x),a,b)
>>> print c
[0.5,  0.33,  0.25]
```

The function map creates a new list by applying Eq. (2.1) for each element of the input lists. The statement lambda creates a small anonymous function at runtime which tells what should be done with the input lists (we discuss this briefly in Sect. 2.10). As you can see, the example contains much lesser code and, obviously, programming is done at a much higher level of abstraction than in case with the usual loops over list elements.

To build a new list, one can also use the "math" module. Let us show a rather practical example based on this module: assume we have made a set of measurements, and, in each measurement, we simply count events with our observations. The statistical error for each measurement is the square root of the number of events, in case of counting experiments like this. Let us generate a list with statistical errors from the list with the numbers of events:

```
>>> data=[4,9,25,100]
>>> import math
```

```
>>> errors= map(lambda x: math.sqrt(x),data)
>>> print errors
[2.0, 3.0, 5.0, 10.0]
```

The above calculation requires one line of the code, excluding the standard "import" statement and the "print" command.

Yet, you may not be totally satisfied with the "lambda" function: sometimes one needs to create a rather complicated function operating on lists. Then one can use the standard Python functions:

```
>>> a=[1.,2.,3.]
>>> b=[3.,4.,5.]
>>> def calc(x,y):
>>> ... return (y-x)/(x+y)
>>> c= map(calc,a,b)
>>> print c
[0.5, 0.33, 0.25]
```

The functionality of this code is totally identical to that of the previous example. But this time, the function calc() is the so-called "named" Python function. This function can contain rather complicated logic which may not fit to a single-line "lambda" statement.

One can also create a new list by selecting certain elements. In this case, use the statement filter() which accepts a function with single argument. Such a function must return the logical true if the element should be selected. In the example below we create a new list by taking only positive values:

```
>>> a=[-1,-2,0,1,2]
>>> print "filtered:",filter(lambda x: x>0, a)
filtered: [1, 2]
```

As before, the statement "lambda" may not be enough for more complicated logic for element selection. In this case, one can define an external (or named) function as in the example below:

```
>>> a=[-1,-2,0,1,2]
>>> def posi(x):
>>> ... return x > 0
>>> print "filtered:",filter(posi, a)
filtered: [1, 2]
```

Again the advantage of this approach is clear: we define a function posi(), which can arbitrary be complicated, but the price to pay is more coding.

Finally, one can use the function reduce() that applies a certain function to each pair of items. The results are accumulated as shown below:

```
>>> print "accumulate:",reduce(lambda x, y: x+y,[1,2,3])
>>> accumulate: 6
```

The same functional programming methods can be applied to the tuples.

2.7   Java Collections in Jython

It was already said that the concept of collections is very important for numeric
computation and data analysis, since "packing" multiple records with information
into a single unit is a very common task.

There are many situations when it is imperative to go beyond the standard Python-
type collections implemented in Jython. The strength of Jython is in its complete
integration with Java, thus one can call Java collections to store data. The approach
of accessing Java collection is particularly useful when working with other scripting
language, such as Groovy, JRuby, and BeanShell. Yes, the power of Java is in your
hands!

To access Java collections, first you need to import the classes which implement
Java collections from the standard Java package java.util. Java collections usually have the class names starting with capital letters, since this is the standard
convention for class names in the Java programming language. With this observation
in mind, there is little chance for mixing Python collections with Java classes during
the code development. In this section, we will consider several collections from the
Java platform.

2.7.1   List. An Ordered Collection

To build an ordered collection which contains duplicates, use the class List from
the Java package java.util. Since we are talking about Java, one can check what
is inside of this Java package as:

```
>>> from java.util import *
>>> dir()
[ .. "ArrayList","Currency","Date",List,Set,Map]
```

Here we printed only a few Java classes to fit the long list of classes to the page
width. One can easily identify the class ArrayList, a class usually used to keep
elements in a list. Check the type of this class and its methods using either dir()
or the JythonShell code assist:

```
>>> from java.util import *
>>> jlist=ArrayList()
>>> type(jlist)
<type "java.util.ArrayList">
>>> dir(jlist):
[... methods ...]
>>> jlist. #  [Ctrl]+[Space] to show help
```

As you can see, the type() method indicates that this is a Java instance, so we have
to use the Java methods of this instance for further manipulation.

It was already discussed how to access a complete Java API using the DMelt java
libraries: one can use help.doc(obj) to start a browser with Java documentation
of the class that corresponds to the object obj. Let us add elements to this list and
print them:

```
>>> from java.util import *
>>> jlist=ArrayList()
>>> from jhplot import *
>>> help().doc(jlist) # a web browser with Java API
```

Now let us add elements to this list and print them:

```
>>> e=jlist.add("test")
>>> e=jlist.add(1)
>>> jlist.add(0,"new test")
>>> e=jlist.add(2)
>>> print jlist
[new test, test, 1, 2]
>>> print jlist.get(0)
new test
>>> print jlist.toArray()
array(java.lang.Object,["new test", "test", 1, 2])
```

You may notice that when we append an element to the end of this list, we assign the
result to the variable "e". In Jython, it returns "1" for success (or true for Java).
We also can add an object obj at the position characterized with the index i using
the method add(i,obj). Analogously, one can access elements by their integer
positions. For example, one can retrieve an object back using the method get(i).
The list of elements can be retrieved in a loop exactly as we usually do for the Jython
lists. Let us show a more complete example:

```
from java.util import *

jlist=ArrayList()
for i in range(100): # append 100 integers
    jlist.add(i)
print jlist.size()
```

```
jlist.set(0,100) # replace at 0 position
s=jlist
print type(s)

newlist=jlist.subList(0,50) # range 0-50
for j in newlist:
    print j
```

Listing 2.3 Java list example

Run the above code and make sense of its output.

Probably, there are no strong reasons to use the "List" class while working with
Jython, since the Python lists discussed in the previous section should be sufficient
for almost any data analysis task. However, it is possible that you will need to use Java
classes in order to integrate your application into a Java application, i.e., converting
your code into a pure Java codding, or when you need to convert your program into
other scripting languages on the Java platform, such as Groovy or JRuby.

2.7.1.1 Sorting Java Lists

One can do several manipulations with the List using the Java Collection class.
Below we show how to sort a list using the natural ordering of its elements, and how
to reverse the order:

```
>>> from java.util import *
>>> jlist=ArrayList()
>>> jlist.add("zero"); jlist.add("one"); jlist.add("two")
>>> Collections.sort(jlist)
>>> print jlist
>>> [one, two, zero]
>>> Collections.reverse(jlist)
>>> print jlist
>>> [zero, two, one]
```

The next question is how to sort a list with more complicated objects, using some
object attribute for sorting. Consider a list containing a sequence of other lists as in
the case shown below:

```
>>> from java.util import *
>>> jlist=ArrayList()
>>> jlist.add([2,2]); jlist.add([3,4]); jlist.add([1,1])
>>> print jlist
[[2, 2], [3, 4], [1, 1]]
```

Here there is a small problem: How can we tell to the method sort() that we want
to perform a sorting using a first (or second) item in each element list? Or, more
generally, if each element is an instance of some class, how can we change ordering
objects instantiated by the same class?

One can do this by creating a small class which implements the `Comparator` interface. We will consider Jython classes in Sect. 2.11, so at this moment just accept this construction as a simple prescription that performs a comparison of two objects. The method `compare(obj1,obj2)` of this class compares objects and returns a negative value, zero, or a positive integer value depending on whether the object is less than, equal to, or greater than the specified object. Of course, it is up to you to define how to perform such object comparison. For the above example, each object is a list with two integers, so one can easily prototype a function for object comparison. Let us write a script that orders the list in increasing order using the first element of each list:

```
from java.util import *

jlist=ArrayList()
jlist.add([2,2]); jlist.add([3,4]); jlist.add([1,1])

class cmt(Comparator):
    def compare(self, i1,i2):
        if i1[0]>i2[0]: return 1
        return 0

Collections.sort(jlist,cmt())
print jlist
```

Listing 2.4 Sorting Java lists

After running this script, all elements will be ordered and the print method displays `[[1, 1],[2, 2],[3, 4]]`.

We will leave the reader here. One can always find further information about Java lists in any textbook dedicated to the Java programming language.

2.7.2 Set. A Collection Without Duplicate Elements

The `Set` container from the package `java.util` is a Java collection that cannot contain duplicate elements. Such a container can be created using general-purpose implementations based on the `HashSet` class:

```
>>> from java.util import *
>>> s=HashSet()
>>> e=s.add("test")
>>> e=s.add("test")
>>> e=s.add(1)
>>> e=s.add(2)
>>> print s
[1, 2, test]
```

As you can see from this example, the string `"test"` is automatically removed from the collection. Operations with the Java sets are exactly the same as those with the `ArrayList`. One can loop over all elements of the set collection using the same method as that used for the `ArrayList` class, or one can use a method by calling each element by its index:

```
>>> for i in range(s.size()):
>>> ...print s[i]
```

As in the case with Java lists, you may face a problem when you go beyond simple items in the collection. If you want to store complicated objects with certain attributes, what method should be used to remove duplicates? You can do this as well but make sure that instances of the class used as elements inside the Java set use hash tables (most of them do). In the case of the example shown in Sect. 2.7.1.1, you cannot use `HashSet` since lists are unhashinable. But with tuples, it is different: tuples have hash tables, so the code snippet below should be quite healthy:

```
>>> from java.util import *
>>> s=HashSet()
>>> e=s.add( (1,2) )
>>> e=s.add( (2,4) )
>>> e=s.add( (1,2) )
>>> print s
[(2, 4), (1, 2)]
```

As you can see, the duplicate entry `(1,2)` is gone from the container. If you need to do the same with Python lists, convert them first into tuples as shown in Sect. 2.6.2.

2.7.3 SortedSet. Sorted Unique Elements

Next, why not to keep all our elements in the Java data container in a sorted order, without calling an additional sorting method each time we add a new element? The example below shows the usage of the `SortedSet` Java class:

```
>>> from java.util import *
>>> s=TreeSet()
>>> e=s.add(1)
>>> e=s.add(4)
>>> e=s.add(4)
>>> e=s.add(2)
>>> print s
[1, 2, 4]
```

as you can see, the second value "4" is automatically removed and the collection appears in the sorted order.

2.7.4   Map. Mapping Keys to Values

As it is clear from the title, now we will consider the Java `Map` collection which maps keys to specific objects. This collection is analogous to the Python dictionary, see Sect. 2.6.3. Thus, a map cannot contain duplicate keys as we have learned from the Python/Jython dictionaries.

Let us build a map collection based on the `HashMap` Java class:

```
>>> from java.util import *
>>> m=HashMap()
>>> m.put("a", 1)
>>> m.put("b", 2)
>>> m.put("c", 3)
>>> print m
{b=2, c=3, a=1}
```

Now you can see that Java maps have the same functionality as the Jython dictionaries. As for any Java collection, the size of the `Map` is given by the method `size()`. One can access the map values using the key:

```
>>> print m["a"]
1
```

Similar to the lists, one can print all keys in a loop:

```
>>> for key in m:
>>> ... print key, "corresponds to", m[key]
b corresponds to 2
c corresponds to 3
a corresponds to 1
```

Here we print all keys and also values corresponding to the keys.

2.7.5   Java Map with Sorted Elements

This time we are interested in a map with sorted keys. For this one should use the class `TreeMap` and the same methods as for the `HashMap` class discussed before:

```
>>> from java.util import *
>>> m=TreeMap()
>>> m.put("c", 1)
>>> m.put("a", 2)
>>> m.put("b", 3)
>>> print m
{a=2, b=3, c=1}
```

Compare this result with that given in the previous subsection. Now the map is sorted
using the keys.

2.7.6 Real-Life Example: Sorting and Removing Duplicates

Based on the Java methods discussed above, we can do something more complicated.
In many cases, we need to deal with a sequence of data records. Each record, or event,
can consist of strings, integers, and real numbers. So we are dealing with lists of lists.
For example, assume we record one event and make measurements of this event by
recording a string describing some feature and several numbers characterizing this
feature. Such example was already considered in Sect. 2.6.1.5.

Assume we make many such observations. What we want to do at the end of
our experiment is to remove duplicates based on the string with a description, and
then sort all the records (or observations) based on this description. This looks like
a real project, but not for Jython! The algorithm shown below does everything using
several lines of the code:

```
from java.util import *

data=ArrayList()
data.add( ["star",1.1,30] )
data.add( ["galaxy",2.2,80] )
data.add( ["galaxy",3.3,10] )
data.add( ["moon",4.4,50] )

map=TreeMap()
for row in data:
    map.put(row[0],row[1:])

data.clear()
for i in map:
    row=map[i]
    row.insert(0,i)
    data.add(row)

print data
```

Listing 2.5 Sorting and removing duplicates

Let us give some explanations. Fist, we make a data record based on the list "data"
which holds all our measurements. Then we build the TreeMap object, and use its
first element to keep the description of our measurement in the form of "key-value".
The rest of our record is used to fill the map values (see row[1:]). As you already
know, when we fill the TreeMap object, we remove duplicate elements and sort the
keys automatically. Once the map is ready, we remove all entries from the list and

refill it using a loop over all the keys (which are now ordered). Then we combine the key value to form a complete event record. The output of the script is given below:

```
[["galaxy",3.3,10],["moon",4.4,50],["star",1.1,30]]
```

We do not have extra record with the description "galaxy" and, expectedly, all our records are appropriately sorted.

2.8 Random Numbers

A generation of random numbers is an essential phase in scientific programming. Random numbers are used for estimating integrals, generating data encryption keys, data interpretation, simulation, and modeling complex phenomena. In many examples of this book, we will simulate random data sets for illustrating numerical and statistical techniques.

Let us give a simple example that shows how to generate a random floating point number in the range [0, 1] using the Python `random` module:

```
>>> from random import *
>>> r=Random()
>>> r.randint(1,10) # a random number in range [0.10]
```

Since we do not specify any argument for the `Random()` statement, a random seed from the current system time is used. In this case, every time you execute this script, a new random number will be generated.

In order to generate a random number predictably for debugging purpose, one should pass an integer (or long) value to an instance of the `Random()` class. For the above code, this may look as: `r=Random(100L)`. Now the behavior of the script above will be different: every time when you execute this script, the method `randint(1,10)` will return the same random value, since the seed value is fixed.

Random numbers in Python can be generated using various distributions depending on the applied method:

```
>>> r.betavariate(a,b)      # Beta distribution (a>0,b>0)
>>> r.expovariate(lambda)   # Exponential distribution
>>> r.gammavariate(a, b)    # Gamma distribution.
>>> r.gauss(m,s)            # Gaussian distribution
>>> r.lognormvariate(m,s)   # Log normal distribution
>>> r.normalvariate(m,s)    # Normal distribution
>>> r.randint(min,max)      # int in range [min,max]
>>> r.random()              # in range [0.0, 1.0)
>>> r.uniform(min,max)      # real number in [min,max]
```

In the examples above, `"m"` denotes a mean value and `"s"` represents a standard deviation for the output distributions.

One can reseed the random numbers, and obtain and reset the internal state of random number generator using the methods:

```
>>> r.seed(i)              # set seed (i integer or long)
>>> state=r.getstate()  # current internal state
>>> r.setstate(state)    # restores internal state
```

Note that if "i" in the method `seed(i)` is omitted, the current system time is used to initialize the generator.

Random numbers are also used for manipulations with Python/Jython lists. One can randomly rearrange elements in a list as:

```
>>> list=[1,2,3,4,5,6,7,8,9]
>>> r.shuffle(list)
>>> print list
[3, 4, 2, 7, 6, 5, 9, 8, 1] # random list
```

One can pick up a random value from a list as:

```
>>> list=[1,2,3,4,5,6,7,8,9]
>>> r.choice(list)   #  get a random element
```

That is all it takes. Similarly, one can get a random sample of elements as:

```
>>> list=[1,2,3,4,5,6,7,8,9]
>>> print r.sample(list,4) # random list
>>> [4, 2, 3, 6]
```

Of course, the printed numbers will be different in your case.

This was a small introduction to the topic of random numbers using the Python language. DMelt supports a vast number of Java classes designed to generate random numbers using predefined distributions. These classes can be used with Java, as well as with all scripting languages on the Java platform. We will come back to this topic in Sect. 10.3.

2.9 Time Module

The time module is rather popular due to several reasons. First, it is always a good idea to find our the current time. Second, it is an essential module for more code-related tasks, such as optimization and benchmarking analysis programs or their parts. Let us check the methods of the module `time`:

```
>>> import time
>>> dir(time)    # check what is inside
```

```
["__doc__", "accept2dyear", "altzone", "asctime",
 "classDictInit", "clock", "ctime", "daylight",
 "gmtime", "locale_asctime", "localtime", "mktime",
 "sleep", "strftime", "struct_time", "time",
 "timezone", "tzname"]
```

You may notice that there is a method called __doc__. This looks like a method
to keep the documentation for this module. Indeed, by printing the documentation
of this module as

```
>>> print time.__doc__
```

you will see a rather comprehensive description. Let us give several examples:

```
>>> time.time()   # time in seconds since the Epoch
>>> time.sleep()  # delay for a number of seconds
>>> t=time.time()
>>> print t.strftime("4-digit year: %Y, 2-digit year: \
                      %y, month: %m, day: %d")
```

The last line prints the current year, the month and the day with explanatory annotations.

To find the current day, the easiest is to use the module datetime:

```
>>> import datetime
>>> print "The date is", datetime.date.today()
>>> The date is 2015-11-14
>>> t=datetime.date.today()
>>> print t.strftime("4-digit year: \
    %Y, 2-digit year: %y, month: %m, day: %d")
```

To force a program to sleep a certain number of seconds, use the sleep()
method:

```
>>> seconds = 10
>>> time.sleep(seconds)
```

2.9.1 Benchmarking

For tests involving benchmarking, i.e., when one needs to determine the time spent
by a program or its part on some computational task, one should use a Jython module
returning high-resolution time. The best is to use the module clock() which returns
the current processor time as a floating point number expressed in seconds. The

resolution is rather dependent on the platform used to run this program but, for our
benchmarking tests, this is not too important.

To benchmark a piece of code, enclose it between two `time.clock()` statements as in this example:

```
>>> start = time.clock();  \
    [SOME CODE FOR BENCHMARKING];  \
    end = time.clock()
>>> print "The execution of took (sec) =", end-start
```

Let us give a concrete example: we will benchmark the creation of a list with integer
numbers. For benchmarking in interactive mode, we will use the `exec()` statement.
This code benchmarks the creation of a list with the integer numbers from 0 to 99999:

```
>>> code="range(0,100000)"
>>> start=time.clock();List=exec(code);end=time.clock()
>>> print "Execution of the code took (sec)=",end-start
Execution of the code took (sec) = 0.003
```

Alternatively, one can write this as:

```
>>> List=[]
>>> code="for x in range(0,100000): List.append(x)"
>>> start=time.clock();exec(code);end=time.clock()
>>> print "Execution of the code took (sec)=",end-start
```

2.10 Python Functions and Modules

Jython supports code reuse via functions and classes. The language has many built-in
functions that can be used without calling the `import` statement. For example, the
function `dir()` is a typical built-in function. But how one can find out which functions have already been defined? The `dir()` itself cannot display them. However,
one can use the statement `dir(module)` to get more information about a particular
module. Try to use the lines:

```
>>> import __builtin__
>>> dir(__builtin__)
..."compile", "dict", "dir", "eval" ..
```

This prints a rather long list of the built-in functions available for immediate use (we
show here only a few functions).

Other ("library") functions should explicitly be imported using the `import` statement. For example, the function `sqrt()` is located inside the package `"math"`, thus it should be imported as `"import math"`. One can list all functions of a particular package by using the `dir()` function as shown in Sect. 2.1.3.

It is always a good idea to split your code down into a series of functions, each of which would perform a single logical action. The functions in Jython are declared using the statement `def`. Here is a typical example of a function that returns `(a-b)/(a+b)`:

```
>>>def func(a,b):
>>>   ... "function"
>>>   ... d=(a-b)/(a+b)
>>>   ... return d
>>>print func(3.0,1.0)
0.5
>>> print func.__doc__
function
```

To call the function `func()`, a comma-separated list of argument values is used. The `"return"` statement inside the function definition returns the calculated value back and exits the function block. If no return statement is specified, then `"None"` will be returned. The above function definition contains a string comment `"function"`. A function comment should always be on the first line after the `def` attributed. One can print the documentation comment with the method `__doc__` from a program from which the function is called.

One can also return multiple values from a function. In this case, put a list of values separated by commas; then a function returns a tuple with values as in this example:

```
>>>def func(a,b,c=10):
>>>   ... d1=(a-b)/(a+b)
>>>   ... d2=(a*b*c)
>>>   ... return d1,d2
>>>print func(2,1)
(0, 20)
>>> >print func(2.,1.0)
(0.5, 30.0)
```

The example shows another features of Python functions: the answer from the function totally depends on the type of passed argument values. The statement `func(2,1)` interprets the arguments as integer values, thus the answer for `(a-b)/(a+b)` is zero (not the expected 0.5 as in case of double values). Thus, Python functions are *generic* and any type can be passed in.

One can note another feature of the above example: it is possible to omit the parameter and use default values specified in the function definition. For the above example, we could skip the third argument in the calling statement, assuming `c=10` by default.

All variable names assigned to a function are local to that function and exist only inside the function block. However, you may use the declaration `"global"` to force a variable to be common to all functions.

```
>>>def func1(a,b):
>>>  ... global c
>>>  ... return a+b+c
>>>def func2(a,b):
>>>  ... global c
>>>  ... c=a+b
>>>
>>>print func2(2,1)
None
>>>print func1(2,1)
6
```

Thus, once the global variable `"c"` is assigned a value, this value is propagated to other functions in which the `"global"` statement was included. The second function does not have the `"return"` statement, thus it returns `"None"`.

We should note that a function in Python/Jython can call other functions, including itself.

You can also create an anonymous function at runtime, using a construct called `"lambda"` discussed in Sect. 2.6.4. The example below shows two function declarations with the same functionality. In one case, we define the function using the standard ("named") approach, and the second case uses the `"lambda"` anonymous declaration:

```
>>> def f1 (x): return x*x
>>> print f1(2)
4
>>> f1=lambda x: x*x
>>> print f1(2)
4
```

Both function definitions, `f1` and `f2` do exactly the same operation. However, the "lambda" definition is shorter.

It is very convenient to put functions in files and use them later in your programs. A file containing functions (or any Jython statement!) should have the extension `".py"`. Usually, such a file is called a "module". For example, one can create a file `"Func.py"` and put these lines:

```
def func1(a,b):
    "My function 1"
    global c
    return a+b+c
```

```
def func2(a,b):
    "My function 2"
    global c
    c=a+b
```

Listing 2.6   File "Func.py"

This module can be imported into other modules. Let us call this module from the JythonShell prompt with the following commands:

```
>>> import Func
>>>print Func.func2(2,1)
None
>>>print Func.func1(2,1)
6
```

We can access functions exactly as if they are defined in the same program, since the `import` statement executes the file "`Func.py`" and makes the functions available at runtime of your program.

Probably, we should remind again that one can import all functions with the statement "`from Func import *`", as we usually do in many examples of this book. In this case, one can call the functions directly without typing the module name.

Another question is where such modules should be located? How can we tell Jython to look at particular locations with module files? This can be done by using a predefined list `sys.path` from the "`sys`" module. The list `sys.path` contains strings that specify the location of Jython modules. One can add a module location using the `append()` method: in this example, we added the location "`/home/lib`" and printed out all directories containing Jython modules:

```
>>> import sys
>>> sys.path.append("/home/lib")
>>> print sys.path
```

Here we have assumed that we put new functions in the directory "`/home/lib`".

Now we are equipped to go further. We would recommend to read any book about Python or Jython to find more detail about modules and functions.

2.11   Python Classes

As for any object-oriented language, one can define a Python/Jython class either inside a module file or inside the body of a program. Moreover, one can define many classes inside a single module.

Classes are templates for creation of objects. Class attributes can be hidden, so one can access the class itself only through the methods of the class. Any Python book or tutorial should be fine in helping to go into the depth of this subject.

A Python class is defined as:

```
>>> class ClassName[args]:
>>> ... [code block]
```

where [code block] indicates the class body and bounds its variables and methods.

The example below shows how to create a simple class and how to instantiate it:

```
>>> class Func:
>>> ... "My first class"
>>> ... a="hello"; b=10
>>>
>>> c=Func()
>>> print c.a, c.b
hello 10
```

The class defined above has two public variables, a and b. We create an instance of the class "Func" and print its public attributes, which are the variables of the type string and integer. As you can see, the class instance has its own namespace. The members of the namespace are accessible with the dot. As for functions and modules, classes can (and should) have documentary strings.

The created instance has more attributes that can be shown as a list using the built-in function dic(): Try this line:

```
>>> dir(Func)
["__doc__", "__module__", "a", "b"]
```

The command displays the class attributes and the attributes of its class base. Note that one can also call the method dir(obj), where "obj" is an instance of the class (c in our example), rather than explicitly using the class name.

But what about the attributes that start from the two leading underscores? In the example above, both variables, a and b, are public, so they can be seen by a program that instantiates this class. In many cases, one should have private variables seen by only the class itself. For this, Python has a naming convention: one can declare names in the form __Name (with the two leading underscores). Such convention offers only the so-called "name mangling" which helps to discourage internal variables or methods from being called from outside a class.

In the example above, the methods with two leading underscores are private attributes generated automatically by Python during class creation. The variable __doc__ keeps the comment line which was put right after the class definition, and the second variable __module__ keeps a reference to the module where the class is defined.

```
>>>  print  c.__doc__
My  first  class
>>>  print  c.__module__
None
```

The last call returns `"None"` since we did not put the class in an external module file.

2.11.1 Initializing a Class

The initialization of a Python class can be done with the __init__ method, which takes any number of arguments. The function for initialization is called immediately after creation of the instance:

```
>>>  class  Func():
>>>      "My  class  with  initialization"
>>>      def  __init__(self,  filename=None):
>>>        self.filename=filename
>>>      def  __del__(self):
>>>        #  some  close  statement  goes  here
>>>      def  close(self):
>>>        #  statement  to  release  some  resources
```

Let us take a closer look at this example. You may notice that the first argument of the __init__ call is named as self. You should remember this convention: every class method, including __init__, is always a reference to the current instance of the class.

After a class instance is initialized and the associated resources are allocated, make sure they are released at the end of a program. This is usually done with the __del__ method which is called before Jython garbage collector deallocates the object. This method takes exactly one parameter, self. It is also a good practice to have a direct cleanup method, like close() shown in this example. This method can be used, for example, to close a file or a database. It should be called directly from a program which creates the object. In some cases, you may wish to call close() from the __del__ function, to make sure that a file or database was closed correctly before the object is deallocated.

2.11.2 Classes Inherited from Other Classes

In many cases, classes can be inherited from other classes. For instance, if you have already created a class `"exam1"` located in the file `"exam1.py"`, you can use this class to build a new ("derived") class as:

```
>>> from exam1 import exam1
>>> class exam2(exam1):
>>>    ... [class body]
```

As you can see, first we import the ancestor class "exam1", and then the ancestor of the class is listed in parentheses immediately after the class name. The new class "exam2" inherits all attributes from the "exam1" class. One can change the behavior of the class "exam1" by simply adding new components to "examp2" rather than rewriting the existing ancestor class. In particular, one can overwrite methods of the class "exam1" or even add new methods.

2.11.3 Java Classes in Jython

The power of Jython, a Java implementation of the Python language, becomes clear when we start to call Java classes using Python syntax. Jython was designed as a language that can create instances of Java classes and has access to any method of such Java class.

This is exactly what we are doing to do while working with the DMelt libraries. The example below shows how to create the Java Date object from java.util and use its methods:

```
>>> from java.util import Date
>>> date=Date()
>>> date.toGMTString()
"09 Jun 2009 03:48:17 GMT"
```

One can use the code assist to learn more about the methods of this Java class. Remember, one can simply type the object name followed by a dot and use [Ctrl]+[Space] in JythonShell for help. Similarly, one can call dir(obj), where obj is an object that belongs to the Java platform. When working with the DMelt IDE code editor, use a dot and the key [F4].

In this book, we will use Java-based numerical libraries from DMelt, thus most of the time we will call Java classes of this package. Also, in many cases, we call classes from the native Java platform. For example, the AWT classes "Font" and "Color" are used by many DMelt objects to set fonts and colors. For example, Sect. 3.3.1 shows how to build a Java instance of graphical canvas based on the Java class HPlot.

2.11.4 Not Covered Topics

In this book, we will try to avoid going into the depths of Python classes. We cannot cover here many important topics, such as inheritance (the ability of a class to inherit propertied from another class) and abstract classes. We would recommend any Python or Jython textbook to learn more about classes.

As we have mentioned before, we would recommend to develop Java libraries to be linked with Jython, rather than building numerical libraries using pure-Python classes; for the latter approach, you will be somewhat locked inside the Python language specification, plus this may result in slow overall performance of your application. Of course, you have to be familiar with the Java language in order to develop Java classes.

2.12 Parallel Computing and Threads

A Jython program can perform several tasks at once using the so-called threads. A thread allows programs to be parallelizable, thus one can significantly boost their performance using parallel computing on multi-core processors.

Jython provides a very effective threading compared to CPython, since JAVA platform is designed from the ground up to support multithread programming. A multithreading program has a significant advantage in processing large data sets, since one can break up a single task into pieces that can be executed in parallel. At the end, one can combine the outputs. We will consider one such example in Sect. 12.4.

To start a thread, one should import the Jython module "threading". Typically, one should write a small class to create a thread or threads. The class should contain the code to be executed when the thread is called. One can also put an initialization method for the class to pass necessary arguments. In the example below, we create ten independent threads using Jython. Each thread prints integer numbers. We create instances of the class shown above and start the thread using the method start() which executes the method run() of this class.

```
from threading import Thread

class test(Thread):
  def __init__ (self,fin):
    Thread.__init__(self)
    self.fin = fin
  def run(self):
    print "This is thread No="+str (self.fin)

for x in xrange ( 10 ):
    current=test(x)
    current.start()
    print "done!"
```

Listing 2.7 A thread example

Here we prefer to avoid going into detailed discussion of this topic. Instead, we will illustrate the effectiveness of multithreading programs in the following chapters when we will discuss concrete examples.

2.13 Arrays in Jython

This is an important section: here we will give the basics of objects which can be used for effective numerical calculations and storing consecutive values of the same type.

Unfortunately, the Java containers discussed in Sect. 2.7 cannot be used in all cases. Although they do provide a handy interface for passing arrays to Java and DMelt objects to be discussed later, they do not have sufficient number of built-in methods for manipulations.

The standard Python lists can be used for data storage and manipulation. However, they are best suited for general-purpose tasks, such as storing complex objects, especially if they belong to different types. They are rather heavy and slow for numerical manipulations with numbers.

Here we will discuss Python/Jython arrays that can be used for storing a sequence of values of a certain type, such as integers, long values, floating point numbers, etc. Unlike lists, arrays cannot contain objects with different types.

The Jython arrays are directly mapped to Java arrays. If you have a Java function which returns an array of double values, and is declared as `double []` in a Java code, this array will be seen by Jython as an array.

To start working with the arrays, one should import the module `jarray`. Then, for example, an array with integers can be created as:

```
>>> from jarray import *
>>> a=array([1,2,3,4], "i")
```

This array is initialized from the input list `[1,2,3,4]` and keeps integer values, see the input character `"i"` (integer). To create an array with double values, the character `"i"` should be replaced by `"d"`. Table 2.3 shows different choices for array types. The length of arrays is given by the method `len(a)`.

Arrays can be initialized without invoking lists. To create an array containing, say, ten zeros, one can use this statement:

```
>>> a=zeros(10, "i")
```

here, the first argument represents the length of the array, while the second specifies its type.

A new value `"val"` can be appended to the end of an array using the `append (val)` method if the value has exactly the same type as that used during array creation. A value can be inserted at a particular location given by the index `"i"` by

Table 2.3 Characters used to specify the types of Jython arrays

Jython array types

Typecode	Java type
z	Boolean
c	char
b	byte
h	short
i	int
l	long
f	float
d	double

calling the method `insert(i,val)`. One can also append a list to the array by using the method `fromlist(list)`.

The number of occurrences of a particular value `"val"` in an array can be given by the method `count(val)`. To remove the first occurrence of `"val"` from an array, use the `remove(val)` method.

2.13.1 Array Conversion and Transformations

Many Java methods return Java arrays. Such arrays can be converted to Jython arrays when Java classes are called from a Jython script.

Often it is useful to convert arrays to Jython list for easy manipulation. Use the method `tolist()` as below:

```
>>> from jarray import *
>>> a=array([1,2,3,4], "i")
>>> print a.tolist()
[1, 2, 3, 4]
```

One can reverse all elements in arrays using the `reverse()` method. Finally, one can also transform an array into a string applying the `tostring()` method.

There are not too many transformations for Jython arrays: in the following chapters, we will consider another high-level objects which are rather similar to the Jython arrays but have numerous methods for scientific calculations.

2.13.2 Performance Issues

We have already noted that in order to achieve the best possible performance for numeric calculations, one should use the built-in methods, rather than Python language constructs.

Below we show a simple benchmarking test in which we fill arrays with one
million elements. We will consider two scenarios: in one case, we use a built-in
function. In the second case, we use a Python-type loop. The benchmarking test was
done using the time module discussed in Sect. 2.9. The only new component in this
program is the one in which we format the output number: here we print only four
digits after the decimal point.

```
import time
from  jarray import *

start=time.clock()
a=zeros(1000000, "i")
t=time.clock()-start
print "Build-in method (sec)= %.4f" % t

start=time.clock()
a=array([], "i")
for i in range(0,1000000,1):
    a.append(0)
t=time.clock()-start
print "Python loop  (sec) %.4f" % t
```

Listing 2.8 Benchmarking Jython arrays

Run this small script by loading it in the editor and using the "[run]" button. The
performance of the second part, in which integers are sequentially appended to the
array, is several orders of magnitudes slower than for the case with the built-in array
constructor `zeros()`.

Generally, the performance of Jython loops is not so dramatically slow. For most
examples to be discussed later, loops are several times slower than equivalent loops
implemented in built-in functions.

2.13.3 Used Memory

When working with array and other memory-consuming objects, it is important to
find out how much memory used by the Java virtual machine for an application. This
is important for code debugging and optimization. The amount of memory currently
allocated to a process can be found using the standard Java library as in the example
below:

```
>>> from java.lang import Runtime
>>> r=Runtime.getRuntime()
>>> Used_memory = r.totalMemory() - r.freeMemory()
>>> "Used memory in MB = ", Used_memory/(1024*1024)
```

We will emphasize that this only can be done in Jython, but not in CPython which does not have any knowledge about the Java virtual machine.

We remind that if you use the DMelt IDE, one can look at the memory monitor located below the code editor.

2.14 Exceptions in Python

Exception is an unexpected error during program execution. An exception is raised whenever an error occurs.

The Python language handles exceptions using the "try"-"except"-"else" block. Let us give a short example:

```
>>> b=0
>>> try:
>>> ... a=100/b
>>> except:
>>> ... print "b is zero!"
```

Normally, if you do not enclose the expression a=100/b in the "try"-"except" block, you will see the message such as:

```
>>> a=100/b
Traceback (innermost last):
  File "<input>", line 1, in ?
  ZeroDivisionError: integer division or modulo by zero
```

As you can see, the exception in this case is ZeroDivisionError.

When reading files, it is often important to check whether a string can be converted to a float. For this purpose, there is a special ValueError exception. Here is an example of how to convert a string into float. If this is impossible, this example will raise an exception:

```
a="2.0"; b=0
try:
    b=float(a)
except ValueError:
    print "Not a float"
```

Another example of the exceptions is "a file not found" that happens while attempting to open a non-existing file (see the next chapter describing Jython I/O). This exception can be caught in a similar way:

```
>>> try:
>>> ... f=open("filename")
>>> except IOError, e:
>>> ... print e
```

This time the exception is IOError, which was explicitly specified. The variable e contains the description of the error.

Exceptions can be rather different. For example, NameError means unknown name of the class or a function, TypeError defines an operation for incompatible types and so on. One can find more details about the exceptions in any Python manual.

2.15 Input and Output

2.15.1 User Interaction

A useful feature you should consider for your Jython program is interactivity, i.e., when a program asks questions at runtime and a user can enter desired values or strings. To pass a value, use the Python method input():

```
>>> a=input("Please type a number: ")
>>> print "Entered number=",a
```

In this example, the input() method prints the string "Please type a number:" and waits for the user response.

But what if the entered value is not a number? In this case, we should handle an exception as discussed in Sect. 2.14.

If you want to pass a string, use the method raw_input() instead of input().

The above code example works only for the stand-alone Jython interpreter, outside the DMelt IDE. For the DMelt IDE, this functionality is not supported. In fact, you do not need this feature at all: when working with the IDE, you are already using an interactive mode. However, when you run Jython using the system prompt, the operations input() or raw_input() are certainly very useful.

2.15.2 Reading and Writing Files

File handling in Jython is relatively simple. One can open a file for read or write using the open() statement:

```
>>> f=open(name, option)
```

where "name" represents a file name including the correct path, "option" is a string which could be either "w" (open for writing, old file will be removed), "r" (open for reading) or "a" (file is opened for appending, i.e., data written to it is added on at the end). The file can be closed with the close() statement.

Let us read a file "data.txt" with several numbers, with each number being positioned on a new line:

```
>>> f=open("data.txt","r")
>>> s=f.readline()
>>> x1=float(s)
>>> s=f.readline()
>>> x2=float(s)
>>> f.close()
```

At each step, readline() reads a new line and returns a string with the number, which is converted into either a float or integer.

When several numbers are located on one line, and they are separated by a space, then we should split each line into pieces. For example, if we have two numbers separated by white spaces in one line, such as "100 200", we can read this line and then split it as:

```
>>> f=open("data.txt", "r")
>>> s=f.readline()
>>> x=s.split()
>>> print s
["100","200"]
```

As you can see, the variable "x" is a list which contains the numbers in the form of strings. Next, you will need to convert the elements of this list into either float or integer numbers:

```
>>> x1=float( x[0] )
>>> x2=float( x[1] )
```

In fact, the numbers can also be separated by any string, not necessarily by white spaces. Generally, use the method split(str), where "str" is a string used to split the original string.

There is another powerful method—readlines(). It reads all lines of a file and returns them as a list:

```
>>> f=open("data.txt")
>>> for l in f.readlines():
>>> ... print l
```

To write numbers or strings, use the method write(). Numbers should be coerced into strings using the str() method. Look at the example below:

```
>>> f=open("data.txt", "w")
>>> f.write( str(100)+"\n" )
>>> f.write( str(200) )
>>> f.close()
```

here we added a new line symbol, so the next number will be printed on a new line.

One can also use the statement "print" to redirect the output into a file. This can be done with the help of the >> operator. Note that, by default, this operator prints to a console.

Let us give one example that shows how to print ten numbers from zero to nine:

```
>>> f=open("data.txt", "w")
>>> for i in range(10):
>>> ...    print >> f, i
>>> f.close()
```

One can check the existence of the file using the Jython module "os":

```
>>> import os
>>> b=os.path.exists(name)
```

where name is the file name, "b=0" (false in Java) if the file does not exist, and "b=1" (true in Java) in the opposite case.

2.15.3 Input and Output for Arrays

Jython arrays considered in the previous section can be written into an external (binary) file. Once written, one can read its content back to a new array (or append the values to the existing array).

```
>>> from  jarray import *
>>> a=array([1,2,3,4],"i")
>>> f=open("data.txt","w")
>>> a.tofile(f)          # write values to a file
>>> f.close()
>>> # read values
>>> f=open("data.txt","r")
>>> b=array([],"i")
>>> b.fromfile(f,3) # read 3 values from the file
>>> print b
array("i",[1, 2, 3])
```

It should be noted that the method fromfile() takes two arguments: the file object and the number of items (as machine values).

2.15.4 Working with CSV Python Module

The CSV ("Comma-Separated Value") file format is often used to store data structured in a table. It is used for import and export in spreadsheets and databases and to exchange data between different applications. Data in such files are either separated by commas, tabs, or some custom delimiters.

Let as write a table consisting of several rows. We will import the Python module csv and write several lists with values using the code below:

```
import csv
w=csv.writer(open("test.csv", "w"),delimiter=",")
w.writerow(["London", "Moscow", "Hamburg"])
w.writerow([1,2,3])
w.writerow([10,20,30])
```

Listing 2.9 Writing a CSV file

Execute this script and look at the current directory. You will see the file "test.csv" with the lines:

```
London,Moscow,Hamburg
1,2,3
10,20,30
```

This is the expected output: each file entry is separated by a comma as given in the delimiter attribute specified in our script. One can put any symbol as a delimiter to separate the values. The most popular delimiter is a space, tab, semi-column, and the symbol " | ". The module also works for quoted values and line endings, so you can write files that contain arbitrary strings (including strings that contain commas). For example, one can specify the attribute quotechar=" | " to separate fields containing quotes.

In the example below we read a CSV file and, in the case of problems, we print an error message using the Python exception mechanism discussed in Sect. 2.14:

```
import csv

r = csv.reader(open("test.csv", "rb"), delimiter=",")
try:
  for row in r:
    print row
except csv.Error, e:
    print "line %d: %s" % (reader.line_num,e)
```

Listing 2.10 Reading a CSV file

Let us convert our example into a different format. This time we will use a double quote (useful when string contains comma inside!) for each value and tab for value separations. The conversion script is based on the same Python csv module and will look as:

```
import csv
reader=csv.reader(open("test.csv","rb"),delimiter=","
    )
```

```
writer=csv.writer(open("newtest.csv","wb"),\
                  delimiter="\t",\
                  quotechar="'", quoting=csv.QUOTE_ALL
    )
for row in reader:
    writer.writerow(row)
```

Listing 2.11 Converting CSV file

The output file will look as:

```
"London"    "Moscow"    "Hamburg"
"1"         "2"         "3"
"10"        "20"        "30"
```

But what if we do not know which format was used for the file you want to read in? First of all, one can always open this file in an editor to see what it looks like, since CSV files are human readable. One can use the DMelt editor by printing this line in the JythonShell prompt:

```
>>> view.open("newtest.csv", 0   )
```

which opens the file "newtest.csv" in the IDE. Alternatively, one can determine the file format automatically using the Sniffer method for safe opening of any CSV file:

```
import csv
f=open("newtest.csv")
dialect = csv.Sniffer().sniff(f.read(1024))
f.seek(0)
reader = csv.reader(f, dialect)
for row in csv.reader(f, dialect):
        print row
```

Listing 2.12 Reading a CSV file using sniffer

This time we do not use the exception mechanism, since it is very likely that your file will be correctly processed.

We will come back to the CSV file format in the following chapters when we discuss Java libraries designed to read the CSV files.

2.15.5 Saving Objects in a Serialized File

If you are dealing with an object from the Python language specification, you may want to store this object in a file persistently (i.e., permanently), so another application can read it later. In the Python language, one can save and load objects using the pickle module. In Jython on the Java platform, you can do this too. Let us write a list with numbers to the file "data.pic":

```
>>> import pickle
>>> f=open("data.pic","w")
>>> a=[1,2,3]
>>> pickle.dump(a,f)
>>> f.close()
```

One can restore the list back as:

```
>>> import pickle
>>> f=open("data.pic","r")
>>> a=pickle.load(f)
>>> f.close()
```

Note that one cannot save Java objects using the same approach. Also, any object which has a reference to a Java class cannot be saved. We will consider how to deal with such special situations in the following chapters.

2.15.6 Storing Multiple Objects

To store one object per file is not a very useful feature. In many cases, we are dealing with multiple objects that need to be saved in files and restored. Multiple objects can be stored in one serialized file using the shelve module. This Jython module can be used to store anything that the pickle module can handle.

Let us give one example in which we store two Jython objects, a string and a list:

```
>>> import shelve
>>> sh=shelve.open("data.shelf")
>>> sh["describe"]="My data"
>>> sh["data"]=[1,2,3,4]
>>> sh.close()
```

Despite the fact that we specified the file "data.shelf", the above example creates two files, "data.shelf.dir" and "data.shelf.dat" . The first file contains a "directory" with the persistent data. This file is in a human-readable form, so if you want to learn what is stored inside the data file, one can open it and read its keys. For the above example, the file contains the following lines:

```
"describe", (0, 15)
"data", (512, 22)
```

The second file, "data.shelf.dat", contains the actual data in binary form.

One can add new objects to the "shelf" file. In the example below, we add a Jython map to the existing file:

```
>>> import shelve
>>> sh=shelve.open("data.shelf")
>>> sh["map"]={"x1":100,"x2":200}
>>> sh.close()
```

Let us retrieve the information from the shelve storage and print out all saved objects:

```
>>> import shelve
>>> sh=shelve.open("data.shelf")
>>> for i in sh.keys():
>>>   ...print i, " = ",sh[i]
>>> sh.close()
```

The output of this code is:

```
describe  =  My data
data   =   [1, 2, 3, 4]
map   =   {"x2": 200, "x1": 100}
```

Finally, one can remove elements using the usual del method.

As you can see, the "shelve" module is very useful since now one can create a small persistent database to hold different Jython objects.

2.15.7 Using Java for I/O

This section shows how to write and read data by calling Java classes from the java.io package primarily designed for input and output to files. Let us give an example of how to write a list of values into a binary file using the DataOutput Stream Java class. In the example below we also use the Java class Buffered OutputStream to make the output operations to be more efficient. In this approach, data are accumulated in the computer memory buffer, and are only written when the memory buffer is full.

```
from java.io import *
fo=FileOutputStream("test.d")
out=DataOutputStream(BufferedOutputStream( fo ))
list=[1.,2.,3.,4]
for a in list:
    out.writeFloat(a)
out.close()
fo.close()
```

Listing 2.13 Writing data using the java.io package

The output of this example is binary data. The `DataOutputStream` class can write any object with data using appropriate methods. For numeric calculations, the most popular methods are:

- `writeBoolean(val)`—for Boolean values;
- `writeInt(val)`—for integers;
- `writeLong(val)`—for long integers;
- `writeDouble(val)`—for double values;
- `writeFloat(val)`—for float values,

where `val` is an appropriate numeric (or Boolean) value.

Now let us read the stored float numbers sequentially. We will do this in an infinite loop using the `"while"` statement until we reach the end of the file (i.e., until the "end-line" exception is thrown). Then, the `break` statement exits the infinite loop. Since we know that our data are a sequence of float numbers, we use the method `readFloat()`. Generally, one can use these methods to read numeric values:

- `readInt()`—read an integer value;
- `readDouble()`—read a double value (64-bit floating point);
- `readFloat()`—read a float value (32-bit floating point).

Let us consider the following simple example:

```
from java.io import *
fo=FileInputStream("test.d")
inf=DataInputStream(BufferedInputStream(fo))
while 1:
  try:
      f=inf.readFloat()
      print f
  except:
      print "end of file"
      break
inf.close()
fo.close()
```

Listing 2.14 Reading data using the `java.io` package

We will continue the discussion of high-level Java classes for I/O which allows us to store objects or sequences of objects in Chap. 9.

2.15.8   Reading Data from the Network

Files with data may not be available from a local file storage, but they can exist in network-accessible locations. In this case, one should use the module `"urllib2"` which reads data from URLs using HTTP, HTTPS, FTP network protocols. Here is an example of how to read the HTML web page with Jython news:

```
>>> from urllib2 import *
>>> f = urlopen("http://www.jython.org/Project/news.html")
>>> s=f.read()
>>> f.close()
>>> print s
```

This code snippet is very similar to the I/O examples shown above, with only one exception: now we open a file using the `urlopen` statement. The web access should be without authentication. One can always check the response headers as `f.info()`, while the actual URL can be printed using the string `f.geturl()`. As usual, one can also use the method `readlines()` to put all HTML page lines into a Python list.

One can also use a DMelt module for downloading files from the Web. It has one advantage: it shows a progress bar during file retrievals. This will be discussed in Sect. 15.1.1.

If authentication is required during file access, a client should retry the request with the appropriate name and password. The module `"urllib2"` also provides such functionality, but we will refrain from further discussion of this advanced topic.

2.16 Real-Life Example. Collecting Data Files

Here we will consider a rather common data analysis task: we collect all files located in a file system, assuming that all such files have the extension ".dat". The files will be located in the root directory "/home", which is the usual user-home directory on the Linux/UNIX platform. Our files contain numbers, each of which is positioned on a new line. We will persuade the following task: we will try to sum up all numbers in the files and calculate the sum of all numbers inside these files.

A snippet of a module "walker.py" which returns a list of files is given below. The module accepts two arguments: the root directory for scanning and the extension of the files we are interested in. The function builds a list of files with the appended full path. We will call the function `walk()` recursively until all directories are identified:

```
import os
def walker (dir,extension):
  files=[]
  def walk( dir, process):
   for f in os.listdir( dir ):
    fpath = os.path.join( dir, f)
    if os.path.isdir(fpath) and not os.path.islink(fpath):
       walk( fpath, process )
    if os.path.isfile( fpath ):
       if fpath.endswith(extension):
          files.append(fpath)
  walk(dir,files)
  return files
```

Listing 2.15 File "walker.py"

Let us test this module. For this, we will write a small program which: (1) imports the module "walker.py"; (2) lists all descendant files and subdirectories under the specified directory and fills the file list with all files which have the extension ".dat"; (3) then it loops over all files in the list and reads the numbers positioned on every new line; (4) Finally, all numbers are summed up. The code which does all of this is given below:

```
import os
from walker import *
files= walker("/home/",".dat")
sum=0
lines=[]
for file in files:
    ifile = open(file,"r")
    lines=lines+ifile.readlines()
    ifile.close()
    for i in range(len(lines)):
        sum=sum+float(lines[i])
print "Sum of all numbers=", sum
```

Listing 2.16 File collector

The described approach is not the only one. The module which lists all files recursively can look much sorter using the os.walk function:

```
def getFileList(rootdir):
  fileList = []
  for root, subFolders, files in os.walk(rootdir):
    for f in files:
        fileList.append(os.path.join(root,f))
  return fileList

print getFileList("/home/")
```

Listing 2.17 Building a file list

This code builds a list of files in the directory "/home/".

In Sect. 9.12.1 we will show another efficient code based on the DMelt Java class that can also be used in pure Java applications. As in the example above, it builds a list of files recursing into all subdirectories.

The above code can be significantly simplified if we know that all input files are located inside a single directory, thus there is no need to transverse all subdirectories.

```
>>> list=[]
>>> for f in os.listdir("/home/"):
>>>    if not file.endswith(".dat"):   continue
>>>    list.append(f)
```

Finally, there is a simpler approach: import the module "`glob`" and scan all files:

```
>>> import glob
>>> list=glob.glob("/home/*.dat")
```

The asterisk (*) in this code indicates that we are searching for a pattern match, so every file or directory with the extension ".dat" will be put into a list, without recursing further into subdirectories. One can specify other wildcard characters, such as "`/home/data?.dat`", that matches any single character in that position in the name starting from "`data`", Another example: "`/home/*[0-9].dat`" string considers all files that have a digit in their names before the extension ".dat".

Often, in order to process data stored in many files, it is useful to divide a list with file names into several lists with equal number of files in each list. In this way, one can process files in parallel using multiple computers or multiple processors. This task can easily be achieved with the code given below:

```
def splitlist(seq, size):
   newlist = []
   splitsize = 1.0/size*len(seq)
   for i in range(size):
     k1=round(i*splitsize)
     k2=round((i+1)*splitsize)
     newlist.append(seq[int(k1):int(k2)])
     newlist.append(seq[k])
   return newlist
```

Listing 2.18 File list splitter

The code accepts a list of files and an integer value `size` which specifies how many lists need to be generated. The function returns a new list in which each entry represents a list of files. The number of entries in each sublist is roughly equal.

2.17 Using Java for GUI Programming

Undoubtedly, the major strength of Jython is in its natural integration with Java, a language used to build Jython. This opens infinite opportunities for a programmer. Assuming that you had already a chance to look at one of these Java books [2–6], you can start immediately to use Java libraries to write a Jython code.

Below we show a small example of how to write a graphical user interface which consists of a frame, a button, and a text area. While the code still uses the Python syntax, it calls classes from the Java platform.

```
from java.awt import *
from javax.swing import *

fr = JFrame("Hello!")
pa1 = JPanel()
pa2 = JTextArea("text",6,20)
def act(event):
   pa2.setText("Hello, data")

bu=JButton("Hello", actionPerformed=act)
pa1.add(bu)
fr.add(pa1,BorderLayout.SOUTH)
fr.add(pa2,BorderLayout.NORTH)
fr.setDefaultCloseOperation(JFrame.DISPOSE_ON_CLOSE)
fr.pack()
fr.setVisible(True)
```

Listing 2.19 Swing GUI using Jython

In this example, we call Java `swing` components directly, like they are usual Python classes. The main difference with Python is in the class names: Java classes always have names starting with capital letters. This feature can be used to figure out whether the object comes from a Python module, or it was implemented as a Java class.

When comparing this code with Java, one should note several important differences: there is no need to use the `"new"` statement when creating Java objects. Also, there is no need to put a semicolon at the end of each Java method or class declaration. We should also recall that Java Boolean values in Jython can be interpreted as either "1" (true) or "0" (false).

Let us continue with our example. Create a file with the name, "gui.py", copy the lines from the example below, and run this file in the DMelt editor. You will see a frame as shown in Fig. 2.1. By clicking on the button, the message "Hello, data" should be shown.

In the following chapters, we try to follow our general concept: We will avoid a detailed discussion of GUI program development, since a Jython macro already provides sufficiently interactive interface for object manipulations. In this book, we aim to show how to develop data analysis programs for which GUI type of features are less frequent, compared to "macro"-type of programming. Since Jython macros allow manipulations with objects without dealing with low-level features of programming

Fig. 2.1 A Java Swing
frame with a button "Hello"

languages, in a sense, they are already some sort of "user-interfaces". In addition, Jython macros have much greater flexibility than any GUI-driven application, since they can quickly be altered and rerun.

Yet, GUI is an important aspect of our life and we will discuss how to add GUI features to scientific applications in the appropriate chapters.

2.18 Concluding Remarks

This concludes our introduction to the world of Python, Jython, and Java. There is one message I have tried to convey here—a combination of Python and Java gives you an extremely powerful and flexible tool for your research. There are dozens of books written for each language and I would recommend to have some of them on your table if you want to study the topic in-depth. To learn about Jython, you can always pick up a Python book. In several cases, you may look at Jython and Java programming books, especially if you will need to do something very specific and nonstandard using Java libraries. But, I almost guarantee, such situations will be infrequent if you learn how to use the DMelt libraries to be discussed in the following chapters.

References

1. The Apache Common-Math library. http://commons.apache.org/proper/commons-math/
2. Richardson C, Avondolio D, Vitale J, Schrager S, Mitchell M, Scanlon J (2005) Professional Java, JDK 5 edn. Wrox, Birmingham
3. Arnold K, Gosling J, Holmes D (2005) Java(TM) programming language, 4th edn., Java SeriesAddison-Wesley Professional, Boston
4. Flanagan D (2005) Java in a nutshell, 5th edn. O'Reilly Media Inc, Sebastopol
5. Eckel B (2006) Thinking in Java, 4th edn. Prentice Hall PTR, Prentice
6. Bloch J (2008) Effective Java, 2nd edn. The Java Series. Prentice Hall PTR, Prentice

Chapter 3
Mathematical Functions

Functions allow programming code reuse, and thus are essential in any programming language. We have discussed earlier in Sect. 2.10 how to define general-purpose functions in the Python programming language. But this book is about numerical and statistical computations, therefore, we will turn to the question of how to build *mathematical* functions.

In this chapter, we will remind how to construct mathematical functions in Python and then we will discuss how to extend this approach to call Java-implemented mathematical functions using Jython. The latter topic describes the flagship concept of numerical computations used in this book. In addition to simplicity, this approach also allows easy transformation of code examples into Java programs, or other scripting languages, such as Groovy and JRuby, which can directly access Java numeric libraries.

3.1 Python Functions

Before going into the description of Java classes that can be used to build mathematical functions, let us first remind how to create functions using the Python language. A named function is declared with the heading statement def(), the name of the function, parentheses, and finally a colon. In the case of mathematical functions, it is likely you will need to import the Python module "math" before or during creation of functions. The general form of a Python function definition is

```
>>> def FunctionName ( arg1, arg2, .. ) :
>>> ... import math
>>> ... -- math statements --
>>> ...   return value
```

© Springer International Publishing Switzerland 2016   85
S.V. Chekanov, *Numeric Computation and Statistical Data Analysis
on the Java Platform*, Advanced Information and Knowledge Processing,
DOI 10.1007/978-3-319-28531-3_3

In this example, `arg1`, `arg2`, etc., is a comma-separated list of arguments. The arguments could be values, lists, tuples, strings, or Jython/Java objects.

We have discussed earlier in Sect. 2.10 how to build functions in Jython. Many mathematical functions are already defined in the module "`math`", and one can list all such functions using the `dir(math)` statement (after importing this module first). Below we will consider several useful examples illustrating how to construct an arbitrary function:

Absolute value:

```
>>> def abs(x) :
>>> ... "absolute value"
>>> ... if x<0: return -x
>>> ... return x
>>>
>>> print abs(-200)
200
```

Factorial:

```
>>> def factor(x) :
>>> ... "calculate factorial"
>>> ... if x<0:
>>> ...     raise ValueError, "Negative number!"
>>> ... if x<2: return 1
>>> ... return long(x)*factor(x-1)
>>>
>>> print factor(10)
3628800
```

One could build complicated mathematical functions, as shown in this example with two arguments, x and y:

```
>>> def myfun(x,y) :
>>> ... "calculate complicated function"
>>> ... from math import *
>>> ... return cos(x)*sin(y)+(2**x)+y**4
>>>
>>> print myfun(0.3,0.8)
2.3260608626777355
```

Mathematical functions in the Python language can be rather inefficient when they have to be displayed, since any drawing involves loops with multiple calls to the same function but with different arguments. Below we will discuss implementations of mathematical functions in Java numerical libraries included into DMelt, and show how to use them with Jython. Being more efficient and flexible, they are also tightly integrated into the DMelt graphical canvas used for object visualization.

3.2 Functions in DMelt

Now let us turn to functions that are implemented in Java. For functions with one independent variable, we deal with a dependence of one value, say y, on another, usually called x. DMelt includes the F1D class to describe, evaluate, and display such functions. Such a function can be instantiated in Jython as

```
>>> from jhplot import *
>>> f1=F1D("definition")
```

where the string "definition" should be replaced by a mathematical formula. The only independent variable should be specified as x. The definition string can contain any combinations of "+", "−", "*," and "/" operations, parenthesis "()", and predefined mathematical functions. For numerical values, scientific notations with "E" can be used, i.e., a number is split into a mantissa s and exponent p of the form sEp, which is evaluated as $s * 10^p$. The program calculates the formula from left to right, according to the order of operators as shown below (from highest to lowest):

- Expressions within parentheses
- Predefined functions, such as log, cos etc.;
- Power and square root;
- Multiplication and division;
- Addition and subtraction.

In order to change the order of evaluation, the part of the formula to be calculated first should be enclosed in parentheses.

The function definition can contain predefined mathematical functions listed in Table 3.1. The function definition also recognizes the constant π.

To evaluate a function at a fixed point, use the eval(x) method. For evaluation of a list of numerical values, pass a Jython list as one single argument to the eval(x) method. In this case, the eval(x) method returns an array of values y calculated at specific values x from the input list.

Let us give an example that makes this feature clear:

```
from jhplot import *
f1=F1D("x^2+pi*sqrt(x)")
print f1.eval(20)
a=f1.eval([10,20,30])
print a.tolist()
```

Listing 3.1 Creating a F1D function

The execution of this script gives:

```
414.049629462
[109.934, 414.049, 917.207]
```

Table 3.1 Predefined
mathematical functions to be
used together with the `F1D`
Java class

Function	Defined as
x^y	Power
**	Power
%	Modulo (remainder)
abs(x)	Absolute value
acosh(x)	Hyperbolic arc cosine
acos(x)	Arc cosine
asinh(x)	Hyperbolic arc sine
asin(x)	Arc sine
atan2(x, y)	Arc tangent (2 parameters)
atanh(x)	Hyperbolic arc tangent
atan(x)	Arc tangent
ceil(x)	Nearest upper integer
cosh(x)	Hyperbolic cosine
cos(x)	Cosine
exp(x)	Exponential
floor(x)	Nearest lower integer
log10(x)	Logarithm base 10
log(x)	Natural logarithm
sinh(x)	Hyperbolic sine
sin(x)	Sine
sqrt(x)	$\sqrt{x}$
tanh(x)	Hyperbolic tangent
tan(x)	Tangent

You may note that since `eval(d)` returns an array, we converted this array into a list using the `tolist()` method for shorter printout (still we have truncated the output numbers to fit them to the page width).

Functions may have arbitrary parameters. In this case, you should not create "parsed" object of this function during the initialization step. A parsed function means that an object created using the `F1D` class is prepared for fast and efficient numeric evaluation, plotting, and integration, without the need to interpret the string with a function definition. This example shows a function with a parameter "alpha",

```
from jhplot import *
f1=F1D("(2*alpha)/alpha+x",False) # no parsing
f1.setPar("alpha",100)             # substitute alpha=100
f1.parse()                         # create parsed object
print f1.eval([1,2,3])             # calculates f(x=1,2,3)
```

Listing 3.2 Initiation of a `F1D` function without parsing

As you can see, we call `eval(x)` after parsing the function. The same applies for plotting functions, a topic which will be discussed in the following sections. Use sufficiently complex names for the free parameters, otherwise, the substitution may fail.

So far we have considered functions that do not include ranges associated with the abscissa. One can also initialize functions that include a specific range to be used for plotting and numeric evaluations:

```
>>> from jhplot import *
>>> f1=F1D("x^3",1,100)  # range [1,100]
```

When the *x* range is explicitly defined, it will be used for integration and plotting as shown in the following sections. In contrast, functions without defined range will be plotted in the range given by the external methods, such as plotting canvases or methods used for integration.

Creating an F1D function using a string with formula definition is fast, but it lacks flexibility to build complex functions. We will consider how to construct custom functions using Java scripting in Sect. 3.7.

3.2.1 Java Implementation of F1D

The object created in the above examples is an instance of the F1D Java class, thus it does not belong to the Python programming language, although we call it using Python syntax. When we instantiate the object of the class F1D, we call the Java library package `jhplot` stored in the file "jhplot.jar" located in the "lib/system" directory. How can we check this? As usual, call the method `type()`:

```
>>> from jhplot import *
>>> f=F1D("x*exp(x)")
>>> type(f)
<type "jhplot.F1D">
```

Even more: one can also use the Java methods to access the class instance and its name:

```
>>> c=f.getClass()
>>> print c
<type "jhplot.F1D">
>>> print c.getCanonicalName()
jhplot.F1D
>>> print c.getMethods()
... list of all method
```

As you can see, one can use several "get" methods to access the name of this object. This cannot be done if an object belongs to Jython. We can also print all methods associated with this object using the method `getMethods()`. We do not print a very long output from this method to save space.

One can find all the methods associated with this class as explained in Sect. 1.4.10. We remind that one can find help on the `jhplot.F1D` class using several methods:

```
>>> from jhplot import *
>>> f1=F1D("x*x")
>>> f1.              # + press   [F4]  (internal help)
>>> help().doc(f1)  # web-based Java API
```

Since we are dealing with the Java object, one can work with the `F1D` function using either Java code or BeanShell. The only difference is in the syntax of your program. We will discuss this topic later.

You can learn about attributes of functions using the usual `toString()` method that returns a string:

```
>>> from jhplot import *
>>> f1=F1D("x*x")
>>> print f1.toString()
```

This command shows internal structure of the `F1D` function, such as its name, title, number of evaluation points, and the ranges used to show this function.

One can also convert a function in the standard Mathematical Markup Language (MathML) string that can be included for displaying on Web pages. Try this:

```
>>> from jhplot import *
>>> f1=F1D("x*x*cos(x)")
>>> print f1.toMathML()
```

In the following sections, we will learn now to build mathematical functions using several independent variables and how to create functions using Jython classes which will allow the usage of rather complicated logic in the function definition. But before we will give some elements of calculus based on the `F1D` class, focusing mainly on numerical integration and differentiation.

3.2.2 Manipulations with 1D Functions

3.2.2.1 Integration

An `F1D` function can be numerically integrated in a region between `min` and `max` using the `integral(N,min,max)` method, where `N` is the number of points for integration. This method assumes the trapezoid rule for integration.

There are more options in which a particular integration method can be specified. Assuming that a function `f1` is created, one can integrate the function between a minimum value min and a maximum value max using the N number of integration points as

```
>>> f1d.integral(type,N,min,max)
```

where type is a string which defines the type of integration. This string can take the following values:

"gauss4"   using the Gaussian integration (4 points)
"gauss8"   using the Gaussian integration (8 points)
"richardson" using the Richardson extrapolation
"simpson"   using the Simpson's rule.
"trapezium"   using the trapezium rule.

The code below tests different integration algorithms using the methods shown above. We also benchmark the code (see Sect. 2.9.1) and print the time spent by each algorithm (in milliseconds):

```
from jhplot import *
import time

f1=F1D("sin(1.0/x)*x^2")
methods=["gauss4", "gauss8", "richardson",\
         "simpson","trapezium"]
for m in methods:
    start = time.clock()
    d=f1.integral(m,10000,1,10)
    t = time.clock()-start
    print m+" =",d," time (ms)=",t*1000
```

Listing 3.3 Integration of a function

The result of code execution is given below:

```
gauss4      = 49.1203116758  time (ms)= 42.245
gauss8      = 49.1203116758  time (ms)= 32.382
richardson  = 49.1203116758  time (ms)= 12.162
simpson     = 49.1203116758  time (ms)= 14.173
trapezium   = 49.1203116663  time (ms)=  4.234
```

While the time needed for integration could be different for your tests (clearly, this depends on many factors), it appears that the fastest algorithm is that based on the trapezium method.

If a function was initialized in a specific range, one can integrate it in the range as in this example:

```
>>> from jhplot import *
>>> f1d=F1D("x*3",1,100)
>>> print f1d.integral()
25000097.7502
```

As you can see, you do not need to worry about integration ranges when using functions initiated with a specific x range.

3.2.2.2 Differentiation

A function can numerically be differentiated using the method:

```
>>> a=f1d.differentiate(N,min,max)
```

The method returns an array with the result of differentiation. The size of this array is set using an integer value N.

We will again discuss the topic of integration and differentiation using symbolic mathematical calculations in Sect. 6.3.

3.2.2.3 Expand and Factorize

A function can be expanded or converted to a factorized form:

```
>>> from jhplot import *
>>> f=F1D("(10+x)^4")
>>> f.expand()
>>> print f.getName()
10000+4000*x+600*x^2+40*x^3+x^4
```

Let the apposite, i.e., factorize it:

```
>>> f.factorize()
>>> print f.getName()
(10+x)^4
```

3.2.2.4 Simplify

Functions can also be simplified. This method can help in certain situations to keep function definition simpler:

```
>>> from jhplot import *
>>> f1=F1D("10+20+300*x+20*x+1-20")
>>> f1.simplify()
>>> print f1.getName()
11+320*x
```

As discussed before, one can also use free parameters in the function definitions. The `simplify()` method works in this case even without parsing the function.

```
from jhplot import *

f1=F1D("(2*alpha)/alpha+2+beta*x",0)
f1.simplify()
print f1.getName()
f1.setPar("alpha",100)
f1.setPar("beta",100)
f1.parse()                # prepare for numeric evaluation
print f1.getName()
print f1.eval([1,2,3])
```

Listing 3.4 A function with parameters

3.3 Plotting 1D Functions

The remaining issue is how to visualize the `F1D` functions. Below we will learn how to build a canvas suited for plotting of `F1D` objects and discuss the main options for their visualization.

3.3.1 Building a Graphical Canvas

To plot the function with one variable, first we need to create a graphical canvas. This can be done by creating a canvas object of the `HPlot` class:

```
>>> from jhplot import *
>>> c1=HPlot()
```

This creates a canvas with the default title "Canvas" and with the default frame size of 600 × 400 screen pixels. To display the canvas in a pop-up frame, execute the statement:

```
>>> c1.visible(1)
```

It should be reminded that "1" means Boolean "True", while "0" means "False". One can also use the shortcut `visible()` instead of `visible(1)`. If you do not want to pop-up the canvas frame and want to keep it in the computer memory, set the argument of the method `visible()` to zero (or "False").

There are more constructors for this class. For example, one can customize the frame size with the method:

```
>>> c1=HPlot("My Canvas",800,600)
```

which creates a canvas with a frame size of 800 by 600 pixels. One can also resize the canvas using the mouse.

The created canvas can be divided into several plot regions (or "pads"). In this case, use the following initialization:

```
>>> c1=HPlot("Canvas",600,400,iX,iY)
```

which creates a canvas of a size 600 by 400 pixels, but, in addition, the last two numbers are used to make two plot regions inside the canvas frame. The first integer number after 400 shows that we need `iX` plot regions in X, and the second number is used to set the number of pads in Y. For example, after setting `iX=2` and `iY=1`, the above constructor creates two pads positioned horizontally.

A function can be plotted inside the canvas using the `draw(obj)` method, where `obj` is an instance of the `F1D` class. One can navigate to the current region using the method `cd(i1,i2)`, where `i1` and `i2` specify the pad in X and Y. For example, if one needs to plot a function inside the first pad, use:

```
>>> c1.cd(1,1)
>>> c1.draw(f1d)
```

where `f1d` represents an object of the `F1D` function. If the function should be shown on the second region, use

```
>>> c1.cd(1,2)
>>> c1.draw(f1d)
```

By default, the `HPlot` canvas has the range between 0 and 1 for the X or Y axis. One should specify the necessary range using the method:

```
>>> c1.setRangeX(min.max)      # range of X (auto-range for Y)
>>> c1.setRange(0,min.max)     # as before
>>> c1.setRange(x1,x2,y1,y2)   # range for X and Y
```

where `x1` is a minimum value in X, `x2` is a maximum value in X, while `y1` and `y2` are the same but for the Y-axis. Alternatively, one can set "auto-range" using the method `c1.setAutoRange()`.

Before plotting a function, one can show a global title of the entire canvas. This can be done using the method setGTitle(str), where str represents a string. In the simplest case, it accepts one argument: a string with the title text. One can customize the text color and/or the font size as will be shown later. As usual, use the code assist (see Sect. 1.4.10) to learn about possible choices.

In addition to the global title, one can set titles for each drawing pad using the method setName(s), with "s" being a string with some text. We can also annotate X and Y axes. The methods for this are setNameX(s) and setNameY(s), where "s" is an annotation string.

Let us give a more concrete example. First, import the Color class from the Java AWT library and then set appropriate annotations:

```
>>> from java.awt import Color
>>> from jhplot import *
>>> c1.setGTitle("GlobalTitle",Color.red)
>>> c1.setNameX("X axis")
>>> c1.setNameY("Y axis")
>>> c1.setName("Pad title")
>>> c1.visible()        # make it visible
>>> c.setAutoRange()    # set autorange for X and Y
```

All the entries above are self-explanatory. One may add a background color for the canvas using:

```
>>> c1.setBackgroundColor(c)
```

where "c" stands for the Java AWT color. Colors can be set using any of the static methods shown in Table 3.2.

Table 3.2 Colors from the Java AWT library

AWT color	Color
Color.black	Black color
Color.blue	Blue color
Color.cyan	Cyan color
Color.darkGray	Dark gray
Color.gray	Gray color
Color.green	Green color
Color.lightGray	Light gray color
Color.magenta	Magenta color
Color.orange	Orange color
Color.pink	Pink color
Color.red	Red color
Color.white	White color
Color.yellow	Yellow color

A color can also be defined in a more flexible way using the constructor
`Color(r,g,b)`, with the specified red, green, and blue values. Each value must
be in the range [0–1]. Alternatively, one can use the same constructor, but specifying red, green, blue (integer) values in the range [0–255]. One can also specify
a transparency level (or "alpha" value) using the fourth argument. There are other
constructors for Java `Color` class, so please refer to any Java textbook.

To check what colors are available and to define your own colors, use the DMelt
dialog:

```
>>> import utils; utils.ShowColors()
```

The above command brings up a color chooser frame which can be used to select a
custom color using the mouse.

Custom fonts for the legends can be specified using the `Font` class from the same
AWT package:

```
>>> from java.awt import Font
>>> font=Font("Lucida Sans",Font.BOLD, 12)
>>> c1.setLegendFont(font)
```

Here we created a font instance from the font collection existing in your computer
environment. A custom font can be created with the statement `Font("name",
style,size)` from the specified font name, style, and font size. The style can be
either `Font.PLAIN` (simple), `Font.ITALIC` (italic) or `Font.BOLD` (bold).

To find the names of fonts is trickier, but possible. Use the methods `get
AvailableFontFamilyNames` or `getAllFonts`. The lengthy statement
below illustrates how to print out the available fonts in your system:

```
>>> from java.awt import *
>>> e = GraphicsEnvironment.getLocalGraphicsEnvironment()
>>> print e.getAllFonts()
>>> print e.getAvailableFontFamilyNames()
```

When using DMelt, call the predefined command which lists the available fonts:

```
>>> import utils; utils.ShowFonts()
```

Use the above approach to find the necessary fonts to be used for the methods
to display labels and titles. In addition, one can use the GUI-dialogs of the `HPlot`
canvas. For example, double-click on the global title to bring up a setup dialog used
to change the global text, fonts, and colors. Similarly, one can edit the margins of
the pads.

3.3.2 Drawing 1D Functions

Once a canvas is ready, one can plot a 1D function discussed in Sect. 3.2 as:

```
>>> from jhplot import *
>>> c1=HPlot()
>>> c1.visible()
>>> c1.setAutoRange()
>>> f1=F1D("2*exp(-x*x/50)+sin(pi*x)/x", -2.0,5.0)
>>> c1.draw(f1)
```

Obviously, -2.0 and 5.0 specify the range for the abscissa. This range is used to draw
the function. There is an important difference with respect to the functions shown in
Sect. 3.2: now we explicitly define the range for evaluation and drawing. One can set
the range for the function using the methods:

```
>>> f1.setMin(min)
>>> f1.setMax(max)
```

where `min` and `max` are the minimum and maximum values for the abscissa. Naturally, the corresponding "getter" methods `getMin()` and `getMax()` return these
values back. It should be noted that the specified abscissa range does not affect the
method `eval(x)` which determines the value of a function at a certain x value.

If you initialize an `F1D` function without ranges in x, the range of the function
will be determined by plotting the canvas, i.e., after `setRangeX(min,max)` was
called.

Finally, you may need to define the plotting resolution. By default, a function is evaluated at 500 points in the specified abscissa range. One can find this
using the `getPoints()` method. One can change this value using the method
`setPoints(i)`, where "i" is an integer number (should be sufficiently large). You
should remember that setting a large number makes your code slower and requires
more memory for plotting.

In order to plot several functions on the same plotting area, simply repeat
the `draw(obj)` statement for each function or, even simpler, use the method
`draw(obj)` to visualize a Jython list of `F1D` functions as in the example below:

```
from jhplot import *
c1=HPlot()
c1.visible(); c1.setAutoRange()
f1=F1D("2*exp(-x*x/50)+sin(pi*x)/x", -2.0,5.0)
f2=F1D("2*sqrt(x)")
c1.draw([f1,f2])
```

Listing 3.5 Plotting two functions on the same canvas.

Note that the first function includes the range in x. The second function does not
have any particular range, so it will be plotted in the range defined by the previous

function. For all these examples, the color used for drawing is the same, thus it
is difficult to separate the plotted functions visually. One can draw the functions
using various colors after importing the `Color` class from the Java AWT package
as discussed before. Moreover, one can define the line width as

```
from java.awt import Color
from jhplot import *
c1=HPlot()
c1.visible(); c1.setAutoRange()
f1=F1D("2*exp(-x*x/50)+sin(pi*x)/x", -2.0,5.0)
f2=F1D("exp(-x*x/10)+cos(pi*x)/x", -2.0,5.0)
f1.setPenWidth(2)
f1.setColor(Color.green)
f2.setColor(Color.red)
c1.draw([f1,f2])
```

Listing 3.6 Plotting two functions using `java.awt.Color`.

where the method `setPenWidth(i)` accepts an integer number for the line width
in terms of the number of pixels. To draw dashed lines, use the `setPenDash()`
method. One can change the dashed line length by specifying an integer value
between 0 and 40. One can also use the `update()` method to redraw the plot.

The example above shows that an `F1D` function contains several methods for
drawing. Some most important graphical options are listed in Table 3.3.

As mentioned before, instead of using all these graphical methods in Jython
scripts, one can also edit function attributes using a GUI-driven dialog: navigate the
mouse to the pad with the graph and select [Edit settings] with the mouse
button. Then select [Y item] and you will see a pop-up window with various
attributes. Similarly, one can edit the global title and margins of the pads.

Table 3.3 The most important methods for graphical representations of the `F1D` class

Methods	Definitions
setColor(c)	Set line color
setPenWidh(i)	Width of the line
setPenDash(i)	Dashed style with "i" being the length
setLegend(b)	Set (b = 1) or not (b = 0) the legend
setTitle(text)	Set title text

For the methods shown in this table, "b" indicates a Boolean value (0 for true and 1 for false),
while "i" is an integer parameter. The notation "d" indicates a float value. The attributes "c" and
"f" correspond to the Color and Font classes of Java AWT. "text" represents a string

3.3.3 Plotting Functions on Different Pads

To plot two or more functions on different plot regions, one should construct an appropriate canvas. Before calling the method draw(obj), the current plotting pad has to be changed using the cd(i1,i2) method. In the example below, we make a canvas with two pads and then navigate to the necessary pad when we need to plot a function:

```
from java.awt import Color
from jhplot import *
c1=HPlot("Canvas",600,400,1,2)
c1.visible()
c1.setAutoRange()
f1=F1D("2*exp(-x*x/50)+sin(pi*x)/x", -2.0, 5.0)
f2=F1D("exp(-x*x/10)+cos(pi*x)/x", -2.0, 5.0)
f1.setColor(Color.green)
c1.draw(f1)

c1.cd(1,2)        # go to the second pad (1,2)
c1.setRange(0,10,0,20)
c1.draw(f2)
```

Listing 3.7 Using 2 pads

3.3.4 Short Summary of HPlot Methods

Table 3.4 shows the major methods of the HPlot class. Note that there are more than 300 methods associated with this class, which are divided into the "getter" (they start from the "get" string) and "setter" (starting from the string "set") groups of methods.

One can learn about the methods of the HPlot canvas using the DMelt code assist, i.e., typing c1. followed by the [F4] key, where c1 is an instance of the HPlot class. One can also use the method dir(c1) to print the methods.

We will discuss the HPlot canvas in Sect. 8.2.1 in more detail. In addition, we can show how to use other graphical tools to display functions in Chap. 8.

3.3.5 Example

Let us show a complete script example for visualization of several 1D functions using two plot regions. Save these lines in a file, say "test.py", open this file into the DMelt editor, and click on the "[run]" button (or press [F8]) to execute the script.

Table 3.4 Most important methods of the `HPlot` class

Methods	Definitions
cd(X, Y)	Go to a current region in X and Y
clearAll()	Remove drawings from all regions
clearAllData()	Clean data from all regions all graph settings are kept
clear()	Clear the current region
clearData()	Remove data from the current region all graph settings are kept
clear(X, Y)	Remove a region given by X and Y
destroy()	Destroys the canvas frame
draw(o)	Draw some object, like F1D, etc.
export(file)	Export to an image (png, eps, eps, svg, svgz). The format is given by the file extension
removeAxes()	Remove all axes
setAntiAlias(b)	Set (b = 1) or not set anti-aliasing for graphics
setAutoRange(a, b)	Set (b = 1) or not set (b = 1) auto-range for axis
setAutoRange(b)	Set auto-range (b = 1) or not (b = 0) for all axes
setAutoRange()	Set auto-range for all axes
setAxesColor(c)	Set color for axes
setBackgroundColor(c)	Background color of the graph
setBox(b)	Set or not a bounding box around the graph
setBoxColor(c)	Color of the bounding box
setBoxFillColor(c)	Fill color of the bounding box
setBoxOffset(d)	Offset of the bounding box
setGrid(a, b)	Show grid (b = 1) or not (b = 1) for axis
setGrid(a, b)	Show grid (b = 1) or not (b = 1) for axis
setGridColor(c)	Grid color
setGridToFront(b)	Grid in front of drawing (b = 1) or behind (b = 0)
setGTitle(text, f, c)	Set attributes for global title
setLegend(b)	Set legend (b = 1) or not (b = 0)
setLegendFont(f)	Set legend font
setLegendPosition(a, pos)	Set legend position given by pos value
setLogScale(a, b)	Set (b = 1) or not set (b = 0) log scale for axis
setRange(a, min, max)	Min and max for axis
setRange(minX, maxX, minY, maxY)	Min and max range for X and Y
setTicsMirror(a, b)	Set (b = 1) or not set (b = 0) mirror ticks for axis
updateAll()	Update plots in all regions
update()	Update current plot defined by cd()
visible(b)	Make canvas visible (b = true) or invisible (b = false)
visible()	Make canvas visible

For the methods shown in this table, "b" indicates a Boolean value (0 or 1), while "a" is an integer parameter indicating axis (a = 0 means X-axis, a = 1 means Y-axis). The notation "d" indicates a float value. The attributes "c" and "f" correspond to the `Color` and `Font` classes of Java AWT. "text" represents a string

```
from java.awt import Color
from jhplot  import *

c1 = HPlot("Canvas",600,400,2,1)
c1.visible()
c1.setGTitle("F1D Functions",Color.red)
c1.setNameX("X axis")
c1.setNameY("Y axis")

c1.cd(1,1)
c1.setAutoRange()
c1.setName("Local title")
f1 = F1D("2*exp(-x*x/50)+sin(pi*x)/x",-2.0,5.0)
f1.setPenDash(4)
c1.draw(f1)
f1 = F1D("exp(-x*x/50)+pi*x",-2.0,5.0)
f1.setColor(Color(10,200,50))
f1.setPenWidth(3)
c1.draw(f1)

c1.cd(2,1)
c1.setAutoRange()
f1 = F1D("20*x*cos(x)",-0,5.0)
f1.setColor(Color.red)
f1.setPenWidth(3)
c1.draw(f1)

f1 = F1D("10*sqrt(x)+20*x",0,5.)
f1.setColor(Color.blue)
f1.setPenWidth(3)
c1.draw(f1);

f1 = F1D("15*sqrt(x)+20*x*x",0,5.)
f1.setColor(Color.blue)
f1.setPenDash(3)
c1.draw(f1)
```

Listing 3.8 Using HPlot canvas

The execution of this script creates a canvas with several plotted functions as shown in Fig. 3.1.

Note again that all functions in this example are defined in the specific ranges. Therefore, the canvas tries to adopt these ranges. If ranges are not given during initialization, you should specify the x range for the canvas.

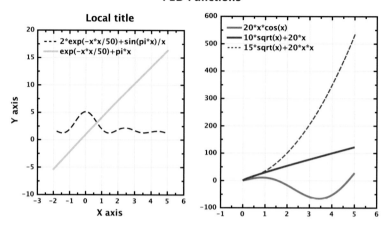

Fig. 3.1 Several F1D functions plotted on two pads of the same HPlot canvas

3.4 2D Functions

3.4.1 Functions in Two Dimensions

By considering functions with more than one independent variable, we are now beginning to venture into high-dimensional spaces. To build a two-dimensional (2D) function, DMelt has the F2D Java class. It is defined in the same way as F1D. The only difference is it takes two independent variables, x and y.

```
>>> from jhplot import *
>>> f1=F2D( "definition")
```

where "definition" is a string that must be replaced by the actual mathematical formula with two independent variables, x and y. As for the F1D class, the function definition can contain any combination of predefined operators, functions, and constants listed in Table 3.1.

To evaluate a 2D function at a fixed point (x, y), use the eval(x,y) method. One can evaluate a function for lists with x and y values using the method eval(x[],y[]) as in the example below:

```
>>> from jhplot import *
>>> f2=F2D("sqrt(x)*sin(y)/x + y^3")
>>> f2.eval(2,0.1)
0.07
>>> f2.eval([1,2],[2,4])
array([D,[array("d",[8.9,63.2]),array("d",[8.6,63.4])])
```

The output numbers below are truncated to fit them to the width of this page.

As discussed before, functions of two variables x and y may have arbitrary parameters. In this case, you should not create "parsed" object of this function during the initialization step, i.e., pass "False" argument during the initialization step. After substitution of a value, one can call parse() method that makes the function ready for numeric evaluation.

The F2D functions can be numerically integrated in a region between minX-maxX (for x) and between minY-maxY (for y) using the method:

```
>>> f2.integral(N1,N2,minX,maxX,minY,maxY)
```

where N1 and N2 are the number of points for integration for X and Y. The method assumes the trapezoid rule for integration. If the function was initiated with the ranges, the method f2.integral() does the same job but using a predefined number of points.

3.4.2 Displaying 2D Functions

The F2D functions can be shown as a 3D surface using HPlot3D canvas. This canvas has similar methods as those for the HPlot canvas, but allows drawing objects in 3D. What we should remember is that we have to prepare the function for drawing beforehand, which means one should specify the ranges for x and y axes during the initialization (or after, but before drawing).

During the initialization step, one can set the ranges as

```
>>> from jhplot import *
>>> f2=F2D("sqrt(x)*sin(y)/x + y^3",minX,maxX,minY,
    maxY)
```

so the function will be plotted for x in the interval [minX-maxX], and in the interval [minY-maxY] for the y independent variable.

After initialization, one can set the range using the setMinX(min), setMinX (max) methods for x, and analogously for y. The number of points for evaluation can be set using setPoints(n) methods (the default number of points is 500).

Here is a typical example that shows how to plot either a single function or two functions on the same lego plot:

```
from java.awt import Color
from jhplot import *

c1=HPlot3D("Canvas",600,700, 2,2)
c1.visible()
c1.setGTitle("F2D examples")

f1=F2D("sin(y)*x");
f2=F2D("sin(x+y)")
```

```
f3 = F2D ( " sqrt (2 * x * y ) + x ^ 2 " )
f4 = F2D ( " x ^ 2 + y ^ 2 " )

c1 . cd ( 1 , 1 )
c1 . setRange ( - 0.5 , 0.5 , - 0.5 , 0.5 )
c1 . setScaling ( 8 ) ;   c1 . setRotationAngle ( 30 )
c1 . draw ( f1 )

c1 . cd ( 2 , 1 )
c1 . setRange ( - 2 , 2 , - 2 , 2 )
c1 . setAxesFontColor ( Color . blue )
c1 . setColorMode ( 3 )
c1 . setScaling ( 8 )
c1 . setElevationAngle ( 30 )
c1 . setRotationAngle ( 35 )
c1 . draw ( f2 )

c1 . cd ( 1 , 2 )
c1 . setRange ( - 2 , 2 , - 2 , 2 )
c1 . setColorMode ( 4 )
c1 . setLabelFontColor ( Color . red )
c1 . draw ( f3 )

c1 . cd ( 2 , 2 )
c1 . setRange ( - 1 , 2 , - 2 , 2 )
c1 . setScaling ( 8 )
c1 . draw ( f4 , f1 )
```

Listing 3.9 Plotting 2D functions

The resulting figure is shown in Fig. 3.2.

This example features one interesting property of the HPlot3D canvas: instead of plotting one function as c1.draw(f1), we plot both functions on the same pad after passing two functions as arguments (see the last line). This is somewhat different from the HPlot behavior, where we could pass any number of functions in the form of a list. For the HPlot3D canvas, one can plot two functions at most. Also, the class does not accept lists of functions. Later, we will show that one can plot 2D histograms or even mixing histograms with functions using the method draw(obj1,obj2).

For surface plots, the presentation style can be changed using the method setColorMode(i), where $i = 0$ for wireframe, $i = 1$ for hidden, $i = 2$ for color spectrum, $i = 3$ for grayscale, and $i = 4$ for dualshades. The methods setScaling(), setElevationAngle(), and setRotationAngle() are self-explanatory. The plots can be rotated with the mouse and the title can be modified using in exactly the same way as for the HPlot canvas. We will return to the HPlot3D canvas in Sect. 8.9.

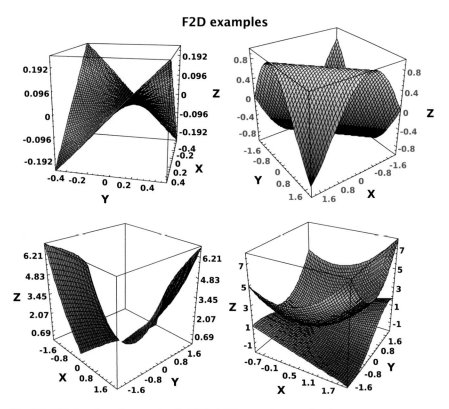

Fig. 3.2 F2D functions shown using the HPlot3D canvas in several regions

3.4.3 Using a Contour Plot

The 2D functions can also be shown using a contour (density) style. This can be done with the help of the same HPlot3D canvas but adding the line c1.setContour() after the definition of the c1 object. This is shown below:

```
from java.awt import Color
from jhplot import *

c1 = HPlot3D("Canvas",600,600)
c1.setNameX("X")
c1.setNameY("Y")
c1.setContour()
c1.visible()
f1=F2D("x^2+y^2", -2.0, 2.0, -2.0, 2.0)
c1.draw(f1)
```

Listing 3.10 Contour plot

Fig. 3.3 A contour representation of the function $x^2 + y^2$ using the HPlot3D canvas

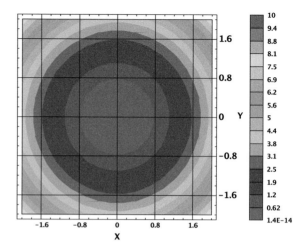

The execution of this script leads to the plot shown in Fig. 3.3. The color bar on the right indicates the density levels used for drawing in color. They can be redefined using several methods of this canvas.

However, it is more practical to show the F2D functions using the canvas based on the class HPlot2D. As we will discuss in Sect. 8.8, this class has significantly more options to display the contour and density plots since it was designed mainly for such types of plots.

Section 3.7 shows how to create and plot 2D functions using Java scripting instead of parsing strings with a mathematical formula.

3.5 3D Functions

3.5.1 Functions in Three Dimensions

You may already have guessed that for three-dimensional (3D) functions, DMelt has the class called F3D. It is defined in the same way as F1D and F2D, with the only difference that it can take up to three independent variables: x, y and z.

```
>>> from jhplot import *
>>> f3 = F3D("definition")
```

where "definition" is a string that should be replaced by the actual formula. As for the F1D and F2D classes, the function definition can be constructed from a combination of operators and predefined mathematical functions given in Table 3.1. To evaluate a 3D function at a fixed value (x, y, z), use the usual method eval(x,y,z):

```
>>> from jhplot import *
>>> f3=F3D("sqrt(x)*sin(y)/x + z^3")
>>> f3.eval(2,0.1,4) # calculate for x=2, y=0.1, z=4
64.0705928859
```

Probably, we can stop here and will not go into the drawing part of this section. We have to be creatures living in four-dimensional space in order to be interested in how to draw such functions (actually, if such a creature is reading this book and is still interested in this option, please contact me—we may discuss how to plot such functions). One can also project 3D functions into a 2D space by fixing some independent variables and then using the F2D and HPlot3D to display such projections.

Below we will discuss more versable classes for dealing with multidimensional functions.

3.6 Functions in Many Dimensions

3.6.1 FND Functions

DMelt supports functions with arbitrary number of independent variables using the FND class. This class is more complicated and has its roots in the known JEP Java package [1].

```
>>> from jhplot import *
>>> fn=FND("definition", "var1,var2,...varN")
```

where the string definition should be replaced by the actual mathematical formula using a combination of predefined functions given in Table 3.1. The second argument tells which characters or strings must be declared to be independent variables in the function definition. Unlike F1D, F2D, and F3D classes, variable names can be any strings (not only x, y, and z). Functions of the class FND can be simplified and differentiated and, obviously, can be evaluated at fixed points. The example below shows this:

```
>>> from jhplot import *

>>> fn=FND("1*x^4+x^2+y+z+h", "x,y,z,h")
>>> fn.simplify()
>>> print "Simplify=",fn.toString()
Simplify= x^4.0+x^2.0+y+z+h
>>>
>>> fn.diff("x") # differentiate using x
>>> print "Differentiate=",fn.toString()
Differentiate= 4.0*x^3.0+2.0*x
>>>
```

```
>>>  fn=FND("1*x^4+x^2+y+z+h",  "x,y,z,h")
>>>  print "Print variables=",fn.getVars()
Print variables=array(java.lang.String,["x","y","z","h"])
>>>
>>>  d=fn.eval("x=4,  y=1,  z=2,  h=0")
>>>  print "Evaluate result=",d
Evaluate results= 275.0
```

In this example, the evaluation of the function happens after fixing all three independent variables.

3.6.2 Drawing FND Functions

The class FND for function representation is flexible. First of all, one can easily deal
with 1D functions using any names for independent variables (remember, F1D can
only accept *x* to define a variable). The only difference you have to keep in mind is
that before drawing an FND function, you should always call the eval() method to
allow for only one independent variable and freezing other variables to fixed values.

Let us consider an example with two independent variables. In this case, we should
set the second variable to some value since we want to plot a 1D function:

```
from jhplot  import  *

c1=HPlot()
f2=FND("sqrt(var1)*sin(var2)","var1,var2")
f2.eval("var1",1,100,"var2=2") # var1 in range 1-100
c1.visible()
c1.setAutoRange()
c1.draw(f2)
```

Listing 3.11 Drawing an FND function

In the example above, first we evaluate the function in the range between 1 and 100
and fixing var2 to 2 before calling the draw(obj) method.

Below is the example for a function with three independent variables. After fixing
two variables, we plot this function in 1D:

```
from jhplot  import  *

c1  =  HPlot("Example",600,400)
c1.setGTitle("FND function")
c1.visible()
c1.setAutoRange()
f2=FND("x*x+y+20+z","x,y,z")
f2.eval("x",-2,10,"y=2,z=100")
c1.draw(f2)
```

Listing 3.12 Drawing an FND function

Now we have fixed two variables, y and z, and plotted the function in terms of the independent variable x between -2 and 10. It should be noted that the variables to be fixed are separated by a comma in the `eval()` statement.

     The FND class requires a string that describes the function. In the next section, we will consider an approach when a function is created without using string definitions.

3.7 Custom Functions

Building a custom mathematical function using Jython scripts makes sense when the logic of the mathematical expression is so complicated that it is better to define it in a separate code block rather than to use a string during the instantiation of the function. In this way, one can build infinitely complicated functions in many dimensions and evaluate them using DMelt Java libraries.

3.7.1 Custom Functions and Their Methods

The most flexible way to define mathematical functions that (1) can be arbitrary complicated and cannot be easily defined by one-line string as shown in Sect. 3.2 and; (2) can call third-party numerical Java libraries is to create a new class. This can be done by extending the FNon class as shown in the example below. Let us assume a noncontinuous function of the form:

$$f(x) = \begin{cases} x^2 * \sqrt{x}, & \text{if } x \geq 3, \\ 0, & \text{otherwise} \end{cases} \tag{3.1}$$

The function "jumps" at $x = 3$, so it can be a problem to define it using a string. Let us program this function as a class that calls the function $\sqrt{x}$ from the standard `java.lang.Math` Java package:

```
from java.lang import Math
from jhplot import *
class MyFunc(FNon):
  def value(self, x):
      y=x[0]*x[0]*Math.sqrt(x[0])
      if (x[0]<3): y=0
      return y
c1=HPlot(); c1.visible(); c1.setAutoRange()
pl=MyFunc("non-continuous",1,0)   # 1 vars, 0 pars
f1=F1D(pl,0,5)                     # plot in range [0,5]
c1.draw(f1)
```

Listing 3.13 Creating a noncontinuous function using the Java class FNon

Fig. 3.4 A noncontinuous function defined by Eq. 3.2 and plotted using a Jython script

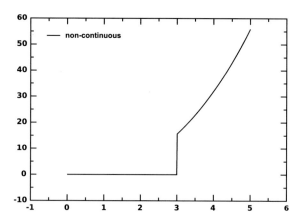

In this example, x[0] indicates a variable (0 means that the dimension of this function is one). Note that we have initiated the class MyFunc using the two arguments: one argument specifies the number of variables, while the second argument tells about the number of free parameters. After this class is initialized, we used its object as an argument to create an F1D function for visualization. Figure 3.4 shows the result of the above script.

Now let us consider a function with two free parameters, say "*a*" and "*b*":

$$f(x) = \begin{cases} x * x * \sqrt{x} * a + b, & \text{if } x \geq 3, \\ 0, & \text{otherwise} \end{cases} \tag{3.2}$$

The implementation of this function is straightforward:

```
from java.lang import Math
from jhplot import *
class MyNewFunc(FNon):
  def value(self, x):
    y=x[0]*x[0]*Math.sqrt(x[0])*self.p[0]+self.p[1]
    if (x[0]<3): y=0
    return y
```

We use self.p[0] (self.p[2]) to specify the parameter "*a*" ("*b*"). You can set the parameters to specific values after the initialization of this class. Append the following lines to the above code:

```
p1 = MyNewFunc("new",1,2)  # one variable, 2 parameters
print p1.numberOfParameters()
print p1.dimension()
print p1.parameterNames().tolist()
p1.setParameter("par0",100)  # a=100
p1.setParameter("par1",20)   # b=20
print p1.parameters()
print "when x=10 f=",p1.value([10])
```

This code instantiates the function object, prints all its parameters, and displays the dimension of the function. Then we set the free parameters to 100 (for "a") and 20 (for "b"). We print such parameters and then calculate the value of this function at the point $x = 10$. Use the same approach as before for displaying this function, i.e., create a "plottable" function in 1D as f1=F1D(pl,0,10), where we assume the x range between 0 and 10.

For completeness, let us show how to create a special function $1 - erf(x)$ using the external Apache math Java library which provides the error function $erf(x)$:

```
from jhplot import *
from org.apache.commons.math3.special.Erf import *
class MyFunc(FNon):
      def value(self, x):
                 return 1-erf(x[0])
pl = MyFunc("1-erf(x)",1,0)
f1=F1D(pl,-4,4)
```

Listing 3.14 Creating $1 - erf(x)$ function using the Apache library

You can try to plot this function using the same approach as in the previous example.

3.7.1.1   Java Class **FNon**

You may start to wonder what exactly has been done in the latter examples when defining functions using the class FNon. The above examples use the concept known as "inherence", a mechanism which simply overwrites the method "value" that belongs to the class FNon. The same concept can be used for other scripting languages which support this concept.

The last example in the previous section can be rewritten using Java. The code below shows how to create a custom mathematical function by extending the Java class FNon.

```
import jhplot.*;

public class Func_erf extends FNon {
public Func_erf(String title, int vars, int pars){
 super(title,vars,pars);
}
public double value(double[] x) { // over-write
  return 1-org.apache.commons.math3.special.Erf.erf(x
     [0]);
} }
```

Put these lines to a file "Func_erf.java" and compile it. Then you can draw this function using the HPlot Java class. Please read Sect. 16.2 which explains in more detail the relationship between Jython and Java codes.

3.7.2   Custom Functions Using Expression Builder

One can define mathematical functions using an alternative approach that does not use the FNon class. Let us show how to create an F1D function using the ExpressionBuilder class [2]:

```
from jhplot   import  *
from jhplot.math.exp4j import ExpressionBuilder

e=ExpressionBuilder("3*sin(x)-2/(x - 2)")
e.variables("x")
ff=F1D("Test",e.build(),10,100)
print ff.eval(10);
c1  = HPlot()
c1.visible(1)
c1.setAutoRange()
c1.draw(ff)
```

Listing 3.15 Building a function using Expression Builder

This shows that, instead of the string, one can use the class ExpressionBuilder as an argument in defining the F1D function. The ExpressionBuilder class can be used to create an Expression object after calling build() which is capable of function evaluation. In principle, the above approach looks the same as if we would use a string and pass it to F1D directly. But, there is one advantage in using the ExpressionBuilder class: now we can define an arbitrary function in a completely programmable way.

Here is what it looks like when we want to plot a function defined as $y = x^2$ (for $x < 100$) and $y = x$ (for $x \geq 100$):

```
from jhplot   import  *
from jhplot.math.exp4j import *
from jhplot.math.exp4j.function import *

class  cf2(Function): # extend Function Java class
   def   apply(self,val):
      if (val[0]<100):
                     return val[0]*val[0]
      if (val[0]>=100):
                     return  val[0]

e=ExpressionBuilder("cf2(x)")
func=cf2("cf2",1) # name cf2, 1 variable
e.function(func)
e.variables("x")   # set name "x" to variable
c1  = HPlot(); c1.visible(1); c1.setAutoRange()
f=F1D(e.build(), 10,1000)
c1.draw(f)
```

Listing 3.16 Building a custom function using Expression Builder

Let us explain how we did this. We created a class `cf2` extending the Java class `Function`. We used this object as an argument for `ExpressionBuilder`. Then we specified the variable x. Finally, we used the object returned by the method `build()` as input to create an `F1D` function for plotting.

Now let us consider a 2D function of two variables, x and y, and show it in a 3D canvas. The function will be defined as $y = x * y$ (for $x < 50$ or $y < 50$) and $y = x^2 + y^2$ (for $x \geq 50$ and $y \geq 50$). Our code will look similar to the 1D case, only now we will have two variables:

```
from jhplot   import *
from jhplot.math.exp4j import *
from jhplot.math.exp4j.function import *

class cf2d(Function):
  def  apply(self,val):
     if (val[0]<50 or val[1]<50):
                 return val[1]*val[0]
     if (val[0]>=50 and val[1]>=50):
                 return  val[0]*val[0]+val[1]*val[1]

e=ExpressionBuilder("cf2d(x,y)")
func=cf2d("cf2d",2) # 2 variables
e.function(func)
e.variables("x")
e.variables("y")

c1 = HPlot3D(); c1.visible(1); c1.setAutoRange()
f=F2D(e.build(), 0,200, 0,200)
c1.draw(f)
```

Listing 3.17 Showing a 2D function using 3D canvas

The output of this code is shown in Fig. 3.5.

Fig. 3.5 Showing a 2D function defined by a Python script from Listing 3.17 using the `ExpressionBuilder` Java class

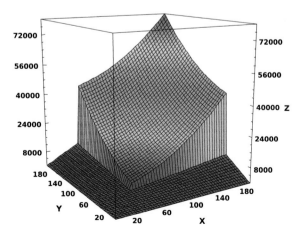

3.7.3 Custom Functions in Jython

So far we have considered a rather general approach to create custom functions that can be used not only in Jython codes, but also in Java and other scripting languages based on Java.

For constructing functions in this section, we will use the package shplot implemented in the Python language. This package still uses Java classes, therefore, shplot is a Jython package and cannot be used in other Java scripting languages, nor in Java programs. This Python module is imported automatically if one uses DMelt. If you are using something else, one can find this module in the directory "system/shplot". In the latter case, we remind that in order to be able to use it, one should import the Jython module sys and append the directory with the shplot module to the Jython variable to the list "sys.path".

```
import sys
sys.path.append(SystemDir+"/macros/system/shplot" )
from shplot import *
```

We assume that you have appended these lines to all the codes shown later. If you run the examples outside the IDE, define the variable SystemDir as the path to the directory where the DMelt is installed.

Let us build a simple second-order polynomial function, $y = c * x^2 + b * x + a$ using a pure-Jython approach. For educational purpose, we will include some complication into the function definition: if $x < 0$, then we assume that the polynomial behavior will vanish and the function will just be a constant $y = a$.

Let us put the following code in a separate file called "p2.py":

```
from shplot import *

class p2(ifunc):
  def value(self, v):
    if v[0]>0:
      d=self.p[2]*(v[0]*v[0])+self.p[1]*v[0]+self.p
    [0]
    else:
      d=self.p[0]
    return d
```

Listing 3.18 File "p2.py"

The example features several important properties: instead of the "jhplot" package we use the "shplot" library which represents a high-level Jython module based on the Java jhplot package.

Second, our class "p2" inherits properties of the "ifunc" class (see the class inheritance topic in Sect. 2.11.2). This is important as it provides the necessary functionality when we will decide to draw such function.

Finally, we specify our mathematical algorithm in the function `value()`, which returns the results of the calculation. Obviously, `self.p[]` list represents our free parameters a, b, c, while `v[0]` corresponds to the variable x.

One can also use any mathematical function provided by the `"math"` package of Jython. In this section, we will restrict ourselves with simple examples that do not require calls to external mathematical functions.

Now let us build a function from the class defined above. We will instantiate the function object as

```
>>> p=function(title,dimension,paramNumber)
```

where `"title"` is a string with the function title, `"dimen"` is a dimension of the function and `"paramNumber"` is the number of free parameters. Names for the variables and parameter names will be set to default values. Alternatively, one can use custom names for the variables and parameters using this constructor:

```
>>> p=function(title,names,pars)
```

where `"name"` and `"pars"` are lists of strings defining the names for independent variables and parameters names (this overwrites the default names).

Let us come back to our example and instantiate the second-order polynomial function given in the module "p2.py". We will create a function by assigning the title `"p2"`, dimension of this function (1) and the number of parameters (3 parameters).

```
from p2 import *

p=p2("p2",1,3)
print p.title()
print p.dimension()
print p.numberOfParameters()
print p.parameterNames()
print p.variableNames()
print p.variableName(0)
```

Listing 3.19 Building a second-degree polynomial function

The execution of this script gives

```
p2
1
3
array(java.lang.String, [u"par0", u"par1", u"par2"])
array(java.lang.String, [u"x0"])
x0
```

Let us discuss the output in more detail: we print out the title of this function with the method `title()` and the dimension with the method `dimension()`. The number

of parameters is given by the method `numberOfParameters()`. The method
`parameterNames()` returns the parameter names. Since we did not assign any
custom names, it prints the default names, `"par0,"` `"par1,"` and `"par2"`.
Then we print the variable name (the default is `"x0"`).

Now let us evaluate this function at several points. But before that, let us assign
some numerical values to our free parameters. One can set a single value by calling
the parameter name, or setting it at once in the form of a list. Then we will evaluate
the function at $x = 10$ and $x = -1$:

```
from p2 import *

p=p2("p2",1,3)
p.setParameter("par0", 10)
p.setParameter("par1", 20)
p.setParameter("par2", 30)
# set all 3 parameters in one go
p.setParameters([10,20,30])
print "Value at x=10 =", p.value([10])
print "Value at x=-1 =", p.value([-1])
```

Listing 3.20 Assigning values to the parameters of a second-order polynomial function ("p2")

Note how this is done: first we pass a list to the function, and then return it with the
method `value()`. As you may guess, this is necessary if you have several variables
to pass to a function. Executing the above script gives:

```
Value at x=10 = 3210.0
Value at x=-1 = 10.0
```

Now let us assign custom names for parameters. This time we will instantiate it using
the second constructor as

```
from p2 import *

p=p2("p2",["x"],["a","b","c"])
print p.parameterNames()
print p.variableNames()
p.setParameter("a", 10)
p.setParameter("b", 20)
p.setParameter("c", 30)

print "Value at x=10 =", p.value([10])
print "Value at x=-1 =", p.value([-1])
```

Listing 3.21 Assigning custom names to a second-order polynomial function ("p2")

The code assigns the constant parameters "*a*," "*b*," and "*c*" to the values 10, 20, and 30, respectively. We also set the variable name to the convenient "*x*". One can see this from the output shown below:

```
array(java.lang.String, [u"a", u"b", u"c"])
array(java.lang.String, [u"x"])
Value at x=10 = 3210.0
Value at x=-1 = 10.0
```

More advanced users can look at the methods of the object `"p"`. They will find that this object is constructed from the AIDA class `IFunction` and contains many other methods that have not been discussed here.

3.7.3.1 Using External Libraries

Previously, we have shown a simple example illustrating how to construct a mathematical function. But we do not restrict ourselves here with constructions from Jython: since our coding is based on Java, one can call a function or library from Java API, or Java-based external library. We can direct you, for example, to the Colt Java library [3] (included into DMelt) which provides a large set of special functions and functions commonly used for probability densities.

To build a function using this approach, you will need first to import necessary classes from third-party libraries and replace parameters with `self.p[]`, while variables should be replaced with `v[]` in the method `value()`. Let us show how to make a simple Jython module containing a Bessel function and Beta function:

```
from shplot import *

class bessel(ifunc):
    def value(self, v):
        from cern.jet.math import Bessel
        return Bessel.i0e(v[0])

class beta(ifunc):
    def value(self, v):
        from cern.jet.stat.Probability import beta
        return beta(self.p[0],self.p[1],v[0])
```

Listing 3.22 Calling Java libraries. Module "special.py"

First, we import the Java package `"cern.jet.math"` from the Colt library mentioned above and use this library to construct the Bessel and Beta functions. Please look at Java API of the Colt library to learn about such functions. We will save these lines in a module called "special.py".

Calling both functions from the module "special.py" is trivial. Just import this module and take care of the number of parameters you pass to such functions:

```
from special import *

b1=beta("Beta function",1,2)
b2=bessel("Bessel order 0",1,0)
```

Listing 3.23 Calling special functions

One can evaluate the functions at any allowed value `"x"` after specifying values for free parameters. In the case of the Bessel function, we do not have free parameters and the evaluation of such function is straightforward.

3.7.3.2 Plotting Custom Functions

Now we know how to build a function object from a script. Next, we will show how to manipulate with such custom functions and plot them.

The first thing you probably will need to do is to convert a function into F1D or F2D objects for plotting. Below, we show how to convert IFunction to the standard F1D, using the function Bessel defined in the previously constructed module "special.py".

```
from special import *
from jhplot import *

p=bessel("Bessel",1,0)
f=F1D(p)
print f.eval([0.1,0.2,0.5])
```

Listing 3.24 Conversion to F1D

so, it looks easy: just pass the custom function to the constructor of the F1D function. In the above example, we evaluate this function at several points using a list of x values. The execution of this script prints the output list of y values, [0.9071, 0.8269, 0.6450].

One can return the object IFunction back as getIFunction(), a handy method of the F1D function. We will see that the IFunction class is important when dealing with curve fitting in Sect. 11.2.

Probably you have already realized how to plot our Bessel function. The script below displays this function in the range [0–100]:

```
from special import *
from jhplot import *

p=bessel("Bessel",1,0)
f1=F1D(p,1,100)
c1 = HPlot()
```

```
c1.visible()
c1.setAutoRange()
c1.draw(f1)
```

Listing 3.25 Conversion to F1D

Plotting 2D functions is as easy as in the 1D case. First build a custom function from the script using two variables. Then pass it to the `F2D` constructor exactly as in the above example. Always pay attention to free parameters: they have to be defined beforehand by passing the function object to the `F2D` constructor.

We will return to the subject of custom functions when we will discuss the class `IFunction` in Sect. 11.2, which will be used for fitting experimental data. You will also learn how to access predefined functions or create functions of the class `IFunction` from a string.

3.8 Parametric Surfaces in 3D

3.8.1 FPR Functions

DMelt has a support for drawing parametric functions or equations in 3D. This feature has its root in the initial version of the 3D Graph Explorer program [4].

We remind that a parametric function is a set of equations with a certain number of independent parameters. To build a parametric function, one should use the `FPR` class ("F" abbreviates the word "function" and "PR" corresponds to "parametric"). The general definition of a parametric function is

```
>>>    f1=FPR("definition"),
```

where `"definition"` is a string representing the equation used for function definition. For parametric surfaces in 3D, we should define x, y and z in terms of two independent variables, `"u"` and `"v"`, which can vary in the range [0, 1]. To construct a string representing the parametric function, one can use the predefined functions given in Table 3.1. In addition, the standard comparison operations `"=="`, `"!="`, `"<"`, `">"`, `"<="` and `">="` can be used.

For example, the equation:

$$u = 2 * Pi * u; \quad x = \cos(u); \quad y = \sin(u); \quad z = v$$

defines a cylinder. Note that `"Pi"` means the predefined π value, and the multiplication sign can be replaced by a space. Each logical unit should be separated by a semicolon, and the entire expression should be passed to the `FPR` constructor as a string.

Let us show several other equations: a torus can be written as

$$u = 2 * Pi * u; \quad v = 2 * Pi * v; \quad r = 0.6 + 0.2 * \cos(u);$$
$$z = 0.2 * \sin(u); \quad x = r * \cos(v); \quad y = r * \sin(v)$$

A cone can be written as

$$u = 2 * Pi * u; \quad z = 2 * (v - 0.5);$$
$$x = z * \cos(u) * 0.8 - 1; \quad y = z * \sin(u) * 0.8 + 0.6$$

A hex cylinder can be written as

$$u = 2 * Pi * u; \quad x = \cos(u) * 0.8 + 0.3;$$
$$y = \sin(u) * 0.8 - 0.6; \quad z = 2 * (v - 0.5)$$

A sphere with a radius "r" is

$$r = 0.7; u = 2 * Pi * u; v = Pi * v; \quad x = r * \cos(u) * \sin(v);$$
$$y = r * \sin(u) * \sin(v);$$
$$z = r * \cos(v)$$

In the above examples, for simplicity, we set the constants defining geometrical sizes to arbitrary values.

In order to vary the parameters u and v over a different range than [0,1], one should scale and shift them by a certain value. For example, if one needs to change the range of u to [5,15], use this transformation: $u = u * (15 - 5) + 5$.

The parametric functions can be shown using the HPlot3DP class (the capital letter "P" in its name means "parametric"). This class provides a canvas for drawing parametric functions and can be used for interactive work with these functions (zooming and rotations). One can draw a surface using the usual draw(obj) method, where obj represents an object of the class FPR. One can make drawing pads as usual and navigate to the pads using the cd(i1,i2) method, or one can draw several parametric functions on the same pad by repeating the draw(obj) method. This can be seen from this example:

```
from java.awt import Color
from jhplot import *

c1 = HPlot3DP("Canvas",700,600,2,1)
c1.setGTitle("Parametric surfaces")
c1.visible()

f1=FPR("r=0.7; u=2 Pi u; v=Pi v; \
       x=r cos(u) sin(v); y=r sin(u) sin(v); z=r cos(v)")
f1.setDivisions(50,50)
```

```
f1.setLineColor(Color.blue);

f2=FPR("u=2 Pi u; v=2 Pi v; r=.7+.2cos(u); \
        z=.2 sin(u)-0.5; x=r cos(v); y=r sin(v)")
f2.setFillColor( Color(20,170,170) )

f3=FPR("ang=atan2(y,x); r2=x*x+y*y;\
        z=sin(5(ang-r2/3))*r2/3")

c1.draw([f1,f2])

c1.cd(2,1)
c1.setCameraPosition(-1.0)
c1.setFog(0)
c1.draw(f3)
```

Listing 3.26 Drawing parametric functions

In this example, the lengthy strings that define the functions were broken to fit the page width. Figure 3.6 shows the resulting plot.

Let us discuss several graphical methods for the FPR functions. First of all, one can set colors for the lines and for the filled area, as well as the line width as

```
>>> f1.setFilled(b)
>>> f1.setFillColor(c)
>>> f1.setLineColor(c)
>>> f1.setPenWidth(i) # set line width i=1,2,3,4 etc.
```

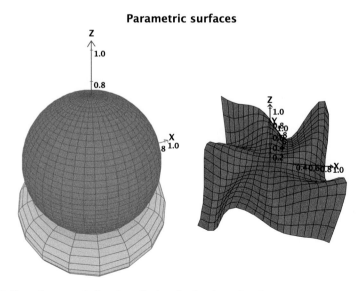

Fig. 3.6 Several parametric functions displayed using the HPlot3DP canvas

where b is ether 0 (not filled area) or 1 (Java version of "True") if the object has to be filled. "c" is the usual Java AWT `Color` class used to fill a surface. One can set the width of the lines as `setPenWidth(i)`, where "i" is an integer value. Finally, all above graphical attributes can be obtained using the corresponding "getter" methods.

One can add a transparency level to the filled color by setting the so-called alpha value for the AWT `Color` class. For example,

```
>>> f1.setFillColor( Color(0.1,0.9,0.7,0.5)
```

sets the transparency level for filled color to 50 % (see the last argument 0.5).

The numbers of divisions for "u" and "v" independent variables are given by the method

```
>>> f1.setDivisions(divU,  divV)
```

where `divU` and `divV` are the numbers of divisions (21 is the default value). The larger the number of divisions, the smoother the surface is. However, this requires more computer resources, especially during interactive work such as rotations or zooming. One can also include the number of divisions during the construction of a parametric function:

```
>>> f1=FPR("definition",divU,divV)
```

We will discuss several methods associated with the class `HPlot3DP` in Chap. 8. One can also look at the corresponding API documentation of this class.

3.8.2 3D Mathematical Objects

In addition to the parametric functions discussed before, DMelt contains the 3D-XplorMathJ package [5] as a third-party library which allows viewing various parametric functions in 3D. A user can specify and draw any function using the menu of this program. The package contains an impressive catalog of useful mathematical objects, ranging from planar and space curves to polyhedra and surfaces to differential equations and fractals.

The 3D-XplorMathJ can be started using [Tools]→[3D surfaces] from the Toolbar menu of the DMelt IDE. How to code with Python scripting using this package is described in Sect. 8.9.

3.9 Function Minimization

Minimization methods have large importance in data analysis; often, we need to determine how well a data set is described by a given function. To answer this question, we need to build a function that reflects a degree of difference between data and a function, and then optimize its parameters such that the best possible description of the data under study can be obtained. We will study this subject in Chap. 11. Here, we will discuss the numerical aspect of this question.

Numeric minimization of a function belongs to a broader mathematical subject of "optimization problem". In fact, maximizing a real function also belongs to the optimization problem. To transform the maximization problem to a minimization problem in many cases is simple since this requires an inversion of the original function. Therefore, in this section, we will discuss the minimization problem.

We will use the most general form for a description of mathematical functions described in Sect. 3.7. Our goal is to minimize a function of the form $a * (x - 2)^2 * \sqrt{x} + b * x^2$, where a and b are free parameters. The function will be minimized for $a = 5$ and $b = -6$ values. To do this, we will use the "JMinuit" package, which is a Java port of the Minuit program [6]. In basic terms, the program searches for minimum in a user-defined function with respect to free parameters. To achieve fast convergence, supply an initial value for the minimization. The value should be reasonably close to the expected result (in our example, we set it to 10). The minimization procedure can use different techniques and arbitrary parameter constraints.

```
from jhplot.shapes import Line
from java.lang import *
from java.awt  import *
from jhplot import *

class Func(FNon):
  def value(self, x):
    a=self.p[0]; b=self.p[1];
    return a*Math.sqrt(x[0])*(x[0]-2)**2+b*x[0]*x[0]

c1=HPlot()
c1.visible();   c1.setAutoRange()
pl=Func("a(x-2)^{2}#sqrt{x}+bx^2",1,2) # 1 var, 2 pars
pl.setParameter("par0", 5) # define two parameters
pl.setParameter("par1",-6)
print pl.parameters()

from hep.aida.ref.optimizer.jminuit import *
op=JMinuitOptimizerFactory().create()
op.setFunction(pl)
op.variableSettings("x0").setValue(10) # set initial value
op.optimize()                          # optimization
res=op.result()

ff=F1D(pl,0,10) # show the function and min value
xmin=res.parameters()[0]
ymin=ff.eval(xmin)
pl=Line(xmin,ymin-10,xmin,ymin+10,BasicStroke(4),Color.red)
c1.add(pl)
c1.draw(ff)
print res.parameters()," status=",res.optimizationStatus()
```

Listing 3.27 Numeric minimization of a 2D function

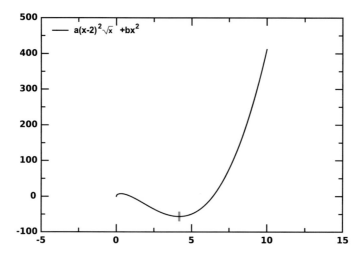

Fig. 3.7 Numeric minimization of a function. The found minimum is shown with the small *red line* (color figure online)

This code outputs the x value at which the function takes minimum value. Figure 3.7 shows the function and the result of minimization. Use the method `optimization Status()` to check the status of the optimization. For a converged minimization, it should return the integer value 3. The results are not converged when the status is 6. Please refer the Java API of the `IOptimizerResult` class. If the minimization fails, try another value for the initial guess.

By default, the optimization algorithm uses the "Migrad" method. It is fairly reliable steepest descent method that evaluates function gradient by measuring derivatives of with respect to parameters. We set $a = 50$ and $b = -60$ for free parameters. In order to make the optimization faster, we set $x = 10$ as the initial guess for the minimum value. You can learn about the input configuration to "JMinuit" using the following lines that should be appended to the above example:

```
co=op.configuration()  # create JMinuitConfiguration object
print co.method()      # print some useful methods
print co.precision()
print co.strategy()
print co.tolerance()
```

where `co` is an instantiation of the `JMinuitConfiguration` class from the `hep.aida.ref.optimizer` package.

The output will tell about the used method and initial precision. We will not go into the details of other optimization methods used during initialization; an advanced user is welcome to look at the Java API of the `JMinuitOptimizerFactory()` class.

If the minimization fails, consider using other minimization methods. In the above example, you can set a new minimization technique right after the `create()`, but before the `optimize()` call:

```
co=op.configuration()
co.setMethod(method)
op.setConfiguration(co)
```

where "method" is a string that can take one of the following values:

- "MIG"—Migrad method. This is the default, commonly used, method. However, it depends on the knowledge of the first derivative and may fail if such knowledge is inaccurate.
- "SIMP"—"simplex" method. This is a multidimensional minimization that does not use first derivatives, so it should not be insensitive to the precision with which first derivative is calculated. But it is rather a slow method.
- "MINI"—minimize method is equivalent to Migrad. If Migrad fails, it reverts to "SIMP" and then calls Migrad again.
- "IMP"—improved Migrad method that has improvements for searches of a global minimum.
- "SEE"—the seek method uses a Monte Carlo method with random jumps to find a minimum.

Similarly, one can define the minimization strategies, precision, tolerance, and a maximum number of iterations. See the `JMinuitConfiguration` class for detail.

3.9.1 Minimization of Multidimensional Functions

Minimization of functions in several dimensions is easily done using the technique discussed above, but one should correctly define the number of variable and free parameters during the initialization. Let us consider a function of the form $a * (x - 2)^2 + (y + 1)^2$, and we want to find its minimum when $a = 2$. Here is a small example that creates a class FuncXY, initialize it with two variables and one free parameter. Then we do the minimization assuming the starting point for the search is $(x = 10, y = -10)$:

```
from java.lang import Math
from jhplot import *

class FuncXY(FNon):
  def value(self, x):
    return (x[0]-2)*(x[0]-2)*self.p[0]+(x[1]+1)*(x[1]+1)
```

```
p1=FuncXY("a*(x-2)^2 + (y+1)^2",2,1)
p1.setParameter("par0",2)
from hep.aida.ref.optimizer.jminuit import *
op=JMinuitOptimizerFactory().create()
op.setFunction(p1)
op.variableSettings("x0").setValue(10)
op.variableSettings("x1").setValue(-10)
op.optimize()
res=op.result()
print res.parameters()
```

Listing 3.28 Minimization of a 2D function

For completeness, we show here the Java equivalent of the code:

```
import jhplot.*;
import hep.aida.ref.optimizer.jminuit.*;
import hep.aida.ext.*;

public class Default {
        public static void main(String[] args) {
FNon p1=new FNon("a*(x-2)^2 + (y+1)^2",2,1){
public double value(double[] x) {
  return (x[0]-2)*(x[0]-2)*p[0]+(x[1]+1)*(x[1]+1);
  }
};
p1.setParameter("par0",2);
JMinuitOptimizerFactory mm= new
    JMinuitOptimizerFactory();
IOptimizer op=mm.create();
op.setFunction(p1);
op.variableSettings("x0").setValue(10);
op.variableSettings("x1").setValue(-10);
op.optimize();
IOptimizerResult res=op.result();
}    }
```

Listing 3.29 Minimization of a 2D function n Java

3.9.2   Calling Migrad Directly

Now we will turn to a different approach. Here we will show how to use the Migrad minimization using a mathematical function defined through the Java scripting approach. This approach becomes attractive if one needs better control over the minimization procedure.

Let us show how to minimize a simple function $1 + x^2$ with very detailed output:

```
from org.freehep.math.minuit import *

class func(FCNBase):
   def valueOf(self, par):
      return 1+par[0]*par[0]
Par = MnUserParameters()
Par.add("x", 1., 0.1)
migrad = MnMigrad(func(), Par)
vmin = migrad.minimize()
state=vmin.userState()
print "Min value=",vmin.fval(), " calls=",vmin.nfcn()
print "Parameters=",state.params()
print "Information=",vmin.toString()
```

Listing 3.30 Using Migrad for a 1D function

Now the user has a full control over the parameters using the class `MnUser Parameters`. In this example, we add a parameter with the name x, with the initial value 1 and its error 0.1. One can check if the minimization is successful or not. If not, redefine the minimization strategy as

```
if vmin.isValid()==False:
   migrad = MnMigrad(func(), upar, 2)
   min = migrad.minimize()
   ....
```

where "upar" are input parameters (some of them can be fixed during minimization). This example merely shows how to use the interface of this function; you would need to read the Java API of the `MnMigrad` class to learn more.

Analogously, we can minimize a more complicated 2D function:

```
from org.freehep.math.minuit import *

class func(FCNBase):
  def valueOf(self, par):
    y=par[1]-par[0]*par[0]
    return 10*y*y + (1-par[0])*(1-par[0])
Par = MnUserParameters()
Par.add("x", 1., 0.1)
Par.add("y", 3., 0.1)
migrad = MnMigrad(func(), Par)
vmin = migrad.minimize()
state=vmin.userState()
print "Min value=",vmin.fval()," calls=",vmin.nfcn()
print state.params()
```

Listing 3.31 Using Migrad for a 2D function

In all such examples, you may be interested in plotting minimized functions. This
can be done by using an additional method for the class "func()". Here is the example
showing how to add the method "getPlot()" that returns a plottable object:

```
from jhplot import *
class func(FCNBase):
   def valueOf(self, par):
       return 1+par[0]*par[0]
   def getPlot(self,min,max):
      f=P1D("function")
      step=(max-min)/100.0 # define step
      for i in range(100):
             x=min+step*i
             y=self.valueOf([x])
             f.add(x,y)
      return f
```

After this, one can plot this function as

```
f=func()
p=f.getPlot(min,max)
c1.draw(p)
```

where "c1" is an instance of the HPlot class, and min and max are the minimum
and maximum values for the plotted ranges. You can use the line style to draw a
continuous function.

3.10 File Input and Output

The best way to save DMelt functions in a file is to use the Java serialization mechanism. The Python serialization mechanism, such as pickle or shelve discussed
in Sect. 2.15, will not work as expected when dealing with pure-Java objects that can
only be saved using the native Java serialization. However, you can use the pickle
or shelve modules when you define a Jython function or when using a string with
the function definition.

One can save a function using the Java serialization mechanism. For this, the
IO class included to the jhplot package can be useful. It takes a Java object and
saves it into a file using the method write(obj). Then one can restore this object
back using the method obj=read(). By default, all objects will be saved in a
compressed form using the GZip format.

In the case of functions, we will write only a *proxy* of functions, i.e., objects
that do not contain any graphical attributes. Such objects can be easily serialized to
simple objects. To get such object representing functions, use the get() methods.
In order to restore the function, use the such objects during function initialization.
In the following, we will use file names with the extension ".jser".

Look at the example which shows how to save a function into a file with the name "file.jser":

```
>>> from jhplot   import  *
>>> f1=F1D("2*sin(x)")
>>> print f1.toString()
2*sin(x) (title=2*sin(x), n=500, true)
>>> IO.write(f1.get(),"file.jser") # save function object
>>> s=IO.read("file.jser")         # read the file
>>> f2=F1D(s)                      # restore the function
>>> print "After serialization:",f2.getName()
After serialization: 2*sin(x)
```

Note that we save the function definition, rather than the complete F1D object. The same can be achieved using the standard Java API, so the method write() from the IO class is equivalent to:

```
>>> from java.io import  *
>>> f=FileOutputStream("file.jser")
>>> out=ObjectOutputStream(f)
>>> out.writeObject(f1.get())
>>> out.close()
```

Note that the method read() from the IO class is equivalent to:

```
>>> file = File("file.jser")
>>> fin = ObjectInputStream(FileInputStream(file))
>>> f2 =fin.readObject()
>>> fin.close()
```

One can also write a list of functions, instead of a single function:

```
from jhplot   import  *

f1=F1D("x*x+2")
f2=F2D("x*y+10")
a=[f1.get(),f2.get()]        # list with functions
IO.write(a,"file.jser")      # write list
list=IO.read("file.jser")    # read functions
f1=F1D(list[0])
f2=F2D(list[1])
print "After serialization:"
print f1.toString(), f2.toString()
```

Listing 3.32 Serialization of functions

Generally, the DMelt functions are relatively simple entities; therefore, one can just serialize or write a string with the function definition into a file and then use it to instantiate a new object of this function.

In Chap. 9, we will discuss the input–output (I/O) issues in more detail. We will learn how to save an arbitrary mix of various DMelt and Java objects, including lists of functions, histograms, and arrays. For example, one can write a large sequence of functions persistently using the `HFile` class to be discussed later.

References

1. Funk N, A Java library for mathematical expressions. http://sourceforge.net/projects/jep/
2. Asseg F et al, Exp4j. A mathematical expression evaluator. http://www.objecthunter.net/exp4j/
3. The Colt Development Team, The Colt project. https://dst.lbl.gov/software/colt/
4. Bose A, 3D graph explorer
5. T. D.-X. Consortium, The 3d-xplormathj project. http://3d-xplormath.org/
6. James F (2013) MINUIT. Function minimization and error analysis, reference manual. http://wwwasdoc.web.cern.ch/wwwasdoc/minuit/

Chapter 4
Data Arrays

Numeric computations based on repetitive tasks (or loops) are not too efficient in scripting languages, such as Jython/Python. Loops offer an easy way to do something repeatedly, but creating objects in loops is slow. This has already been illustrated using Jython arrays in Sect. 2.13.2. In most cases, what we really want is to manipulate with primitive data types, such as floats or integers, rather than with immutable objects used for representation of numbers in Jython. Therefore, our strategy for this book will be the following: Jython will be viewed as *interface* language, i.e., a language designed to link and manipulate with high-level Java classes that implement repetitive operations with primitive types.

In the following chapters, we discuss objects used for data storage and manipulation—building blocks from which a typical data analysis program is constructed. Unlike Jython classes, the objects to be discussed below are derived from pure Java classes and imported from the Java libraries of DMelt.

4.1 1D Arrays

In this chapter, we discuss 1D data, i.e., a data set represented by a sequence of double-precision floating-point real values, $N_1, N_2, \ldots N_{max}$. Each number can represent, for example, a value obtained from a single measurement. Below we discuss how to build objects that can be used for: (1) generation of 1D data sets; (2) storing numerical values in containers; (3) writing to or reading from files; (4) performing numerical analyses and producing statistical summaries; (5) visualization and comparisons with a theory or with other data.

As we progress deeper into analysis of one-dimensional (1D) data, you may be unsatisfied with typing code snippets in the console. We will remind that, in this case, one should write code in files with the extension ".py" and execute such macros using the key [F8] or the [run] button as described in Sect. 1.4.

© Springer International Publishing Switzerland 2016   131
S.V. Chekanov, *Numeric Computation and Statistical Data Analysis*
on the Java Platform, Advanced Information and Knowledge Processing,
DOI 10.1007/978-3-319-28531-3_4

4.1.1 P0D Data Container

The P0D class is among the simplest classes of DMelt: it does not have any graphical
attributes. This class is designed to keep a sequence of real numbers. It is similar to
the Java ArrayList class, but keeps only double-precision real values represented
with 64-bit accuracy. To some extent, the class P0D is also similar to the Python list
which keeps a sequence of objects.

For integer values, use the class P0I that has exactly the same methods as P0D.
For the P0I arrays, you may benefit from a lower memory usage and smaller file
size when writing this array into files. Below we discuss only the P0D class.

Let us construct a P0D object to keep a 1D data set. It is advisable to annotate it,
so we can easily obtain its attribute later:

```
>>> from jhplot import *
>>> p0=P0D("measurement")
```

where "measurement" is an optional string describing the data.

Let us remind again how we can learn about the methods associated with DMelt
objects, like the p0 object described above. Here is the list of various methods to
help you learn about the available methods:

- If you are using the JythonShell, type "p0." and press [Ctrl]-[Space];
- In Jython, use the standard Python method dir(p0) to print the methods of the
  object p0;
- If you are using the DMelt source code editor, type "p0." and then press [F4].
  You will get a detailed description of all methods in a pop-up frame. One can
  select a necessary method and check its description using the mouse. In addition,
  one can copy the selected method to the editor area as described in Sect. 1.4.10;
- If you are using Java instead of Jython, and working with Eclipse or NetBeans
  IDE, use the code assist of these IDEs. The description of these IDEs is beyond
  the scope of this book.

In all cases, you might be impressed by the significant number of methods associated
with the P0D class. But let us go slowly and first discuss its most important methods.

One can obtain the title by calling the method getTitle():

```
>>> print p0.getTitle()
measurement
```

One can reassign a new title with the method:

```
>>> p0=P0D()
>>> p0.setTitle("new measurement")
```

Once the p0 object is created, one can add numbers to this container. In practice, 1D data can be filled from a file or some external sources (this will be discussed later). Here, we show how to add a value to the container:

```
>>> p0.add(1)        # add an integer
>>> p0.add(-4.0)     # add a float
>>> p0.add(-4E12)    # add a float
>>> p0.set(2,200)    # insert the value at the 2nd position
```

One can obtain the value at a specific index "i" using the method get(i).

It should be noted that all values are converted into the double representation by the P0D class. The size of the container can be found using the method size(). One can remove all elements from the container using the method clear():

```
>>> print p0.size()  # size of the data
3
>>> p0.clear()       # clean up the container
>>> print p0.size()  # again print the size
0
```

Data from the p0 instance can be obtained in the form of Python/Jython arrays discussed in Sect. 2.13:

```
>>> array=p0.getArray()     # array with double  numbers
>>> array=p0.getArrayInt()  # array with integer numbers
```

If you want to return a list, instead of arrays, you can call the method getArrayList() method. One can convert the array into a list using the method tolist().

A P0D container can be initialized from a Python list as:

```
>>> p0.setArray([1,2,3,4])
```

It should be noted again that, for scripting with Jython, one should use the setArray() and getArray() methods that are optimized for speed. They are much faster than pieces of codes with the methods add() or set() called inside Jython loops.

One can fill the current P0D from a file (see below), generate a sequence of numbers, or fill the P0D with random numbers. For example, to fill a P0D container with a sequence of real numbers between 0 and 100, use:

```
>>> from jhplot  import *
>>> p0=P0D("sequence")
>>> p0.fill(101,0,100) # numbers from 0-100 (step 1)
```

A POD container can be instantiated from the Python list:

```
>>> p0=POD("title",[1,2,3,4]) # with title
>>> p0.POD([1,2,3,4])          # without title
```

POD objects can also be filled with random numbers using a few built-in methods. For example, uniformly distributed numbers can be filled as:

```
>>> p0= POD("Uniform distribution")
>>> p0.randomUniform(1000,0.0,1.0)
```

The first argument is the total number of values to be filled, while the second and third specify the range for random numbers. One can also generate an array with random numbers in accordance with a Gaussian (normal) distribution with the mean 0 and the standard deviation equals one as:

```
>>> p0= POD("Normal distribution")
>>> p0.randomNormal(1000, 0.0, 1.0)
```

As before, 1000 is the total number of entries.

We will consider how to fill the POD class with many other random distributions in Sect. 10.4. In particular, we will explain how to fill a POD with random numbers using various functional forms of probability distributions in Sect. 10.4.3.

To print a POD on the screen, use the method toString() that converts the POD object into a string representation:

```
>>> p0=POD("measurement")
>>> p0.add(1); p0.add(2); p0.add(3);
>>> print p0.toString()
1.0
2.0
3.0
```

In some cases, it is more convenient to show data as a table in a separate frame, so one can sort and search for a particular value. In this case, use

```
>>> p0.toTable()
```

This line brings up a table with filled values. Finally, one can use the print() method for printing using Java System.out.

4.1.2 P0D Transformations

One can add, subtract, multiply, or divide two P0D objects. Having created two
P0D objects, say p0 and p1, one can apply several mathematical transformations
resulting in new P0D objects:

```
>>> p0=p0.oper(p1,"NewTitle","+") # add
>>> p0=p0.oper(p1,"NewTitle","-") # subtract
>>> p0=p0.oper(p1,"NewTitle","*") # multiply
>>> p0=p0.oper(p1,"NewTitle","/") # divide
```

To sort values, use the method:

```
>>> p0.sort()     # sort in the natural order
```

Other useful methods are given below:

```
>>> m=p0.search(val) # 1st occurrence of "val"
```

Here are some other useful methods:

```
>>> p0=p0.merge(p1)        # merge with p0 and p1
>>> p0.range(min,max)      # get range between min and max
```

All values inside the P0D container can be transformed into a new set using an
analytical function and an F1D function discussed in Sect. 3.2. Functions can either
be defined from a string or using Jython functions as shown in Sect. 3.7. We remind
that, in order to define a function, a user should use the variable x and the functions
listed in Table 3.1. As usual, +, −, *, or / and parenthesis ( ) can be used in the
definition.

Let us give a simple example. We will create a P0D, initialize it with the list
[1,2,3,4] and transform this list using the function $x^2 + 1$:

```
from jhplot import *

f1=F1D("x^2+1") # transformation
p0=P0D("numbers",[1,2,3,4,5])
print p0.func("x^2+1", f1)
```

Listing 4.1 Transforming an array using the P0D class

The output of this code is

```
P0D x^2+1
2.0
5.0
10.0
17.0
26.0
```

Note that, in all cases, the objects remain to be the same, only the values will be modified. If one needs a copy of a POD object, use the method `copy()`.

One can obtain a new POD object with all elements smaller, larger, or equal to a specified value. Assuming that p0 is an object with filled values, this can be done using the following method:

```
>>> p1=p0.get(d,str)
```

The method returns a new array with values smaller, larger, or equal to an input value d. The type of operation is specified using a string (shown as str in the above example), which can be equal to " < ", " > ", or "=", respectively.

4.1.3   Statistical Summary

The POD class contains a number of useful methods to access descriptive characteristics of 1D arrays. First of all, let us consider the most simple methods that return double values with certain characteristics of a p0 object of the class POD:

```
>>> m=p0.size()             # size of POD
>>> m=p0.getMin()           # min value
>>> m=p0.getMax()           # max value
>>> m=p0.getMinIndex()      # index of min value
>>> m=p0.getMaxIndex()      # index of max value
>>> m=p0.mean()             # mean value
>>> m=p0.correlation(p1)    # correlation coefficient p1
>>> m=p0.covariance(p1)     # covariance
>>> m=p0.getSum()           # sum of all values
```

In the above examples, p1 is a second POD object used to find correlations between p0 and p1 arrays.

One can also perform a search for a specific value inside the POD arrays. The method `contains(val)` returns true if a value val is found. One can also find an index of the first occurrence of the specified element inside a POD array using the method `find(d)`.

The second set of methods is more elaborate, but requires execution of the method `getStat()`. This method returns a string with characterization of the entire data set. Such a summary contains a very comprehensive statistical characteristics of 1D data set:

```
>>> from jhplot import *
>>> p0=POD([1,2,3,3])
>>> print p0.getStat()   # evaluates statistics
```

The execution of the line with `getStat()` prints a rather long list with the summary statistics:

```
Size: 4
Sum: 10.0
SumOfSquares: 30.0
Min: 1.0
Max: 4.0
Mean: 2.5
RMS: 2.7386127875258306
Variance: 1.6666666666666667
Standard deviation: 1.2909944487358056
Standard error: 0.6454972243679028
Geometric mean: 2.213363839400643
Product: 23.99999999999999
Harmonic mean: 1.9200000000000004
Sum of inversions: 2.083333333333333
Skew: 0.0
Kurtosis: -2.0774999999999997
Sum of powers(3): 100.0
Sum of powers(4): 354.0
Sum of powers(5): 1300.0
Sum of powers(6): 4890.0
Moment(0,0): 1.0
Moment(1,0): 2.5
Moment(2,0): 7.5
Moment(3,0): 25.0
Moment(4,0): 88.5
Moment(5,0): 325.0
Moment(6,0): 1222.5
Moment(0,mean()): 1.0
Moment(1,mean()): 0.0
Moment(2,mean()): 1.25
Moment(3,mean()): 0.0
Moment(4,mean()): 2.5625
Moment(5,mean()): 0.0
Moment(6,mean()): 5.703125
25%, 50%, 75% Quantiles: 1.75, 2.5, 3.25
quantileInverse(median): 0.625
Distinct elements: [1.0, 2.0, 3.0, 4.0]
Frequencies: [1, 1, 1, 1]
```

Once the method `getStat()` is called, one can access the following characteristics:

```
>>> m=p0.variance()      # variance
>>> m=p0.stddeviation()  # standard deviation
>>> m=standardError()    # standard error
>>> m=p0.kurtosis()      # kurtosis
>>> m=p0.skew()          # skewness
>>> m=p0.median()        # median
>>> m=moment(k,c)        # k-th order moment
```

The above methods are quite self-explanatory. The last method returns k-th order moment of the distribution defined as $\sum_{i=0}((x[i] - \text{mean})^k)/\text{size}()$.

4.1.4   Displaying P0D Data

There is only one way to show a P0D array: project it into a *histogram*. This topic is extensively covered in Sect. 7.1. Here, we briefly point out that a histogram is a chart of rectangles drawn on the x-axis whose areas are proportional to the frequency of a range of variables.

One can build a histogram from a P0D using the two methods:

```
>>> h=p0.getH1D(bins)
>>> h=p0.getH1D(bins, min, max)
```

The first method creates a histogram with a given number of bins (bins is an integer number). The minimum and the maximum values of the X range are determined automatically. In the second case, one can explicitly specify the number of bins and the minimum (min) and maximum (max) values.

One question you may ask is: Assume we have two P0D objects. One object represents x values, the second represents y values. We already know how to check correlations between these objects—use the method correlation() as shown in Sect. 4.1.3. But how can one display pairs (x, y) on X-Y plots? Below we show how to do this:

```
from jhplot   import *

c1 = HPlot("Canvas",600,400)
c1.setGTitle("X Y plot")
c1.visible(); c1.setAutoRange()

p1=P0D("numbers",[1,2,3,4,5])
p2=p1.copy()
f1=F1D("x^2")
p2.func("squared", f1)
pp=P1D(p1,p2)
c1.draw(pp)
```

Listing 4.2   Plotting a 1D array using the P0D Java class

You may notice that we made an extra step by creating an object P1D from two P0D arrays. We will discuss this object in Sect. 4.2.

Below is an example that shows how to generate random numbers and transform them using the F1D class. The output is shown as a histogram on the HPlot canvas:

```
from jhplot   import *

c1 = HPlot()
c1.setGTitle("1D data array")
c1.visible(); c1.setAutoRange()
p0= P0D("Normal distribution")
p0.randomNormal(200, 0, 1)
```

```
f1="x*cos(x)+2"
p01=p0.copy(f1)      # transform a copy to a function
p01.func(F1D(f1))

f2="exp(x)-2"
p02=p0.copy(f2)
p02.func(F1D(f2))    # make a new copy and transform
h1=p0.getH1D(20)     # histogram with 20 bins
h2=p01.getH1D(50)    # histogram with 50 bins
c1.draw([h1,h2])     # plot both histograms
```

Listing 4.3   Creating a 1D array using the F1D function

Run this script and try to make sense of it. If it is not easy, skip this section since we will return to a very detailed discussion of the histograms in Sect. 7.1.

4.1.5   File Input and Output

To fill a P0D object from an ASCII file, assuming that each number is on a new line, the method read() can be applied. For such files, use # or * for comments at the beginning of each line.

```
>>> p0=P0D("data from ASCII file")
>>> p0.read(file)
```

where file is a string with the file name (the full path should be included). Data can also be read from a compressed ("zipped") file as:

```
>>> p0=P0D("data from a ZIP file")
>>> p0.readZip(file)
```

or from a gzipped format:

```
>>> p0=P0D("data from a GZIP file")
>>> p0.readGZip(file)
```

In all these cases, the methods read(), readZip(), and readGZip() return zero in case of success. Error codes 1-2 mean that the file is not found and 3 indicates a parse error.

To build a P0D object from an ASCII file, use this constructor:

```
>>> p0=P0D("measurement", file)
```

(in this case, GZIP and ZIP form are not supported).

To write a p0 back to an ASCII file, use:

```
>>> p0.toFile(file)
```

Data can be written to a binary file using the so-called big-endian format as:

```
>>> p0.writeBinary(file)
```

To read data from an existing binary file, use

```
>>> p0.readBinary(file)
```

In this case, the old content of the p0 object will be erased.

4.1.5.1   Serialization

One can save and restore a POD object containing data and other attributes using the Java object serialization. DMelt has a short command for this:

```
>>> p0.writeSerialized(file)
```

The object p0 will be saved in a file with the name "file" using a compressed format. The method returns zero in case of success. One can restore the object from the file as:

```
>>> p1=p0.readSerialized(file)
```

where p1 is a new object restored from the input file.

Now let us show an example of how to write a POD object in an external file in a serialized form and then how to restore the object back using a handy static method write(obj) of the class IO. To read a POD object from the file, use the method read() of the same class. The example below shows how to do this:

```
from jhplot  import *

p0=POD("test")
p0.add(20); p0.add(12)
print p0.toString()
IO.write(p0,"file.jser") #  write to a file
p0s=IO.read("file.jser") #  deserialize POD
print "After serialization:",p0s.toString()
```

Listing 4.4   Serialized I/O

The method `write(obj)` in this example writes a compressed object `obj` in a file. In Sect. 9.2, we will show how to deal with the cases when no compression is used. Often, a program can benefit from the use of uncompressed objects since, in this case, the CPU time wasted for uncompressing files is avoided.

4.1.5.2 XML Format

In some cases, it is convenient to write an object to a human-readable XML file, so one can open it using any editor and look at its structure as well as written data. For the XML format, one should use the following methods from the class `jhplot.IO`:

`writeXML(obj,file)`   writes a Java object (`obj`) into a file with the name `file`;

`readXML(file)`   reads a file with the name `file` and returns a stored object.

You can replace the corresponding lines in the the code of Listing 4.4 and check how the output was written.

It should be noted that the XML style is not recommended if there is a lot of data to be stored, since XML tags increase the file size dramatically. In addition, this approach may not work for all DMelt classes.

For your convenience, the package `"jhplot"` provides two other important methods. An object can easily be converted into an XML string using the methods `IO.toXML(obj)` and `IO.fromXML(str)`. Below we illustrate how to print an object in an XML form and then read it back:

```
>>> from jhplot  import *
>>> p0=P0D("data")
>>> str=IO.toXML(p0)
>>> print str
>>> p0=IO.fromXML(str)
```

It should be noted that the serialization can be done using the native Java API. Below we rewrite the example discussed above using the standard Java serialization class:

```
from jhplot  import *
from java.io import *

p0=P0D("test")
p0.add(20)
p0.add(12)
print p0.toString()

# serialize P0D into a file
f=FileOutputStream("filename.jser")
```

```
out =ObjectOutputStream(f)
out.writeObject(p0)
out.close()

# Deserialize POD from the file
file=File("filename.jser")
fin = ObjectInputStream(FileInputStream(file))
p0s =fin.readObject()
fin.close()
print "After serialization:\n",p0s.toString()
```

Listing 4.5   Java serialized I/O for the POD class

4.1.5.3   Dealing with Object Collections

The next question is how to write many POD objects to a single file. This can easily
be done using Python/Jython lists. The example below writes two POD objects into
a single file:

```
from jhplot   import *

p1=POD("p1")
p1.add(10);  p1.add(12)

p2=POD("p2")
p2.add(1000);  p2.add(2000)
p2.add(2000)

a=[p1,p2] # create a list
IO.write(a,"file.jser") #  write to a file
```

Listing 4.6   Writing multiple 1D arrays

We can restore all objects from the file as:

```
from jhplot   import *

f=IO.read("file.jser") # deserialize list
p1=f[0]
p2=f[1]
print "After serialization:\n"
print p1.toString()
print p2.toString()
```

Listing 4.7   Reading multiple 1D arrays

Similarly, one can write objects and read them back using the methods
writeXML(obj) and readXML() discussed before. Finally, one can use Python
dictionaries for convenient access to the data using the keys.

The serialization can be used for almost any DMelt class. In Sect. 4.2.8 we will show how to use Jython dictionaries to store various objects in a serialized file and fetch them later using the keys.

To write huge sequences of POD arrays, the best approach would be to use the class HFile. This topic will be discussed in Chap. 9 dedicated to input and output of Java objects.

4.2   Arrays for Two Dimensions

The previous section considered a holder with 1D arrays of numbers. More frequently, we need to show data in 2D space. Such bivariate data are necessary to consider when each data point on the (X, Y) plane is represented by two variables, x and y. Needless to say such arrays are ideal to show the relationship between two sets of measurements.

The P1D class is among the central classes designed for data manipulations in two dimensions (2D). The name of the P1D class is similar to that of the class F1D used to display functions on the 2D plane.

The declaration of a P1D object is rather similar to that considered for the POD class. However, this time, the container should be filled with at least two numbers using the method add(x,y):

```
>>> from jhplot import *
>>> p1=P1D("x-y points")
>>> p1.add(10,20)
>>> p1.add(20.,40.)
```

In this brief example, we declare the object P1D with the title "x-y points" and then fill it with two points using the method add(x,y). Each point is represented by two numbers (either integer or float).

As stated before, the method add(x,y) is not the most optimal, especially when it is used inside loops implemented in Python. To avoid performance penalties, use the high-level methods of this class for passing containers with numbers, rather than numbers themselves. In the example shown below, we fill a P1D from two POD arrays which, in turn, can also be filled using high-level methods as discussed in the previous chapter:

```
>>> from jhplot import *
>>> p1=P1D("x-y points")
>>> p01=POD("px"); p02=POD("py")
>>> ax=p01.randomNormal(100,1.0,0.5)
>>> ay=p02.randomNormal(100,10.0,1.0)
>>> p1.fill(ax,ay)
```

In the above example, we fill two POD arrays with random numbers in accordance with the normal distributions and use these arrays to fill a P1D object. The above code can be shortened since the POD arrays can also be passed to the P1D constructor as:

```
>>> p1=P1D("x-y points",p01,p02)
```

Of course, make sure that the sizes of the input arrays are the same.

Having filled the array P1D, one can replace data points using the method set(i,x,y), where "i" denotes a position (index) of the (x, y) point inside the array. The array can be emptied up with the method clear(), while the array size can be found using the method size().

Finally, one can extract data points using the methods getX(i) and getY(i) $(0 \leq i < \text{size}())$. Or, one can get arrays for x and y values without looping over all elements inside the container as:

```
>>> from jhplot import *
>>> p1= P1D("x-y points")
>>> p1.add(10,20)
>>> p1.add(20,50)
>>> ax=p1.getArrayX() # get all X values as array
>>> ay=p1.getArrayY() # get all Y values as array
```

One can also initialize a P1D from an external file as:

```
>>> p1= P1D("data from file", "data.d")
```

The format of the input file is rather simple: each pair of numbers should be on a new line. To include comments, use "#" at the beginning of each line. Later, we will discuss the I/O topics in more detail.

4.2.1   Data with Errors

A P1D object is not as simple as it first may look: it can also hold information on errors on x and y values. Therefore, it is particularly well suited for keeping measurements.

There are two different types of errors: one type comes from the fact that there is always inherent statistical uncertainty in counting random events. Usually, such uncertainties are called statistical or random errors. Also, we will call such errors "1st-level errors" when discussing various technical aspects.

The second type of uncertainties are called "systematic errors," which mainly originate from instrumental mistakes or other (not random) causes. For the P1D, we will call such uncertainties "2nd-level errors."

Thus, a single point at the position (x, y) is characterized by ten double-precision numbers: two numbers give the central position in (x, y) and the other eight numbers

Fig. 4.1 An illustration of a P1D array characterized by ten numbers: two numbers represent positions of a data point on the (x, y) plane, while other eight represent 1st- and 2nd-level uncertainties (see the text). The figure displays several methods to access the information on a single data point defined by an index i

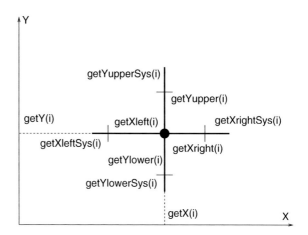

specify its errors. A single point with 1st and 2nd errors is illustrated in Fig. 4.1, together with the "getter" methods used to access all characteristics of a P1D object.

Why do we need this complication when dealing with 2D data? Is it not enough to represent a relationship between two sets of data with two arrays? For example, a P1D can contain information on a set of points representing positions of particles in two dimensions: Each particle will be characterized by the position on the (x, y) surface, while the total number of particles will be given by the size of the P1D array. Well, in this case, you do not need extra information to be stored in this data holder. Probably, you would not need even such a data holder at all, since one can use the usual Jython or Java arrays or the PND class to be discussed in the following chapters. However, there are many situations when each data point represents many measurements. In this case, one should store possible uncertainties associated with the measurements, which can have either statistical or systematical nature (or both). And this is why P1D becomes really handy, since this class is optimized exactly for such tasks.

Let us give a simple example. Assume we measure the average number of cars parked per weekday on some parking spot. If the measurements are done during one month, there should be seven measurements for each day of the week. We calculate the average for each weekday, and represent the entire measurement with seven numbers. Each number has a statistical uncertainty, which is assumed to be symmetric. The measurements can be represented by a single P1D container which, for example, can be filled for two weekdays as:

```
>>> from jhplot import *
>>> p1= P1D("average number of cars with errors")
>>> p1.add(1,av1,err1) # average for Monday
>>> p1.add(2,av2,err2) # average for Tuesday
```

where `av1` and `av2` are the averages for each weekday, and `err1` and `err2` represent their statistical uncertainties.

Let us continue with our hypothetical example and estimate systematic uncertainties for such measurements which reflect inaccuracies of our apparatus. Of course, for this particular example, we do not have any apparatus, but our measurements still may suffer from the uncertainties related to inaccuracies of our observations. We again assume that the systematic uncertainties are symmetric. In the case of the 1st- and 2nd-level errors, we can fill the `P1D` as:

```
>>> from jhplot import *
>>> p1= P1D("average number of cars with errors")
>>> p1.add(1,av1,0,0,err1,err1,0,0,sys_err1,sys_err1)
>>> p1.add(2,av2,0,0,err2,err2,0,0,sys_err2,sys_err2)
```

where `av1` and `av2` are the averages of two different measurements, `"err1"` and `"err2"` their 1st-level errors (i.e., statistical errors). The 2nd-level errors (systematic errors) are added using the `"sys_err1"` and `"sys_err2"` values.

You may wonder why `err` and `sys_err` have been passed to the `add()` twice? The reason is simple: we have used a rather general method, which can also be applied for adding asymmetrical errors. In addition, we had to type "0" which tells that there are no errors attributed for the X-axis. The last point can be clear if we give the most general form of the `add()` method:

```
>>> p1.add(x,y,xLeft,xRight,         # 1st errors on X
              yUpper,yLower,          # 1st errors on Y
              xLeftSys, xRightSys,    # 2nd errors on X
              yUpperSys,yLowerSys)    # 2nd errors on Y
```

where `xLeft` (`xRight`) represents a lower (upper) 1st-level error on the X-axis. For the $(x-y)$ plots, this is represented by a line started at the central position and extended to the left (right) from the central point (see Fig. 4.1). Analogously, `yUpper` (`yLower`) is used to indicate the upper (lower) uncertainty for the Y-axis. Next, four other numbers are used to show the 2nd-level uncertainties for the X and Y axes. As discussed before, such uncertainties are usually due to some systematic effects reflecting instrumental or methodological uncertainties not related to statistical nature.

In many cases, we do not care about systematic uncertainties, so one can use several shortcuts. We have already shown that if one needs to specify only symmetrical statistical uncertainties on the Y-axis then one can use this method:

```
>>> p1.add(x,y,err) # fills X,Y and symmetric error on Y
```

where `err` is a statistical error on the y value, assuming that it is equal to yUpper=yLower. All other errors are set to zero. If the error on Y is asymmetric, use this method:

```
>>> p1.add(x,y,err_up, err_down)
```

where `err_up` and `err_down` are symmetric upper and lower errors on y. If there are only 1st-level errors, then one can fill a P1D as:

```
>>> p1.add(x,y,xLeft,xRight,yUpper,yLower)
```

while the 2nd-level errors are set to zero. Table 4.1 lists the main "setter" methods associated with the P1D class.

Occasionally, it is convenient to reset all errors to zero. For this, use the method:

```
>>> p1.setErrToZero(a)      # set 1st-level errors to 0
>>> p1.setErrAllToZero(a)  # set 1st and 2nd errors to 0
>>> p1.setErrSysToZero(a)  # set 2nd-level error to 0
```

where $a = 0$ for x values and $a = 1$ for y values.

In addition, you may need to generate new errors from the numbers of counted events, when statistical uncertainty for each y value is the squared root of counted numbers [1]. If y represents the counted number of events, its statistical error is $\sqrt{y}$ (upper and lower). One can build a new P1D by assigning the 1st-level errors for x and y separately:

```
>>> p1.setErrSqrt(a)      # set 1st-level errors
```

Errors are set on x (y) when a=0 (a=1).

Table 4.2 gives "getter" methods used to access characteristics of this object.

To obtain an index with either minimum or maximum value, use:

```
>>> p1=p1.getMinIndex(a)  #   index with min
>>> p1=p1.getMaxIndex(a)  #   index with max
```

where a=0 for x values and a=1 for y values.

To find all major statistical characteristics of the P1D array, use the method `getStat()`. Let us consider an example that shows how to access statistics of x and y arrays. In this example, we use the standard Python dictionary to extract and print the mean, variance, and standard deviation of these arrays.

Table 4.1 Some methods used to fill a `P1D` container

Methods	Definitions
add(x, y)	Add (X, Y) (all errors are 0)
add(x, y, err)	Add (X, Y) and 1st-level error on Y
add(x, y, up, down)	Add (X, Y) and asymmetric 1st-level errors on Y
add(x, y, left, right, up, down)	Add (X, Y) and asymmetric 1st-level errors on X and Y
add(x, y, left, right, upper, lower, leftSys, rightSys, upperSys, lowerSys)	Add (X, Y) and asymmetric 1st- and 2nd-level errors on X and Y
set(i, x, y)	Insert (X, Y) at position i (all errors are 0)
set(i, x, y, err)	Insert (X, Y) and 1st-level error on Y
set(i, x, y, up, down)	Add (X, Y) and asymmetric 1st-level errors on Y
set(i, x, y, left, right, up, down)	Insert (X, Y) and asymmetric 1st-level errors on X and Y
set(i, x, y, left, right, upper, lower, leftSys, rightSys, upperSys, lowerSys)	Insert (X, Y) and asymmetric 1st- and 2nd-level errors on X and Y
replace(i, x, y)	Replace (X, Y) at position i (all errors are 0)
replace(i, x, y, err)	Replace (X, Y) and 1st-level error on Y
replace(i, x, y, up, down)	Replace(X, Y) and asymmetric 1st-level errors on Y
replace(i, x, y, left, right, up, down)	Replace (X, Y) and asymmetric 1st-level errors on X and Y
replace(i, x, y, left, right, upper, lower, leftSys, rightSys, upperSys, lowerSys)	Replace (X, Y) and asymmetric 1st- and 2nd-level errors on X and Y
fill(arrayX, arrayY)	Fills from 2 arrays
fill(arrayX, arrayY, arrayE)	Fills from 3 arrays with errors
fill(arrayX, arrayY, arrayUP, arrayDOWN)	Fills from 4 arrays with errors
fill(arrayX, arrayY ...+ 8 arrays with errors)	Fills from 10 arrays with full sets of errors
fill(P0D, P0D)	Fills from 2 P0D arrays
fill(P0D, P0D, P0D)	Fills from 3 P0D arrays with errors
setTitle("text")	Set a title

In the methods shown in this table, "b" indicates a Boolean value (1 for true and 0 for false), while "i" is an integer parameter

Table 4.2   Some methods for accessing information about a P1D object

Methods	Definitions
copy()	Copy to new P1D
getArrayX()	Get array with all X values
getArrayXleft()	Get array with all left X errors
getArrayXleft()	Get array with all left X errors
getArrayXleftSys()	Get array of left 2nd-level errors
getArrayXright()	Get array with all right X errors
getArrayXrightSys()	Get array with right 2nd-level errors
getArrayY()	Get array with all Y values
getArrayYlower()	Get array with all lower Y errors
getArrayYlowerSys()	Get array with all lower 2nd-level Y errors
getArrayYupper()	Get array with all upper Y errors
getArrayYupperSys()	Get array with all upper 2nd-level Y errors
getMax(a)	Max value for axis
getMaxIndex(a)	Get index of max value for axis
getMin(a)	Min value for axis
getMinIndex(a)	Get index of min value for axis
getTitle()	Get title
getX(i)	X value at index i
getXleft(i)	Left error on X at index i
getXleftSys(i)	Left 2nd error on X at index i
getXright(i)	Right error on X at index i
getXrightSys(i)	Right 2nd error on X at index i
getY(i)	Y value at index i
getYlower(i)	Lower error on y at index i
getYlowerSys(i)	Lower 2nd error on Y at index i
getYupper(i)	Upper error on Y at index i
getYupperSys(i)	Upper 2nd error on Y at index i
integral(i1, i2)	Sum up all Y values between i1 and i2
mean()	Mean value
size()	Size of the array
updateSummary()	Update summaries after adding a value

In the methods shown in this table, "b" indicates a Boolean value ("1" is false, "0" is true), while "a" is an integer parameter indicating the axis (a = 0 means X-axis, a = 1 means Y-axis). The notation "d" denotes a float value, while "text" represents a string

```
from jhplot import *

p1= P1D()
p1.add(12,100); p1.add(22,80);
p1.add(28,110); p1.add(54,50);
```

```
statX=p1.getStat(0) # statistics of X values
for keys,values in statX.items():
    print keys, values

statY=p1.getStat(1) # statistics of Y values
for keys,values in statX.items():
    print keys, values
```

Listing 4.8 Statistical characteristics arrays

Finally, one can use the method `integral(i1,i2)` that returns the integral between two indexes, `i1` and `i2`, of a `P1D` data. For this operation, the integration is a sum over all *y* values, so it is rather similar to the integration of 1D functions.

4.2.2 Viewing P1D Data

In addition to the Java serialization mechanism used to store the `P1D` containers, which will be discussed in detail later, one can store entries from a `P1D` in a human-readable format. For example, one can write data in ASCII files using the method:

```
>>> p1.toFile("newfile.d")
```

where `"newfile.d"` is the name of the output file. One can also export data into a LaTeX table as

```
>>> from java.text import DecimalFormat
>>> format=DecimalFormat("##.####E00")
>>> p1.toFileAsLatex("Output.tex",xformat,xformat)
```

One should specify an appropriate format for the numbers to be stored in the LaTeX table. We remind that the class `DecimalFormat` is used to format decimal numbers in the Java platform. The pound sign (`"#"`) denotes a digit, and the dot is a placeholder for the decimal separator. Please refer to the Java API documentations of this class.

Also, one can print a `P1D` container on the console as

```
>>> p1.print()
```

To print the stored values in the shell, convert the `p1` object into a string and print it using the standard `print` method:

```
>>> print p1.toString()
```

In some cases, it is convenient to show data as a table in a separate frame, so one can sort and search for a particular value. In this case, use the statement:

```
>>> p1.toTable()
```

which brings up a frame showing data inside a table. The method calls the class `HTable` to be discussed in Sect. 15.1.3. Analogously, data can be exported to a spreadsheet as discussed in Sect. 15.1.3.

4.2.3 Plotting P1D Data

In order to display a `P1D` representing a set of bivariate data, the usual `HPlot` canvas discussed in Sect. 3.3 can be used. To plot data, follow the same steps as for drawing `F1D` functions: first, create a canvas and then use the `draw(obj)` method to display a `P1D` object on a *scatter* plot:

```
>>> from jhplot import *
>>> c1=HPlot()
>>> c1.visible(); c1.setAutoRange()
>>> p1=P1D("x-y points", "data.d")
>>> c1.draw(p1)
```

In this example, we first create a `P1D` object from an ASCII input file, and then we display data as a collection of points on the (x, y) plane.

Table 4.3 shows the most important graphical attributes associated with the `P1D` class.

It should be noted that one can edit the plot using a pop-up menu `Edit` which allows changing some attributes. Click on the right mouse button to access a GUI-driven dialog with the menu.

There are many methods which come together with the `P1D` data holder to display error bars, assuming that error values have been filled before. To display error bars use the methods:

```
>>> p1.setErr(1)     # show  1st-level errors
>>> p1.setErrSys(1)  # show  2nd-level errors
```

which should be set before drawing the `p1` object on the canvas.

Table 4.4 shows various methods for controlling the attributes of the error bars. As you can see, there are two separate methods to modify the horizontal and vertical error bars. The "getter" methods are similar, but start with the "get" string instead of "set".

Table 4.3 The most important methods for graphical representation of a P1D

Methods	Definitions
setStyle("text")	Set as symbols ("p") or line ("l")
setSymbolSize(i)	Symbol size
setSymbol(i)	Symbol type i = 0–12:
	0: not filled circle
	1: not filled square
	2: not filed diamond
	3: not filled triangle
	4: filled circle
	5: filled square
	6: filed diamond
	7: filled triangle
	8: plus (+)
	9: cross as (x)
	10: star (*)
	11: small dot (.)
	12: bold plus (+)
setColor(c)	Set the line color
setPenWidh(i)	Width of the line
setPenDash(i)	Dashed style with "i" being the length
setLegend(b)	Set (b = 1) or not (b = 0) the legend
setTitle("text")	Set a title

"b" indicates a Boolean value (1 for true and 0 for false), while "i" is an integer parameter. The notation "d" indicates a float value. The attributes "c" and "f" correspond to the Color and Font classes of Java AWT, while "text" represents a Jython string

4.2.4   Contour Plots

The HPlot canvas can be used to display P1D data as contour plots. For such type of plots, we draw colored regions which show the density population, instead of showing separate data points. It is required to bin a (x, y) plane in x and y: the smaller the bin size is, the more chances to resolve a fine structure for the plotted density distribution.

To set up the canvas HPlot for showing contour plots, use the method setContour(1). Table 4.5 shows some methods for the contour plots.

It should be noted that there is a special canvas, called HPlot2D, which is designed to show the contour plots and has more options for manipulations. Please read Sect. 8.8 for more details.

Table 4.4 The most important methods for a graphical representation of the P1D errors

Methods	Definitions
setErrAll(b)	Set all errors (1st- and 2nd-level)
setErrX(b)	Set error on X or not
setErrY(b)	Set error on Y or not
setErrColorX(c)	Color used for X error bars
setErrColorY(c)	Color used for Y error bars
setPenWidthErr(i)	Line width for 1st-level errors
setPenWidthErrSys(i)	Line width for 2nd-level errors
setErrSysX(b)	Set or not 2nd-level error on X
setErrSysY(b)	Set or not 2nd-level error on Y
setErrFill(b)	Fill or nor the area covering errors
setErrFillColor(c)	Fill color
setErrFillColor(c, d)	Fill color + transparency level "d"
setErrSysFill(b)	Fill or not 2nd-level errors
setErrSysFillColor(c)	Fill color
setErrSysColor(c, d)	As before + transparency level "d"

"b" indicates a Boolean value (1 for true and 0 for false), while "i" is an integer parameter. The notation "d" indicates a float value. The attributes "c" and "f" correspond to the Color and Font classes of Java AWT, while "text" represents a Jython string

Table 4.5 HPlot methods for displaying contour plots

Methods	Definitions
setContour(b)	Sets (or not) the contour style
setContourLevels(i)	The number of color levels
setContourBins(iX, iY)	The number of bins in X and Y
setContourBar(b)	Set (or not) a color line showing levels
setContourGray(b)	Set (or not) white–black style

"b" indicates a Boolean value (1 for true and 0 for false), while "i" is an integer parameter

4.2.5 Manipulations with P1D Data

The P1D containers are designed from the ground to support numerous mathematical operations. The operations do not create new objects, but just modify the original containers. To create a new P1D object, use the method copy(). For example:

```
... p1 is created above...
>>> p2=p1.copy() # now p2 is different object
>>> c1.draw(p1)  # draw two different objects
>>> c1.draw(p2)
```

One can merge two P1D containers into one using the method merge(). One can also add, divide, subtract, and multiply the P1D objects. Let us read two P1D containers from files and perform such operations:

```
>>> p1=P1D("first", "data1.d")
>>> p2=P1D("second", "data2.d")
>>> p3=p1.merge(p2)            # merge 2 P1D's into one
>>> p1.oper(p2,"NewTitle","+") # add  p1 and p2
>>> p1.oper(p2,"NewTitle","-") # subtract p2 from p1
>>> p1.oper(p2,"NewTitle","*") # multiply p1 by p2
>>> p1.oper(p2,"NewTitle","/") # divide   p1 by p2
```

The execution speed of these operations is significantly faster compared to equivalent Jython codes based on loops, since all such methods are implemented in the Java libraries. One can skip the string with a new title if you want to keep the same title as for the original P1D. For example, in this case, the additive operation will be p1.oper(p2,"+"). All graphical attributes are preserved during such data manipulations.

For the above operations, the errors on p3 will be propagated accordingly assuming that all 1st- and 2nd-level errors associated with p1 and p2 are independent of each other. For the error propagation, we use a rather standard prescription [1].

To scale a P1D with a number, use the statement:

```
>>> p1.operScale(a, scaleFactor)
```

where scaleFactor is a double or an integer number. If a=0 (Java Boolean "false"), the scaling is applied for x, if a=1, the scaling is applied for y. The title is optional for this operation. It is important to know that the factor "scaleFactor" scales the 1st- and 2nd-level uncertainties as well. If you need to scale only errors, use:

```
>>> p1.operScaleErr(a, scaleFactor)
```

which scales the 1st-level errors for either $X- (a = 0)$ or $Y- (a = 1)$ axis. If one needs to scale also the 2nd-level errors, use:

```
>>> p1.operScaleErrSys(a, scaleFactor)
```

which works exactly as the method operScaleErr(), but this time it is applied for the 2nd-level errors.

Finally, to extract a range of P1D points from the original data container, use

```
>>> p1=p1.range(min,max)
```

where min and max are integer numbers denoting the range.

4.2.6 Advanced P1D Operations

4.2.6.1 Operations with Correlations

The operations considered above assume that there are no correlations between two data holders. In reality, data from different measurements can correlate, so do P1D containers corresponding to such measurements. In this case, one can also specify a correlation coefficient and use it for mathematical operations. A correlation coefficient should be represented with an additional P1D container used for the actual mathematical manipulations. The correlation coefficients should be added using the add() method and included at the positions of errors (statistical or systematical). Look at the example below where we assume that there is a 50 % correlation between two data sets:

```
from jhplot import *

p1= P1D("data1")    # data set 1 points with 2 point
p1.add(10,100,5,5) # 1st-level errors on Y only
p1.add(20,50,5,5)

p2= P1D("data2") # data set 2 with 2 points
p2.add(10,40,5,5) # 1st-level errors on Y only
p2.add(20,40,5,5)

# add with 50% correlations
corr=P1D("correlation coefficients")
corr.add(0,0,0.5,0.5)
corr.add(0,0,0.5,0.5)
p50=p1.copy() # add arrays
p0=p1.copy()
p50.oper(p2,"added with 50% corr.","+","Y",corr)
print p50.toString()
p0.oper(p2,"added with 0% corr.","+")
print p0.toString()
```

Listing 4.9 Adding data with correlations

The output of this script shows that the (x, y) values remain to be the same, but the statistical errors are different for the case with 50 % correlations and without any correlations.

Analogously, one can include correlations for the 2nd-level errors (see the API description for this method). The same feature is supported for any operation, such as subtraction, multiplication, and division.

4.2.6.2   Functional Transformation

One can also transform a `P1D` data using a mathematical function. The error propagation is done for x or y components (or for both). The following functions are supported: "inverse" $(1/y)$, square $(y*y)$ "sqrt(y)" (square root), "exp(y)" (exponential), "log(y)" $(\log_{10}(y))$, and all trigonometrical functions. Let us show an example that illustrates a generic usage of such functional transformations:

```
>>> p1=p1.move("function", "a" )
```

where `"function"` is a string defining a function used for the transformation, and `"a"` is a string indicating the axis, which can either be "X" (apply the transformation for x values), "Y" (apply the transformation for y values) or "XY" (transform both x and y). For example:

```
>>> p1=p1.move("log", "Y" )
```

transforms all y values to $\log_{10}(y)$. Errors for the `p1` container will be transformed appropriately.

4.2.6.3   Smoothing

In some situations you may be interested in smoothing `P1D` values. This technique, which is rather general and can be applied to any data container, will be described in Sect. 14.3. In this section, we can briefly mention that a smoothing can be done by averaging over a moving window of a size specified by the method parameter: if the value of the parameter is "k" then the width of the window is $2*k+1$. If the window runs off the end of the `P1D` only those values that intersect are taken into account. The smoothing may optionally be weighted to favor the central value using a "triangular" weighting. For example, for a value of "k" equal to 2 the central bin would have weight 1/3, the adjacent bins 2/9, and the next adjacent bins 1/9. Errors are kept to be the same.

All of this can be achieved using the command:

```
>>> p2=p1.operSmooth(a,b,k)
```

where `"a"` defines the axis to which the smoothing is applied, i.e., it can be either a=0 (for the "X"-axis) or a=1 (for the "Y" values). When b=1 (Boolean "true") then x or y values are weighted using a triangular weighting scheme favoring bins near the central bin, and "k" is the smoothing parameter which must be nonnegative. If zero, the original `P1D` object will be returned with no smoothing applied.

One can also convert a `P1D` into a Gaussian smoothed container in which each band of the original `P1D` is smoothed by discrete convolution with a kernel approximating a Gaussian impulse response with the specified standard deviation. This can be done using the command:

```
>>> p2=p1.operSmoothGauss(a,sDev)
```

where "sDev" is the standard deviation of the Gaussian smoothing kernel (must be nonnegative).

4.2.7 Weighted Average and Systematical Uncertainties

When there are several measurements with different values x_i and known errors σ_i (or variance, σ_i^2), then it is reasonable to combine the measurements using the so-called weighted average method. In this case, the value x for best estimate and its variance are given by

$$x = \sum (x_i/\sigma_i^2) / \sum (1/\sigma_i^2)$$
$$\sigma^2 = 1 / \sum (1/\sigma_i^2)$$

This calculation can be done in one line as:

```
>>> p1.addAndAverage(P1D[])
```

Let us explain this method. Assume there is a measurement represented by a `p1` and additional measurements defined by an array of `P1D` objects. Then, after calling the above method, `p1` will represent a weighted average with the corresponding 1st-level errors.

The class `P1D` is also very useful when one needs to evaluate systematic uncertainties of many measurements. If we have several measurements with different resulting outcomes, and each measurement is represented by a `P1D` data holder, one can obtain a `P1D` representing the final measurement using the method `getSys()` which returns the final `P1D` object with systematic uncertainties. Let us illustrate this using the following example:

```
>>> p0= P1D("default") # original measurement
>>> pp=[]
>>> p1= P1D("data1")    # other measurements
>>> pp.append(p1)
>>> p2= P1D("data2")
>>> pp.append(p2)
>>> p3= P1D("data3")
>>> pp.append(p3)
>>> ....
>>> psys=p0.getSys(pp) # build uncertainties
>>> psys.setErr(1)
>>> c1.draw(psys)
```

It should be noted that the systematic uncertainties are added in quadrature, thus they are assumed to be independent of each other.

One can also face with the situation like this: there are three P1D objects, the first contains an array with the central values and two others represent lower and upper deviations. One can build a new P1D with the 2nd-level errors that represent the differences between the central P1D values and the upper and the lower P1D using the method operErrSys ("title", a, p1,p2), where "a" represents the axis (0 for X, 1 for Y), and p1 and p2 are objects for the lower and upper errors. The example below illustrates this for two data points:

```
from java.awt import Color
from jhplot  import *

c1 = HPlot()
c1.visible(); c1.setRange(5,25,70,120)
c1.setGTitle("Uncertainties",Color.blue)

p1= P1D("Central")
p1.setColor(Color.blue)
p1.add(12,100); p1.add(22,80)

p2= P1D("Lower")
p2.add(10,90); p2.add(20,75)
c1.draw(p2)

p3= P1D("Upper")
p3.add(10,110); p3.add(20,96)
c1.draw(p3)

p0=p1.operErrSys("Data",1,p2,p3)
p0.setErrSys(1)
p0.setPenWidthErrSys(2)
c1.draw(p0)
```

Listing 4.10 Errors from two input P1D arrays

Fig. 4.2 Displaying a new P1D object with the 2nd-level errors given by two other P1D containers

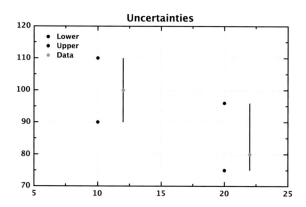

This script plots a p0 object with new 2nd-level errors given by the difference between the input P1D containers. The output plot is shown in Fig. 4.2. The container with the central values can contain statistical errors, while statistical errors in the input P1D containers are ignored.

4.2.8 File Input and Output

The P1D containers can be saved into a file and restored later. As for the POD class, one can fill a P1D container from an ASCII file using the method read(). We remind that each (x, y) pair should be on a separate line, and x and y should be separated by a space. One can use the characters # or * at the beginning of each line for comments:

```
>>> p0=P1D("data from ASCII file")
>>> p0.read(file)
```

In this example, file is a string with the file name (the full path should be included). One can use a shorter version of the code above by passing a file name directly to the constructor:

```
>>> p0=P1D("data from ASCII file", file)
```

as illustrated before. Data can also be read from a compressed (Zip) file:

```
>>> p0=P1D("data from a ZIP file")
>>> p0.readZip(file)
```

or from a file in the Gzip format:

```
>>> p0=P1D("data from a GZIP file")
>>> p0.readGZip(file)
```

In all cases, if the methods read(), readZip(), and readGZip() return zero, then the operation was successful. Error code 1 or 2 tells that the file is not found. If a parse error occurs, the reaGZip() method returns 3.

To write a P1D data into an ASCII file, use the following method:

```
>>> p0.toFile(file)
```

There is another handy method: one can store data in the LaTeX format as:

```
>>> from java.text import DecimalFormat
>>> format=DecimalFormat("##.####E00")
>>> p0.toFileAsLatex(file, format, format)
```

where "format" is an instance of the DecimalFormat class from the Java
package java.text. We passed the object "format" twice, one for *x* values
and one for *y* values.

It should be pointed out that, for the above examples, we could only write (read)
data. All graphical attributes are completely lost after saving the data in ASCII files.
But there is another more elegant way to save a P1D object: As for any DMelt
object, one can save and restore P1D data and other attributes (including those used
for graphical representation) using the Java serialization mechanism.

To serialize the entire P1D object into a file, use the method below:

```
>>> p0.writeSerialized(file)
```

which writes the object p0 including their graphical attributes into a file with the
name file. The method returns zero in case of no I/O problem occurs. One can
restore the object from this file as:

```
>>> p1=p0.readSerialized(file)
```

where p1 is a new object from the file.

The example below shows how to write a P1D into an external file in a serialized
form and then restore the object back:

```
from jhplot   import *

p1=P1D("x-y data")
p1.add(10,20)
p1.add(12,40)
print p1.toString()
IO.write(p1,"file.jser") #  write to a file
p1s=IO.read("file.jser") # read fro the file
print "After serialization:",p1s.toString()
```

Listing 4.11   P1D serialization

The serialization to the XML format can be achieved using the writeXML()
and readXML() methods of the same jhplot.IO class. This class also allows
conversion of the container into XML strings using the toXML() method.

4.2.8.1   Reading and Writing Collections

One can write any number of P1D objects into a single file using the serialization
mechanism. The idea is simple: put all P1D objects into a list and then serialize the
entire list in one step. Of course, one can use Python/Jython tuples or dictionaries
instead of using lists.

Below we show how to save two different objects, P0D and P1D, in one file and then restore them later:

```
from jhplot  import *

p1=P1D("p1 data")
p1.add(10,20);  p1.add(12,40)
print p1.toString()

p2=P0D("p2 data")
p2.add(1000);  p2.add(2000)
print p2.toString()

a=[p1,p2]                     # make a list
IO.write(a,"file.jser")    # write to a file
list=IO.read("file.jser")  # read from the file

p1,p2 = list[0], list[1]
print "After serialization:\n"
print p1.toString(),"\n", p2.toString()
```

Listing 4.12 Serialization of multiple containers

We should note again that one can also use the standard Java for the serialization as shown in Sect. 4.1.5.

It is more convenient to use Python dictionaries to store different objects, since there will be no need to memorize the order of objects in the Python list or tuple holding other objects. We remind that, in the case of dictionaries, we have a one-to-one relationship between the keys and the corresponding values. Below we write three objects into a file, P1D, P0D, and F2D, using the keys for easy retrieval:

```
from jhplot  import *

hold = {}      # create dictionary
p1=P1D("p1dobject")
p1.add(10,20);   p1.add(12,40)
hold[p1.getTitle()] = p1

p2=P0D("p0dobject")
p2.add(1000); p2.add(2000)
hold[p2.getTitle()]=p2

f1=F2D("2*x*sqrt(2*y)")
hold["f2dobject"] = f1.get() # function object

IO.write(hold,"file.jser") # write the dictionary
hold=IO.read("file.jser")  # read the dictionary
print hold.keys() # print all keys

print "After serialization:\n" # fetching objects using keys
print "P1D = ",hold["p1dobject"].toString()
print "P2D = ",hold["p0dobject"].toString()
```

```
f2=F2D(hold["f2dobject"])
print "F2D = ",f2.toString()
```

Listing 4.13 Serialization using dictionaries

In this example, we create a dictionary `hold` to store different objects using the keys (strings). We used the title strings as the keys for the P1D and P0D objects. Then we write this dictionary into a file. In the second part of this code, we read the dictionary from the file and restore all the objects back using the corresponding keys. For example, the object `newhold["p1dobject"]` gives access to the P1D object stored in the dictionary. It should be noted that the keys can be any Jython objects, not only strings.

Finally, one can store and retrieve data using the `HFile` class which is designed to work with large sequences of arbitrary Java objects, including P1D (see Chap. 9).

4.2.9 Example I: Henon Attractor

We will illustrate a typical program based on the P1D class. For our example, we will consider a Henon map (or Henon attractor) [2, 3]. The Henon map can be written as the coupled equations:

$$x_{n+1} = y_n + 1 - a * x_n^2$$
$$y_{n+1} = b * x_n$$

The parameters are usually set to the canonical values, $a = 1.4$ and $b = 0.3$. This simple equation is known to exhibit properties of an attractor with a fractal structure of its trajectories (the so-called "strange attractor").

Let us show how to program this attractor using the P1D class. To visualize the attractor on the (x, y) plane, the SPlot canvas will be used as a lightweight alternative to the HPlot class considered before. The SPlot class will be discussed in Sect. 8.7. Here, we will note that this canvas has a low memory footprint compared to HPlot which is more interactive, but also requires more memory.

The code snippet that implements the Henon map with 10,000 iterations are shown below:

```
from jhplot  import *

a,b=1.4,0.3 # initial values
p=P1D("Henon attractor")
p.setSymbol("Dot")
x,y=0,0
for i in range(10000):
    x1=x
    x=1+y-a*x*x
    y=b*x1
```

```
      p.add(x,y)
c1 = SPlot()
c1.setGTitle("Henon attractor")
c1.visible(); c1.setAutoRange()
c1.draw(p)
```

Listing 4.14   Henon attractor

Figure 4.3 shows the resulting image for $a = 1.4$ and $b = 0.3$. One can examine the fine structures of this attractor by zooming into a specific rectangular area of the plot by clicking and dragging the mouse to draw a rectangle where desired. One can also replace the method `setAutoRange()` with the method `setRange(xmin,xmax,ymin,ymax)` and rerunning the script (the arguments of this method define the zooming rectangle).

In Sect. 8.7, we will rewrite this example using another approach in which we will directly populate the canvas with (x, y) points without the intermediate step based on the `P1D` class.

4.2.10   Example II. Weighted Average

In this section, we will consider how to find a weighted average of several measurements and plot it together with the original measurements. The weighted average of a list with `P1D` objects was already considered in Sect. 4.2.7. In addition, this example shows a somewhat more technical issue: how to mix Java with Jython classes.

For an educational purpose, we will diverge from our original concept which stated that all CPU intensive calculations should be managed inside Java libraries. For this example, we create a custom Jython class "measurement" mixing Jython with Java class `P1D`. The latter will be used mainly for graphical representation.

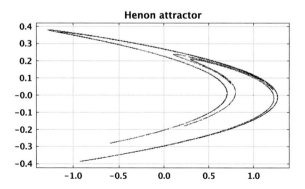

Fig. 4.3   The Henon attractor. Use the mouse for zooming in order to discover its fine structure

Let us create a file `"measurement.py"` with the lines:

```
from jhplot import *

class measurement:
  def __init__(self, number, value, error):
    "A single measurement"
    self.number = number
    self.v  = value
    self.err =  error
    self.p1=P1D(str(self.number))
    self.p1.setSymbolSize(8)
    self.p1.setSymbol(4)
    self.p1.setPenWidthErr(4)
    self.p1.setPenWidthErrSys(2)
    self.p1.add(self.v,self.number,self.err,self.err,0,0)
  def echo(self):
    print  self.number,self.v,self.err
  def getValue(self):
    return self.v
  def getError(self):
    return self.err
  def getNumber(self):
    return self.number
  def getPoint(self):
    return self.p1
```

Listing 4.15 Module "measurement.py"

An object of this class keeps information about a single measurement characterized by an integer number (which defines the type of measurement), measured value (can be accessed with the method `getValue()`), its statistical error (accessed as `getError()`). Finally, we will return the measurement represented in the form of a P1D object. Note the way to fill this object: unlike the previous example, we assign errors in x-direction, rather than for the y-axis. This is mainly done for better representation of the final result.

Let us test this module. We will assume that two measurements have been performed. We fill a list with these two measurements and then print the filled values. Assuming a "counting" experiment in which the statistical error is the square root of the counted numbers of events, our Jython module to add the measurements can look as:

```
from measurement import *
import math

data=[]
data.append( measurement(1,100, math.sqrt(100)))
data.append( measurement(2,120, math.sqrt(120)))

for m in data:
   m.echo()
```

Listing 4.16 Adding data from two measurements

Now let us calculate a weighted average of an arbitrary number of measurements
with errors. The weighted average and its error for two measurements with a common error σ_C are:

$$v3 = \frac{v_1/(\sigma_1^2 - \sigma_c^2) + v_2/(\sigma_2^2 - \sigma_C^2)}{1/(\sigma_1^2 - \sigma_c^2) + 1/(\sigma_2^2 - \sigma_C^2)},$$

$$\sigma_3^2 = \frac{1}{1/(\sigma_1^2 - \sigma_c^2) + 1/(\sigma_2^2 - \sigma_c^2)} + \sigma_C^2,$$

where v_1 and v_2 are measured values and σ_1 and σ_2 are their statistical errors, respectively. We have introduced the common error for generality; in most cases, the com-
mon error $\sigma_C = 0$. The module that calculates the weighted average from a list of
measurements and a common error can be written as:

```
from measurement import *
import math

def average(meas,c):
  "Calculate weigthted average"
  s1,s2=0,0
  for m in meas:
    e=m.getError()*m.getError()-c*c
    w1= 1.0/e
    w2= m.getValue()*w1
    s1=s1+w1
    s2=s2+w2
  err=math.sqrt((1.0/s1)+c*c)
  return measurement(len(meas)+1,s2/s1,err)
```

Listing 4.17 Module "average.py"

Put these lines into a file with the name `"average.py"`.

Now let us use this module and plot the original measurements as well as the
weighted average (for which we used a different symbol). This time, we will use the
class `HPlotJa` which has very similar methods as for the `HPlot` canvas, but it can
be used for an interactive drawing with the mouse, as discussed in Sect. 8.6.

```
from measurement import *
from average import *
import math

c1 = HPlotJa("Canvas")
c1.removeAxes()
c1.showAxis(0)
c1.setGridAll(0,0); c1.setGridAll(1,0)
c1.setRange(50,150,0,5); c1.visible()
c1.setNameX("Measurements")

data=[]
```

```
data.append( measurement(1,100, math.sqrt(100)))
data.append( measurement(2,120, math.sqrt(120)))

p=average(data,1.)
for m in data:
   c1.draw(m.getPoint())

p1=p.getPoint()
p1.setSymbol(5)
c1.draw(p1)
```

Listing 4.18   A weighted average

Figure 4.4 presents the results. Two filled circles show the original measurements, while the third point shows their weighted average calculated by the module "average.py".

4.3   Other Arrays

4.3.1   P2D Data Container

Now let us discuss how to deal with data used to work with the number of dimensions larger than two. A natural extension of P1D is the class P2D used to show data in three dimensions. It is similar to the P1D. The only difference is that it keeps data in a 3D phase space (x, y, z). Also, it has less options for drawing and, in addition, statistical and systematical errors are not supported. In Jython, one can add values to this container in the same way as for P1D, only this time the method add(x,y,z) takes three arguments. In the example below, we create a P2D object and append a single point with the components (1, 2, 3):

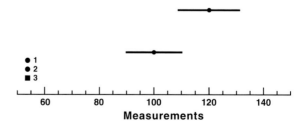

Fig. 4.4   A weighted average (marked by the *filled square* and labeled as "3") of two independent measurements (indicated using the keys "1" and "2")

Table 4.6 The main P2D methods

Methods	Definitions
add(x, y, z)	Add (x, y, z)
clear()	Clean from data
copy()	Copy to a new P2D
fill(arrayX, arrayY, arrayZ)	Fills from 3 arrays
getArrayX()	Get arrays with X
getArrayY()	Get array with Y
getArrayZ()	Get array with Z
getMax(a)	Get max for axis
getMin(a)	Get min for axis
getTitle()	Get title
getX(i), getY(i), getZ(i)	Get points at index i
mean(a)	Get mean for axis
merge(p2d)	Merge with another P2D
set(i, x, y, z)	Insert (x, y, z) at position i
setTitle("text")	Set title
size()	Get size
toTable()	Show in a table

In this table, "b" denotes a Boolean value (1 for Java "true" and 0 for "false"), while "i" is an integer parameter. "a" indicates the axis (a = 0, 1, 2)

```
>>> from jhplot import *
>>> p2= P2D("x-y-z points")
>>> p1.add(1,2,3)
```

Table 4.6 shows the main methods of this class. We will not discuss its details since the P2D arrays are similar to the P1D Java class discussed before. You can always look at the Java API documentation to learn more about this Java class.

4.3.1.1   Drawing P2D Data in 3D

To draw a P2D, one should use the 3D canvas based on the HPlot3D class. This class was discussed in Sect. 3.4.2 and used to draw F2D functions. In the example below, we draw two data sets shown in blue and red:

```
from jhplot import *
from java.awt import Color

c1 = HPlot3D()
c1.visible(); c1.setRange(-5,10,-5,5,-10,20)
c1.setNameX("X"); c1.setNameY("Y")
```

```
h1= P2D("blue data")
h1.setSymbolSize(6)
h1.setSymbolColor(Color.blue)
h1.add(1,2,3)
h1.add(4,4,5)
h1.add(3,2,0)
c1.draw(h1)

h1= P2D("red data")
h1.setSymbolSize(6)
h1.setSymbolColor(Color.red)
for i in range(10):
   h1.add(0.1*i, 0.2*i, 0.5*i)
c1.draw(h1)
```

Listing 4.19 Drawing P2D data in 3D

Table 4.7 lists several most important methods used to draw P2D data on the 3D canvas. The number of methods is not very large, since many drawing methods belong to the actual 3D canvas and are not attributed to the P2D object itself. We will discuss the HPlot3D methods in Sect. 8.9.

We remind that the HPlot3D canvas is similar to HPlot: one can display several plots on the same canvas and change the plotted regions using the cd(i1,i2) method:

```
>>> from jhplot import *
>>> c1=HPlot3D("Canvas",600,400,2,2)  #   2x2 pads
>>> c1.visible()              # set visible
>>> c1.cd(1,1)                # go to the 1st pad
... draw some object ..
>>> c1.cd(1,2)                # go to the 2nd pad
```

As for the HPlot canvas, the first two integers in the constructor HPlot3D define the size of the canvas (600 × 400 pixels for this example), while the two other integers define how many drawing regions (pads) should be shown (2 regions in X and 2 regions in Y).

Let us give a more concrete example of how to work with the P2D:

Table 4.7 Graphical methods for displaying P2D data

Methods	Definitions
setSymbolColor(c)	Set symbol color
setSymbolSize(i)	Set symbol size
getSymbolColor()	Get symbol color
getSymbolSize()	Get symbol size

In the methods shown in this table, "c" denotes the Java Color class, while "i" is used to indicate an integer parameter

```
from java.util import Random
from java.awt import Color
from jhplot import *

c1 = HPlot3D()
c1.visible(); c1.setRange(-5,10,-5,5,-10,30)
c1.setGTitle("Interactive 3D")
c1.setNameX("X"); c1.setNameY("Y")

p1= P2D("3D Gaussian 1")
p1.setSymbolSize(6)
p1.setSymbolColor(Color.blue)

rand = Random()
for i in range(200):
   x=1+rand.nextGaussian()
   y=1+0.5*rand.nextGaussian()
   z=10+4.5*rand.nextGaussian()
   p1.add(x,y,z)

p2= P2D("3D Gaussian 2")
p2.setSymbolSize(10)
p2.setSymbolColor(Color.red)
for i in range(50):
   x=2+2*rand.nextGaussian()
   y=4+0.5*rand.nextGaussian()
   z=6+1.5*rand.nextGaussian()
   p2.add(x,y,z)

c1.draw(p1)
c1.draw(p2)
```

Listing 4.20   Showing P2D data in 3D

In this script, we fill two P2D containers with the Gaussian numbers and plot them on the same canvas. The result of this script is shown in Fig. 4.5. Note that the most efficient way to fill the containers is to fill P0D with Gaussian numbers, and use three P0D containers for the input of the P2D constructor.

4.3.2   P3D Data Container

You may wonder what could be shown with the object called P3D since, by analogy, it must contain points to be shown in a 4D space. The P3D container, by design, still can be used to show 3D data. This time, however, this container can show objects in the 3D space, instead of points. All methods of the P3D are similar to that of P2D, the only difference is that each data point in x, y, z, has an additional parameter representing an extension of the point in the corresponding directions. As before, the canvas HPlot3D should be used for drawing such objects. To fill a

Fig. 4.5 Two arrays defined
by the P2D class displayed
using the HPlot3D canvas

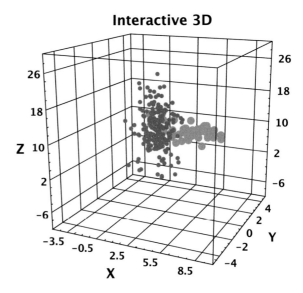

P3D container, one should use the method add(x,dx,y,dy,z,dz) which takes
exactly 6 arguments, with dx, dy, and dz being the extensions in corresponding
directions.

Here, we will stop the discussion of this class since it has limited use for mathematical manipulations. Rather, we will show a simple example of how to use the
P3D class to draw various shapes, such as lines, cubes, and surfaces.

```
from jhplot import *
from java.util import Random
from java.awt import Color

c1 = HPlot3D()
c1.visible(); c1.setRange(-5,10,-4,10,0,20)
c1.setGTitle("P3D objects shown in 3D")
c1.setNameX("X"); c1.setNameY("Y")
c1.setNameY("Y")

h1 = P3D("3D form in blue")
h1.setPenColor(Color.blue)
h1.setPenWidth(3)
h1.add(4.0,1.0,8.0,2.0,3.0,4.0) # 3D cube
h1.add(5.0,2.0,3.0,1.0,8.0,0.0) # 2D panel (Z=0)

h2 = P3D("3D form in red")       # create another cube
h2.setPenColor(Color.red)
h2.add(-0.5,3.0,-1.0,2.0,6.0,2.0)

c1.draw([h1,h2])
```

Listing 4.21 Showing objects in 3D using the P3D class

Fig. 4.6 P3D objects displayed using the HPlot3D canvas

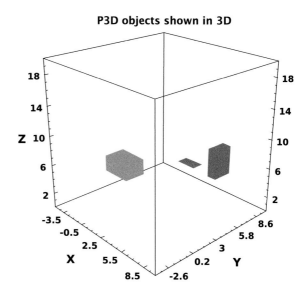

Figure 4.6 shows the resulting plot. We should note that the P3D class is not designed for complicated drawings of geometrical objects in 3D. Generally, it is more appropriate to visualize geometrical shapes using other approaches.

4.3.3   PND Data Container

The PND class is used to manipulate with data in N-dimensional phase space, i.e., this container keeps points represented by N numbers. This explains the appearance of "N" after "P" in the name of this class. As usual, "D" in the name means that the object stores double values. Use the class PNI to store integer values.

Unlike other classes discussed so far, this class does not have any graphical attributes, since it is mainly designed for data manipulations in N-dimensional space. To be able to visualize multidimensional data, one should always project such data into lower dimensions, or use the P1D or P2D classes for drawing.

Below we will discuss the class PND, since its clone for integer values, PNI, has exactly the same methods. As usual, one should first initialize a PND object and then fill it. This can be done by appending lists with numbers using the method add(list). The lists can have any size. The example below shows how to create and fill a PND:

```
>>> from jhplot import *
>>> pn=PND("example")        # build a PND object
>>> pn.add([1,2,3])          # append three values
>>> pn.add([2,3,4])          # another three values
>>> pn.add([2,3,4,3,4])      # append more values
```

```
>>> pn.add([2,3])              # append less values
>>> print pn.getDimension()  # set by the last appended item
2
>>> print pn.toString()
PND: example
1.0 2.0 3.0
2.0 3.0 4.0
2.0 3.0 4.0 3.0 4.0
2.0 3.0
```

It should be noted that the dimension for each row is not fixed, i.e., one can add an array with arbitrary length. The method `getDimension()` returns the dimension of the last appended array. If the dimension is different for each row, one could expect problems for methods based on 2D arrays with fixed number of rows and columns. So, try to avoid the use of rows with different length to avoid problems in future.

As for any other class, one can build a PND from an ASCII file. Once the object is created, one can easily obtain an array from a certain column or row in the form of POD using the following methods:

```
>>> p0=PND("PND from file", file)
>>> p0.getRow(i)     # get a row at index "i" as POD
>>> p0.getColumn(i) # get a column at index "i" as POD
```

where `file` is a string with the file name. Table 4.8 shows the most important methods of the class PND.

To draw a PND object, create a histogram object first. It collects all values stored in the array and project them in one dimension. Let us give a simple example of how to display a PND value:

```
from jhplot import *

c1 = HPlot()
c1.visible(); c1.setAutoRange()
pnd=PND("array")
pnd.add([1,2,3,4])
pnd.add([5,6,7,8])
pnd.add([1,3,2,1])
h1=pnd.getH1D(10) # 10 bins in the range [min,max]
c1.draw(h1)
```

Listing 4.22 Showing PND data as a histogram

We should note that the minimum and maximum values of the histogram are determined automatically. We have only specified the number of bins for data projection. Alternatively, one can make a histogram from a given column of the PND array.

Table 4.8 Some methods of the PND class

Methods	Definitions
add(array[])	Add array
add(P0D)	Add P0D as row
clear()	Remove all entries
copy("text")	New copy with a title "text"
getArray()	Get double array
getArrayList()	Get data in form of array list
getColumn(i)	Get P0D from column at index "i"
getDimension()	Get dimension (last added entry)
getH1D(bins)	Get H1D histogram with "bins"
get(i, j)	Returns value at row "i" and column "j"
get(i)	Returns row "i" as array
getMax()	Get minimum value
getMin()	Get minimum value
getRaw(i)	Get P0D from row at index "i"
remove(i)	Remove a row at index i
rescale(i)	Rescale each column to [0, 1] (i = 0) or [−1, 1] (i = 1)
setArray(array[][])	Set from double array
set(i, array[])	Set array at index i
set(i, P0D)	Set P0D at index i
setTitle()	Set title
size()	Get the size
standardize()	Convert each column to $(x_i - \bar{x})/\sigma$
toString()	Convert data to a string

"i" and "j" denote integer indexes

4.3.3.1 Operations with PND Data

One can perform various operations with the PND data. Below we will discuss the most important methods. Please refer to the Java API documentation of this class.

To scale all data by a constant factor, use this method:

```
>>> p0.operScale(scale)
```

where "scale" is a double value used to scale all elements.

A PND can be rescaled to the range [−1, 1] or [0, 1] with the rescale() method. Another operation is the so-called standardize(), which is useful for the neural network studies to be discussed in Sect. 13.4.1.

One can add, subtract, multiply, and divide two PND arrays. Assuming p0 and p1 are objects of the PND class, this can be done as:

```
>>> p2=p0.oper(p1,"NewTitle",operation)
```

where "operation" is a string which can either be "+", "/", "*", or "/" (they are
self-explanatory). Finally, "NewTitle" is an optimal title which can be dropped.

4.3.4 File Input and Output

All the containers discussed before, P2D, P3D, and PND, can be serialized and
restored back in exactly the same way as discussed in Sect. 4.2.8.

In addition, one can write and read ASCII files with the instances pn of the class
PND using the methods:

```
>>> pn.toFile(file)  # write to a file
>>> pn.write(file)
>>> pn.read(file)     # read from files
```

where file is a string with the file name.

If a multidimensional collection of data must be stored in a single file, the most
convenient way would be to populate a Python/Jython list or a dictionary with the
PND objects and serialize it to a file as shown in Sects. 4.1.5.3 and 4.2.8.

We will come back to the I/O methods in Chap. 9.

4.4 Third-Party Data Containers

DMelt contains many types of arrays which can be used to hold data in the form of
primitive numbers. Typically, they come from third-party Java libraries.

One should not consider such external arrays as being completely independent
of the native DMelt containers which come with the package jhplot. There are
many ways to convert the third-party arrays into the POD, P1D, and PND objects for
manipulation and graphical representation.

4.4.1 Math Arrays

The previous section discussed DMelt arrays, such as P1D, which are "loaded" with
graphical attributes. Such attributes are useful for data visualization, but they may
not be needed in other situations. Plus, a penalty you pay for complex data objects
can be high, since they use more memory and can be slower.

Now we will consider simple arrays which are more optimized for performance, but do not have graphical attributes. Let us import the `ArrayMath` class and see what is inside:

```
>>> from jhplot.math.ArrayMath import *
>>> dir()
```

You will find about 100 static methods of the class `ArrayMath` to be used for manipulations with such arrays. Note that they are not Python lists. The implementation is based on Java, thus you can use this class with other languages on the Java platform. Let us make a list and then perform various manipulations:

```
>>> a=[-1,-2,3,4,5,-6,7,10]      # make a Python list
>>> print mean(a)                # mean value
>>> print standardDeviation(a)   # standard deviation
>>> print sumSquares(a)          # sums the squares
>>> print toString(a)            # print
>>> b=normalize(a)               # renormalizes to 1
```

We will leave the reader here to explore the methods of such arrays. You can use the class `ArrayMath` for rather complex manipulations which are more advanced than the standard Python or Java lists.

4.4.2   Jaida Data Containers

In this section, we will consider the Jaida library which is included to the DMelt package by default. This library provides a number of data containers that are useful for keeping data.

In order to keep data points with uncertainties, use the `Measurement` class. There are several constructors for this class that are best illustrated in this example:

```
>>> from hep.aida.ref.histogram import *
>>> m=Measurement(d)        # d is a value
>>> m=Measurement(d,err)    # d with uncertainty err
>>> m=Measurement(d,errMinus, errPlus)
```

where "d" is a central measurement represented by a double value, "err" is its error. "errMinus" and "errPlus" can be used to add asymmetric uncertainties (i.e., a lower and an upper error).

To retrieve the measurement value and its errors, one should call the following methods:

```
value()           obtain the measurement value
errorMinus()      obtain a lower error
errorPlus()       obtain upper error
```

One should note that this class is conceptually rather similar to the P1D container.

A measurement can also be kept in a more general container, called DataPoint. It has the following constructors:

```
>>> from hep.aida.ref.histogram import *
>>> dp=DataPoint(d[])
>>> dp=DataPoint(d[], err[])
>>> dp=DataPoint(d[], errMinus[], errPlus[])
```

In contrast to the class Measurement, the DataPoint object can hold values and their errors in many dimensions. One can access the values as:

dimension()   get the dimension of data
lowerExtent(i)   the lower value at "i"
upperExtent(i)   the upper value at "i"

Jaida has a special container to hold data represented by the class DataPoint, called DataPointSet. It was designed for holding a set of DataPoint objects. Below we give the two most popular constructors:

```
>>> from hep.aida.ref.histogram import *
>>> dp=DataPointSet(name,title, int dim)
>>> dp=DataPointSet(name,title, int dim, int capacity)
```

The strings "name" and "title" are self-explanatory, dim is the dimension (integer), and capacity is the default capacity (integer). Once the DataPointSet object is created, one can add a DataPoint using the method addPoint(p), where "p" is a data point represented by the DataPoint object.

A DataPointSet object can be translated into the usual P1D object discussed in Sect. 4.2 and used for visualization using the DMelt canvas. What one needs to do is to build a P1D object passing a DataPointSet as argument:

```
>>> from hep.aida.ref.histogram import *
>>> from jhplot import *
>>> #.. create dp using DataPointSet class
>>> p1=P1D(dp) # convert to P1D
```

In this operation, only 1st-level errors of the P1D array will be filled. Table 4.9 lists some methods of this container.

4.4.2.1 Jaida Clouds

Jaida cloud is a set of data points. These objects are rather similar to P0D, P2D, etc., containers of the DMelt library.

To create a 1D cloud, the following constructor should be used:

Table 4.9 Some methods of the `DataPointSet` from the Jaida library

Methods	Definitions
addPoint(p)	Add a DataPoint
clear()	Clear the array
dimension()	Returns the dimension
lowerExtent(i)	Get an upper extend at index "i"
point(i)	Returns DataPoint at index "i"
removePoint(i)	Removes point at index "i"
scale(d)	Scales by a factor "d"
scaleErrors(d)	Scales errors by a factor "d"
scaleValues(d)	Scales values by a factor "d"
upperExtent(i)	Get a lower extent at index "i"

```
>>> from hep.aida.ref.histogram import *
>>> c=Cloud1D()
```

Once the cloud object has been initialized, one can fill it using two methods:

```
>>> c.fill(d)
>>> c.fill(d,w)
```

The second method fills a cloud with a value `"d"` which has a weight `"w"` (both numbers are double). The notion of the weight is similar to that to be discussed in the section devoted to histograms (see the following Chap. 7). In simple words, the number `"w"` represents importance of a data point in the data set. By default, the first method assumes w=1. To display a cloud, one needs to convert it into a histogram. Table 4.10 shows some most important methods of the clouds.

A cloud can be created in 2D using the `Cloud2D` constructor.

```
>>> from hep.aida.ref.histogram import *
>>> c=Cloud2D()
```

which can be filled with two numbers (say, x, y):

```
>>> c.fill(x,y)
>>> c.fill(x,y,w)
```

As before, one can specify a weight `"w"` for each 2D point (x, y) (the default weight is 1). The methods for this cloud are the same as those shown in Table 4.10, but this time there are more methods since we have two values, instead of one. Therefore, each method characterizing the Coud1D class has extra "X" or "Y" string at the

Table 4.10 Some methods of the class `Cloud1D` from the Jaida library

Methods	Definitions
entries()	The number of entries
fill(d)	Fill with data point "d"
fill(d,w)	Fills with "d" and a weight "w"
histogram()	Create a histogram
lowerEdge()	Cloud's lower edge
mean()	The mean value
rms()	The standard deviation
scale(d)	Scale with a factor "d"
upperEdge()	Get the upper edge
value(i)	Get a value at the index "i"
weight(i)	Get a weight at the index "i"

"d" indicates a double value and "i" is an integer index

end of each method name. For example, `meanX()` denotes the mean value in x, `meanY()` is the mean value in y.

Similarly, one can build a cloud in 3D using the constructor `Cloud3D()`. We will leave the reader here as it is rather easy to guess how to fill and how to access values of such cloud, reminding that one single point now is represented by three numbers, (x, y, z).

4.4.3   jMathTools Arrays

Another package incorporated into DMelt is called jMathTool [4]. It contains a collection of Java classes designed for engineering and general scientific computing needs.

Let us explore this package by creating a 1D array with double numbers:

```
>>> from jhplot.math.DoubleArray import *
>>> a=one(N,d)   # N numbers with value d
```

This example instantiates the array with "N" numbers, all of which have the same double value "d". Check the type which corresponds to the object "a" as `type(a)`. You will see that the object "a" is < type "array.array" >. One can find all the methods for manipulation with this object using the code assist.

One can also build an array with numbers incremented by some value:

```
>>> a=increment(N,begin,val)
```

Again, "N" is the total number of values, "begin" is an initial value, and "val" represents a double value used to increment it, x[0]=begin, x[n]=x[n-1]+val. Below we describe some important methods for the 1D arrays (denoted by "a"):

copy(a)	returns a copy
cumProduct(a)	array with the cumulative product, $b_k = \prod_{i=0}^{k} x_i$
cumSum(a)	array with the cumulative sum, $b_k = \sum_{i=0}^{k} x_i$
delete(a,i)	deletes the range starting from "i"
deleteRange(a,i,j)	deletes range from "i" to "j"
increment(N,d,p)	initializes an array with size N, $x[i] = d + i * p$;
insert(a1,i,a2)	inserts an array a2 to array a1 starting from index "i"
max(a)	the maximum values for array
maxIndex(a)	get index of the maximum value
min(a)	the minimum value for array
minIndex(a)	get index of the minimum value
product(a)	product of all values, $\prod_{i=0}^{N} x_i$
random(i)	creates an array of the size "i" with random numbers
random(i,min,max)	creates an array of size "i" with random numbers between min and max
sort(a)	sorts array
sum(a)	sums of all values, $\sum_{i=0}^{N} x_i$

One can print the array using the io.ArrayString method. Here is a complete example of how to create an array, print it, and sum up its values:

```
>>> from   jhplot.math.DoubleArray import *
>>> from   jhplot.math.io.ArrayString import *
>>> a=increment(10,0,1)
>>> print "Array="+printDoubleArray(a)
Array=0.0 1.0 2.0 3.0 4.0 5.0 6.0 7.0 8.0 9.0
>>> sum(a)
45
```

We remind that one can learn more about this type of arrays using the method dir(obj). For the DoubleArray static Java class, call these lines:

```
>>> from   jhplot
>>> dir(jhplot.math.DoubleArray)
```

The DoubleArray class can be used to build 2D arrays almost in the same way as it was done in the 1D case discussed before. Such arrays are ideal for storing matrix values. To initialize a 2D array, the following class can be used:

```
>>> from jhplot.math.DoubleArray import *
>>> a=one(i1,i2,d)
```

where `"d"` is a number to be assigned for the entire matrix with the number `"i1"` of rows and the number `"i2"` of columns.

```
>>> from   jhplot.math.DoubleArray import *
>>> from   jhplot.math.io.ArrayString import *
>>> a=one(2,5,1)
>>> print printDoubleArray(a)
1.0 1.0 1.0 1.0 1.0
1.0 1.0 1.0 1.0 1.0
```

Below we give a list of the most important methods of this class, where `"a"` indicates an input array.

`copy(a)`	returns exact copy
`cumProduct(a)`	array with cumulative product, $b_k = \prod_{i,j=0}^{k} a_{i,j}$
`cumSum(a)`	array with cumulative sum, $b_k = \sum_{i,j=0}^{k} a_{i,j}$
`deleteColumnsRange(a,i1,i2)`	deletes columns between "i1" and "i2"
`deleteRowRange(a,i1,i2)`	deletes rows between "i1" and "i2"
`getColumnCopy(a,i)`	obtain a column "i"
`getColumnDimension(a,i)`	the dimension of column at "i"
`getColumnsRangeCopy(a,i1,i2)`	get columns between "i1–i2"
`getColumnsRangeCopy(a,i1,i2)`	obtain columns between "i1–i2"
`getRowCopy(a,i)`	obtain a row at "i"
`getRowsRangeCopy(a,i1,i2)`	obtain rows between "i1–i2"
`getSubMatrixRangeCopy` `(a,i1,i2,j1,j2)`	get submatrix using ranges i1–i2 (rows) and j1-j2 (columns)
`increment(i,j, b[], p[])`	initialize array $i x j$ as $a[i][j] = b[j] + i * p[j]$.
`insertColumn(a1, a2, i)`	inserts a2[] to a1[][] at column "i"
`insertColumns(a1, a2, i)`	inserts a2[][] to a1[][] at column "i"
`insertRow(a1, a2, i)`	inserts a2[] to a1[][] at row "i"
`insertRows(a1, a2, i)`	inserts a2[][] to a1[][] at row "i"
`max(a)`	maximum values for array
`maxIndex(a)`	an index of the maximum value
`min(a)`	minimum value for array
`minIndex(a)`	an index of the minimum value
`product(a)`	product of all values, $\prod_{i,j=0}^{N} a_{i,j}$
`random(i)`	creates a 2D array of size "i" with random numbers
`random(i,min,max)`	creates a 2D array of the size "i" with random numbers between min and max
`sort(a)`	sorts the array
`sum(a)`	sums of all values, $\sum_{i,j=0}^{N} a_{i,j}$

Table 4.11   Methods for the DoubleArrayList

Methods	Definitions
addAllOf(a)	Appends an array "a"
add(d)	Add a double value
clear()	Clear the array
contains(d)	Returns true (1) if "d" exists
copy()	Returns a DoubleArrayList copy
delete(d)	Deletes 1st element "d"
elements()	Returns elements as list
get(i)	Get double value at index "i"
getQuick(i)	Get double at index "i" without checking
indexOf(d)	Returns index of the 1st occurrence
quickSortFromTo(i1, i2)	Sorts the range [i1–i2] into ascending numerical order
reverse()	Reverses the order
set(i, d)	Set double "d" at index "i"
shuffleFromTo(i1, i2)	Randomly permutes from index "i1" to "i2"
size()	Get the size
toList()	Returns ArrayList

Note that IntegerArrayList and LongArrayList have the same methods

4.4.4   Colt Data Containers

The Colt package [5] provides several enhanced data containers to store primitive values and to perform basic manipulations with them. The arrays support quick access to their elements which is achieved by nonbound checking. There are several classes used to build the Colt arrays:

- `IntArrayList`—used to create integer arrays;
- `DoubleArrayList`—used to create arrays with double values;
- `LongArrayList`—used to create arrays with (long) integer values.

We will consider the `DoubleArrayList` container for further examples since all these classes are very similar.

```
>>> from cern.colt.list import *
>>> a=DoubleArrayList()          # empty list
>>> a=DoubleArrayList(d[])       # with elements d[]
>>> a=DoubleArrayList(maxval)    # maxval is initial capacity
```

The constructors above illustrate various ways to initialize the arrays. Try to look at the methods of this class with the code assist. Table 4.11 lists some of the most important methods of the `DoubleArrayList` class.

The data stored in the `DoubleArrayList` array can be analyzed using the `DynamicBin1D` class from the same package:

```
>>> from cern.colt.list import *
>>> from hep.aida.bin    import *
>>> a=DoubleArrayList()
>>> bin=DynamicBin1D()
>>> bin.addAllOf(a);
>>> print bin.toString()
```

This example prints a comprehensive statistical summary of the array. One should also note that one can build a `POD` from the `DoubleArrayList` object as:

```
>>> from cern.colt.list import *
>>> from jhplot    import *
>>> a=DoubleArrayList()
>>> p0=POD(a)
>>> print p0.getStat()
```

File-based input/output can be achieved through the standard Java built-in serialization mechanism.

The Colt package also includes the class `ObjectArrayList`, which is similar to the Java list discussed in the previous sections.

4.4.5 Lorentz Vector

Containers that hold a group of numbers (arrays, vectors, matrices) are useful abstractions for keeping specific quantities in a structural form. In this section, we will discuss a concrete physics implementation—a representation of a particle using four-momentum which is typically used in relativistic calculations. Such a representation is particularly useful for various transformations involving simulated (or real) relativistic particles.

Before going into the depth of four vectors, let us first take a look at the simplest case when a vector is represented by three coordinates, $\mathbf{p} = (p_x, p_y, p_z)$. This situation requires the `Hep3Vector` class which can be instantiated as:

```
>>> from hephysics.vec import *
>>> v=Hep3Vector(x,y,z)
```

where x, y, z are coordinates of this vector. This class is useful for usual three-vector transformations as illustrated below:

```
>>> from hephysics.vec import *
>>> v1=Hep3Vector(1,1,1)
>>> v2=Hep3Vector(2,2,2)
```

```
>>> v1.add(v2)           # add vector v2
>>> v1.mult(10)          # multiply by  10 (scale)
>>> v1.sub(v2)           # subtract v2
>>> print v1.dot(v2)     # dot product
168.0
>>> print v1.toString()  # print
[28.0, 28.0, 28.0]
```

Next, a four-momentum of a particle can be represented by four numbers: three-momentum $\mathbf{p} = (p_x, p_y, p_z)$ and energy e. Similarly, a position of particle in space and time can be represented with four coordinates, (x, y, z, t), where t is time. This means that either position, in the momentum space or space-time, can be described by four numbers. A class used for such a description is called HepLorentzVector:

```
>>> from hephysics.vec import *
>>> hp=HepLorentzVector(px,py,pz,e)
```

or, in the case of space-time:

```
>>> hp=HepLorentzVector(x,y,z,t)
```

There are a number of useful methods implemented for this class. First of all, one can perform the standard arithmetic operations similar to the usual three-vector. One can access angles and perform transformations as for any other vector representation. Let us give a few examples:

```
>>> from hephysics.vec import *
>>> # (px,py,pz,energy)=(10,20,30,100)
>>> hp=HepLorentzVector(10,20,30,100)
>>> print hp.mag()       # magnitude of three-vector
>>> print hp.phi()       # azimuthal angle
>>> print hp.perp()      # transverse momenta.
>>> print hp.m()         # invariant mass
>>> print hp.theta()     # polar angle
>>> h.add(HepLorentzVector(1,2,3,10))  # add a new vector
```

The transverse momentum is calculated as $(p_x^2 + p_y^2)^{1/2}$, while the invariant mass is $(e^2 - p_x^2 - p_y^2 - p_z^2)^{1/2}$. Please refer to the API of this class to find more methods.

Now let us show how to visualize an object of the HepLorentzVector class. One natural way to do this is to use the 3D canvas HPlot3D, and show a Lorentz vector as a point with a symbol size proportional to the fourth component (either energy or time) as shown below:

```
from jhplot  import *
from  hephysics.vec import *

hp=HepLorentzVector(10,20,30,10)
```

```
c1 = HPlot3D("Lorentz particle",600,400)
c1.setRange(0,100,0,100,0,100)
c1.setNameX("pX"); c1.setNameY("pY"); c1.setNameZ("pZ")
c1.visible()

p= P2D("LorenzParticle")
p.setSymbolSize(int(hp.e()))
p.add(hp.px(),hp.y(),hp.z())
c1.draw(p)
```

Listing 4.23   Visualizing a Lorenz vector

4.4.5.1   Particle Representation

A particle in the package `hephysics` can be represented by the two classes, `LParticle` and `HEParticle`. The first class can be used to keep several characteristics of a particle, while the second class contains rather detailed particle properties, such as spin, parity, and other characteristics which are typically used in particle physics. For simplicity, below we will concentrate on the `LParticle` class.

The class `LParticle` can hold information about four-momentum, particle name, and even another `LParticle` object representing a parent particle. Let us create a particle with the name "proton" and with a known mass (in units called MeV, where one MeV corresponds to 10^6 electron volts):

```
>>> from hephysics.particle import *
>>> p1=LParticle("proton",939.5)
>>> p1.setPxPyPzE(10,20,30,300)
```

The second line of this code sets the three-momentum and energy. One can also set charge of this particle via the method `setCharge(c)`. One can access the information about this particle using various "getter" methods which can be found either using the code assist or the class API.

As for any other Jython or Java object, one can create a list of particles and store them in a file. In the case of Java, use the `ArrayList` class. In the case of Jython, use the Python list as shown below:

```
>>> from hephysics.particle import *
>>> p1=LParticle("proton",939.5)
>>> p1.setPxPyPzE(10,20,30,300)
>>> p2=LParticle("photon",0.0)
>>> p2.setPxPyPzE(1,2,3,30)
>>> list=[p1,p2]
```

The object `list` (as well as the objects of type `LParticle`) can be serialized into a file as any other Java object discussed in Chap. 9. Particles can be visualized in exactly the same way as shown before, since the class `LParticle` is an extension of the class `HepLorentzVector`.

4.5 Multidimensional Arrays

The previous sections discussed arrays used for manipulation and visualization of data in N-dimensional space, or some physics space. For example, Sect. 4.3.3 discussed a container to store points characterized with N numbers. In the strict programming (and mathematical) sense, such arrays are still 2D, i.e., they are characterized by multiple lists of 2D arrays, or by 2D matrix element with two subscripts. In order to run over all elements of such arrays, one must use two nested loops. So, the actual dimension of such arrays is two ($N = 2$).

This is not the same when a position of each element in arrays is characterized by N indices. For example, in the case of 3D arrays ($N = 3$), we will need 3 indices to identify a single element, and three nested loops to run over all elements. Analogously, we will need four indices to refer a single element in 4D arrays, and so on.

2D arrays will be discussed in detail when we come to the description of matrices for linear algebra in Chap. 5. Now we will show how to use Python and Java classes for arrays with arbitrary number of dimensions.

We can create a 3D array using Python lists as in this case:

```
from jhplot.io import *

d=[[[0]*2]*3]*4   # 2x3x4 array with 0 ellements
ArrayReaderWriter.write(d, "3darray.d")
print d
```

Listing 4.24 3D array using the Python approach

You can work now with the object "d" as with any Python lists discussed in the previous section. But now it has three indices. The example also calls a handy Java class `ArrayReaderWriter` for writing and reading 3D, 4D, and 5D arrays to/from files. This approach of mixing Python and Java classes works only with Jython. If you start to work with other languages, you will need to use pure Java classes—this way you can make sure that your code stays in the "Java" domain and can be used with other languages supported by DMelt.

Now we will consider a second approach using the Nd4j library [6] that provides versatile N-dimensional array objects and mimics the semantics of Numpy in CPython. We will work with this library using Jython.

The code shown below uses Nd4j class to build a $3 \times 3 \times 3 \times 3$ array (4D array) with all elements filled with 1. Then we make a transformation of all elements, and add the resulting array to the original array:

```
from org.nd4j.linalg.factory import Nd4j
from org.nd4j.linalg.ops.transforms.Transforms import *
from java.io import File
a=Nd4j.create(Nd4j.ones(81).data(), [3,3,3,3])
a1=exp(a)        # convert all ellements to exp()
a1=a1.add(a)     # add a1+a1
print a1
```

Listing 4.25   4D arrays using Nd4j Java library

The import statements given in this example will help discover more Java classes and methods that can be used to manipulate with multidimensional arrays of the Nd4j library.

References

1. Taylor J (1997) An introduction to error analysis: the study of uncertainties in physical measurements. University Science Books
2. Henon M (1969) Numerical study of quadratic area-preserving mappings. Quart Appl Math 27:291–312
3. Gleick J (1988) Chaos: making a new science. Penguin Books, New York
4. JMATHTOOLS Java Libraries. http://jmathtools.berlios.de
5. The Colt Development Team, The COLT Project. https://dst.lbl.gov/software/colt/
6. ND4J, scientific computing library for the jvm. http://nd4j.org

Chapter 5
Linear Algebra and Equations

Most readers know from graduate textbooks that linear algebra is a branch of mathematics that deals with vector spaces, matrices, determinants, linear transformations, and systems of linear equations. This chapter explains how to put your hands on numeric packages for linear algebra. First of all, we will offer a description of Java classes designed to construct vectors and matrices. Then we will discuss the usual matrix operations. We will finish this chapter with a description of how to solve linear equations using the libraries provided by the DMelt project.

5.1 Vector and Matrix Packages

The construction of vectors and matrices can be done with several third-party Java libraries [1, 2]. They are designed for linear algebra calculations with vectors and two-dimensional matrices of double-precision numbers. Here is a short summary of what such packages can do:

- Creation of vectors and matrices, and basic methods for manipulation with them;
- Calculation of derived quantities, such as condition numbers, determinants, ranks, etc.;
- Elementary operations, such as addition, subtraction, multiplication, scalar multiplication, etc.;
- Various matrix decompositions.

We will start our discussion with the Jama package [1] implemented in Java. Let us first make a vector with the values (1, 2, 3, 4, 5):

```
>>> from Jama import *
>>> m=Matrix([[1,2,3,4,5]])   # create a vector
>>> print m.toString()        # print
1.0,  2.0,  3.0,  4.0,  5.0
```

© Springer International Publishing Switzerland 2016
S.V. Chekanov, *Numeric Computation and Statistical Data Analysis
on the Java Platform*, Advanced Information and Knowledge Processing,
DOI 10.1007/978-3-319-28531-3_5

Analogously, let us build a 2×2 matrix and print its values:

```
>>> from Jama import *
>>> m=Matrix([[1.,2.],[3.,4.]])
>>> print m.toString()
      1.00  2.00
      3.00  4.00
```

The example is rather simple: first we initialize the matrix from a Python list, and then print it out using the toString() method which converts the matrix into a string. This method takes two integer parameters: the column width and the number of digits after the decimal point. One can also pass the DecimalFormat instance from Java API for nice printing, but this requires more typing.

Let us construct a matrix holding a single value, say 100:

```
>>> from Jama import *
>>> m=Matrix(2,3)   # 2 rows and 2 columns
>>> print m.toString()
      0.0   0.0   0.0
      0.0   0.0   0.0
>>> m=Matrix(2,3,100)
>>> print m.toString()
      100.0 100.0 100.0
      100.0 100.0 100.0
```

As you can see from this example, the constructor Matrix takes two arguments: the number of rows and the number of columns. One can obtain a single value using the matrix indices:

```
>>> m=Matrix(2,3,100)
>>> m.set(1,1,200)
>>> print m.get(1,1)
200.0
```

It is relatively easy to construct a matrix from any array. For example, one can build a matrix from two P0D classes discussed in Sect. 4.1 using the method getArray().

```
>>> from jhplot import *
>>> from Jama import *
>>> p1=P0D([2,3])
>>> p2=P0D([4,5])
>>> m=Matrix([p1,p2])
```

Finally, one can fill in a matrix with the method m.random(n,m) using uniformly distributed random elements. In this method, n is the number of rows and m is the number of columns.

One can insert a submatrix of a matrix with the method:

```
>>> m.setMatrix(int[]  r,    int[]  c,  X)
```

where `r` is an array of row indexes and `c` - is an array of column indexes. `X` is the actual sub-matrix defined as `A(r(:),c(:))`.

Now let us learn how to extract the information about methods of the class `Matrix`. First, one can access 2D array of a matrix using the method `getArray()`. To get a single value, use the usual method `get(i1,i2)`, where `i1` is the row index and `i2` is the column index. One can return the entire matrix object, or a submatrix using the `getMatrix(i1,i2,j1,j2)` method, where `i1` is the initial row index, `i2` is the final row index, `j1` is the initial column index and `j2` is the final column index. One can also learn about all methods of this Java class by looking at the corresponding API documentation, or printing the list of methods using the Python method `dir(matrix)`.

A matrix or a group of matrices can be saved into dictionaries or lists, and serialized into a file as any DMelt object discussed in Chap. 9. This is possible because the class `Matrix` implements the Java `Serializable` interface.

5.1.1 Basic Matrix Arithmetic

Assuming that you have two matrices, "A" and "B," one can perform the following operations:

`B.minus(A)`	subtract a matrix A from B
`A.timesEquals(d)`	multiply a matrix by a scalar "d", i.e., $A = d * A$
`B.times(A)`	Linear algebraic matrix multiplication, $A * B$
`B.plus(A)`	$C = A + B$
`A.plusEquals(B)`	$A = A + B$
`A.minusEquals(B)`	$A = A - B$
`A.minus(B)`	$C = A - B$
`A.arrayRightDivide(B)`	Element-by-element right division, $C = A./B$
`A.arrayRightDivideEquals(B)`	Element-by-element right division in place, $A = A./B$
`A.arrayLeftDivide(B)`	Element-by-element left division, $C = A.$ divide B
`A.arrayLeftDivideEquals(B)`	Element-by-element left division in place, $A = A.$ divide B
`A.arrayTimes(B)`	Element-by-element multiplication, $C = A. * B$

`A.arrayTimeEquals(B)`	Element-by-element multiplication, $A = A.*B$
`A.uminus()`	unary minus, i.e., $-A$

One can use the `inverse()` method to inverse all elements. Finally, one can obtain
the basic normalization methods, such as

`norm1()`	the maximum column sum, i.e., summing up absolute values of all column numbers;
`normInf()`	the maximum row sum, i.e., summing up absolute values of all row numbers.

5.1.2 Elements of Linear Algebra

The Jama package provides advanced operations with matrices. For example, the
determinant of a matrix can be obtained with the method `det()`, while the method
`rank()` returns its rank. There are several basic matrix decompositions:

- Cholesky decomposition of symmetric, positive definite matrices;
- QR decomposition of rectangular matrices;
- LU decomposition of rectangular matrices;
- Singular value decomposition of rectangular matrices;
- Eigenvalue decomposition of both symmetric and nonsymmetric square matrices.

The decompositions are accessed by the class `Matrix` to compute solutions of
simultaneous linear equations, determinants and other matrix functions.

Let us consider a small example illustrating the capabilities of the Jama package:

```
from Jama import *

A=Matrix([[2.,2.,3],[4.,5.,6.],[7.,8.,4.]])
print "Determinant",A.det()
B=Matrix([[7.,8.,1],[1.,7.,0.],[4.,1.,9.]])
X = A.solve(B)
print X.toString()
R = A.times(X).minus(B)
print "normInf=",R.normInf()

print "EigenvalueDecomposition:"
Eig=A.eig()
D=Eig.getD()
V=Eig.getV()
print "D=",D.toString(), "V=",V.toString()
```

Listing 5.1 Solving a linear system of equations

The output is shown below:

```
Determinant  -13.0
   14.77    8.85    4.23
  -13.00   -9.00   -2.00
    1.15    2.77   -1.15
normInf=  1.50990331349e-14
EigenvalueDecomposition:
D=
  13.765  0.000    0.000
   0.000  0.307    0.000
   0.000  0.000   -3.072
V=
   0.291    0.761  -0.319
   0.627   -0.675  -0.481
   0.722    0.021   0.860
```

Now let us consider another example: computing eigenvalues and eigenvectors of a real symmetric matrix $A = VDV^T$ with the size of 10×10:

```
from Jama import *
N=10                       # size of the matrix
A=Matrix.random(N,N)  # symmetric positive definite matrix
A=A.transpose().times(A)
e=A.eig()                  # spectral decomposition
V=e.getV()
K=e.getD()
print A.toString()
print V.toString()
print K.toString()
print "||V*V^T-I||=" # check that V is orthogonal
print V.times(V.transpose()).minus(Matrix.identity(N,N)).
      normInf()
print "||AV-DV||="    # check that A V = D V
print A.times(V).minus(V.times(K)).normInf()
```

Listing 5.2 Computing eigenvectors of a real symmetric matrix

Run this example and look at its outputs. We have a number of comments in this code describing each step.

5.1.3 Jampack Matrix Computations

The package Jampack (Java Matrix Package) [2] is a complementary package. Unlike the previous package, it supports complex matrices. In fact, it supports only complex matrices, since the design proceeded from a more general case to a less general one. The package has all basic linear algebra operations and many decomposition methods:

- The pivoted LU decomposition;
- The Cholesky decomposition;
- The QR decomposition;
- The eigendecompostition of a symmetric matrix;
- The singular value decomposition;
- Hessenberg form;
- The Schur decomposition;
- The eigendecompostition of a general matrix.

In this package, a complex number is represented by the class Z(a,b), where a represents a real part while b is an imaginary part of a complex number. One can perform several simple arithmetical operations as shown below:

```
>>> from Jampack import *
>>> z1=Z(10,6)
>>> z2=Z(8,1)
>>> z1.Minus(z2)
>>> print "Real part=",z1.re, " Imaginary part=",z1.im
```

The print statement is used to print real and imaginary parts of the complex number. Below we give an example to help getting started with the complex matrices:

```
from Jampack import *
# initializes real and imaginary parts
A=Zmat([[1,2],[3,4]], [[3,4],[5,6]])
print Print.toString(A,4,3)
nrm = Norm.fro(A)      # Frobenius normalization
A=Times.o(H.o(A), A) # multiply
print Print.toString(A,4,3)
```

Listing 5.3 Working with complex matrices

One should look at the API documentation of this package or use the DMelt help assist to learn about the package.

5.1.4 La4J Library

The La4j library [3] is another open-source Java library for linear algebra. The library provides operations with vectors, matrices (dense matrices as well as sparse), matrix decompositions (SVD, LU, Cholesky, etc.), file input/output and linear system solving.

First let us start with a simple example—creation of vectors using different approaches:

```
from org.la4j import Vector
from java.util import Random

a=Vector.fromCSV("0.1, 0.2,1.0, 2.0")  # from CSV string
b=Vector.fromArray([1,2,3,4])           # build   from a list
c=Vector.random(100,Random())           # random vector
c.shuffle()                             # shuffle elements
print c
```

Listing 5.4 Creating vectors in La4J

Next, here are a few methods to create matrices using this package:

```
from org.la4j import *
from java.util import *

a=Matrix.identity(100)                  # identity matrix
b=Matrix.fromCSV("0.1, 0.2\n1.0, 2.0")  # from CSV string
e=Matrix.from2DArray([[1,2],[2,2]])     # from list
c=Matrix.diagonal(10, 0.4)              # diagonal size 10
d=Matrix.random(10,20,Random())         # 10x00 random matrix
print  d.getRow(1), d.getColumn(2)      # row 1 and column 2
```

Listing 5.5 Creating matrices in La4J

The last two methods extract rows and columns from the matrix. If you need a subset of matrices, such as sparse (in which most of the elements are zero), or dense (in which most of the elements are nonzero), consider using org.la4j.vector.sparse and org.la4j.vector.dense packages.

Now let us show a few common operations on matrices:

```
from org.la4j import *
from java.util import *

d=Matrix.random(10,10,Random())  # 10x10 random  matrix
c=d.multiply(d)                  # multiply d*d
f=c.subtract(d)                  # subtract c-d
f=f.add(f)                       # add f+f
trace=f.trace()                  # find trace
det=f.determinant()              # find determinant
print f.toString()
```

Listing 5.6 Operations on La4J matricies

The package has several dozens of predefined transformations. In addition, it can help to create custom transformations using several convenient methods. For example, let us make a transformation using a user-defined function that selects only diagonal elements:

```
from org.la4j import *
from org.la4j.matrix.functor import *
from java.util import *

a=Matrix.random(10,10,Random())  # 10x10  random matrix

class myfunction(MatrixFunction):
  def evaluate(self, i,j,value):
              if (i==j): return value
              else: return 0
print "Matrix before transform=\n",a.toString()
print "Keep only diagonal elements"
c= a.transform(myfunction())
print c.toString()
```

Listing 5.7 User-defined transformations

As you can see, the package is easy to use. It even provides the needed API to write matrices and vectors to several formats: CSV files, MatrixMarket, and a binary format. Let us show how to write and read back a matrix using the MatrixMarket format—a popular exchange format that provides a mechanism for exchanging data between different programs:

```
from org.la4j import Matrix
from java.text import *

a=Matrix.identity(10)
mm=a.toMatrixMarket(DecimalFormat("#,###,##0.00"))
file=open("out.mm","w")   # write matrix to MatrixMarket
file.write(mm)
file.close()

file=open("out.mm")       # read it back
b = Matrix.fromMatrixMarket( file.read() );
print b
```

Listing 5.8 File output and input using the MatrixMarket format

It is rather simple, but watch out! Here we mixed the Java objects with the Python method "open" to write and read string data. To make your program portable between Java and other scripting languages, one should rewrite this example using pure Java classes.

5.1.5   EJML Matrix Library

Efficient Java Matrix Library (EJML) [4] provides alternative sets of Java classes that can be used for manipulating dense matrices. The library implements all basic matrix operators (addition, multiplication, etc.), decompositions (LU, QR, Cholesky, SVD, eigenvalue, etc.) and matrix features (rank, symmetric, definitiveness). The package can also be used to create random matrices. Matrices can be saved in files using ether CSV files, or using a binary format. Here is a basic example to initiate EJML matrices and perform the most common operations, such as invert, multiply, transpose, finding eigenvalue decomposition

```
from org.ejml.simple import *

S=SimpleMatrix([[1,2,3,4],[3,4,4,5],[3,3,3,1],[3,3,1,1]])
H=SimpleMatrix([[1,0,0,4],[1,4,0,5],[2,3,1,1],[2,3,1,1]])
P=SimpleMatrix([[1,0,0,4],[1,4,0,5],[2,1,1,2],[1,2,2,1]])
SIN=S.invert() # invert matrix "S"
# invert, multiply,transpose
K = P.mult(H.transpose().mult(S.invert()))
K.eig() # eigenvalue decomposition
K.print()
```

Listing 5.9 Operations on EJML matricies

Here are the basic operations for file input and output of matrices:

```
from org.ejml.simple import SimpleMatrix

S=SimpleMatrix([[1,2,3,4],[3,4,4,5],[3,3,3,1],[3,3,1,1]])
S.print("%e")                    # scientific format
S.print("%10.2f")                # custom format
S.saveToFileCSV("file.csv")   # save in CSV output
S.saveToFileBinary("file.d") # save in binary form
B=SimpleMatrix.loadBinary("file.d") # read back
```

Listing 5.10 File input and output

One useful feature of the package is to visualize matrices. Here is an example of creating a random matrix 20×20 with values in the range -10–10. Running this code will show the matrix in a pop-up frame.

```
from java.util import *
from org.ejml.ops import *
# create a random matrix 20x20 with random values -10- 10
A=RandomMatrices.createSymmetric(20,-10,10,Random())
MatrixVisualization.show(A,"Small Matrix")
```

Listing 5.11 Matrix visualization using a 2D density plot

The code creates the image shown in Fig. 5.1 where the color of each cell shows the values of this symmetric matrix.

Fig. 5.1 A random matrix from the EJML package shown as an image. Each cell represents a value. Empty square indicates zero values. The *red color* indicates positive values, while *blue* shows negative values. More intense the color, larger the element's absolute value is (color figure online)

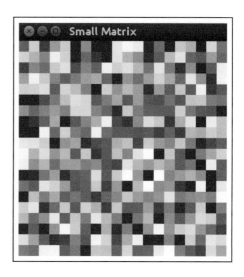

The package can also be used to create dense real matrices using the Java class `DenseMatrix64F`, dense complex matrices (`CDenseMatrix64F` class), as will as fixed-size dense vectors and matrices. Please explore the Java API of this package.

5.1.6 Multithreaded Matrix Computations

Another option for linear algebra matrices is the Colt scientific library [5]. It can be used for the standard matrix operations and decompositions, such as

• Cholesky decomposition of symmetric, positive definite matrices
• LU decomposition (Gaussian elimination) of rectangular matrices
• QR Decomposition of rectangular matrices
• Eigenvalue decomposition of both symmetric and nonsymmetric square matrices
• Singular value decomposition of rectangular matrices.

In addition to these rather standard linear algebra decompositions, Parallel Colt includes trigonometric transforms, such as the discrete Fourier transform (DFT), the discrete Hartley transform (DHT), the discrete cosine transform (DCT) and the discrete sine transform (DST).

Due to recent improvements in the personal computer architecture, linear algebra calculations can be done using multi-core processors. We will consider multithreaded version [6] of the Colt library that can substantially increase the performance on computers with multiple CPUs. The appropriate Java classes are located in the package `cern.colt.matrix`.

First, let us show how to initialize dense and sparse matrices with double values in one, two, and three dimensions:

```
from cern.colt.matrix.tdouble.impl import *

# Dense 1-d matrix (aka vector)
d=DenseDoubleMatrix1D([10,20])

# Dense 2-d matrix
d=DenseDoubleMatrix2D([[10,10],[5,6]])
print d.toString() # let's print it

# Dense 3-d matrix 5x10x3 (not filled)
d=DenseDoubleMatrix3D(5,10,3)

# Sparse hashed 2-d matrix
s=SparseDoubleMatrix2D(2, 2);
print dir(s) # print methods
```

Listing 5.12 Types of matrices of the Colt library

One can give a similar example using integer values and the package
cern.colt.matrix.tint. Let us perform a number of standard linear algebra
operations on matrices:

```
from cern.colt.matrix.tdouble.impl import *

# Dense matrix 2x2
d=DenseDoubleMatrix2D([[10,10],[5,6]])
d.fft2()                # 2D discrete Fourier transform (DFT)
print d.zSum()          # calculate sum of all cells
v=d.vectorize()
print v.toString()  # stacking the columns
```

Listing 5.13 Working with Colt matrices

The example shows only a few possible operations. Please refer the API of this
package.

But what about multithreading discussed above? Let us give a short program that
illustrates the advantage of exploiting modern computer architectures in scientific
computing. We will use the same Colt package, but this time we will specify the
number of processing CPU cores for our computations. The example shown below
can be run on a computer that has more than three processing cores:

```
from cern.colt.matrix import tdouble
from edu.emory.utils  import ConcurrencyUtils
import time

def process(M):
  M.cardinality() # number of cells with non-zero values
  M.dctColumns(0) # discrete cosine transform
  M.dctRows(0)    # inverse discrete cosine transform
  M.dct2(0)       # inverse of the discrete cosine
  M.dht2()        # inverse discrete Hartley transform
```

```
M.dhtColumns()    # inverse discrete Hartley transform
M.dhtRows()       # same for rows
M.dst2(0)         # inverse discrete sine transform
M.dstColumns(0)   # inverse discrete sine transform
M.dstRows(0)      # as before but for rows
M.vectorize()     # stacking the columns of the matrix
M.zSum()          # the sum of all cells
M.normalize()     # normalize

Ncores=4   # number of processing cores
print Ncores," CPU cores. Wait!"
ConcurrencyUtils.setNumberOfThreads(Ncores)
start = time.clock()
# make 2000x2000 random matrix
M=tdouble.DoubleFactory2D.dense.random(2000,2000)
process(M)
print "Time (s)=",time.clock()-start
```

Listing 5.14 Matrix operations using multithreading

The program performs different operations on 2,000 × 2,000 random matrix, and prints the CPU needed to do such calculations. Note that we use a "factory" for a handy creation of the random matrix. You can benchmark this code using a single computational core, i.e., replacing the number 4 with 1 in the above code snippet. After running this code, you will immediately notice substantial decrease in performance (typically the program will run a factor 2–4 slower when using one thread).

Normally, you do not need to worry about setting the number of CPU threads for your program. Parallel Colt attempts to identify the number of CPU threads and will perform the computations that are best optimized for your computer. To check this, remove the statement where you set the number of the cores, i.e., replace the last part of your previous code with this code snippet:

```
print ConcurrencyUtils.getNumberOfThreads()
start = time.clock()
# make 2000x2000 random matrix
M=tdouble.DoubleFactory2D.dense.random(2000,2000)
process(M)
print "Time (s)=",time.clock()-start
```

It should print the actual number of processing cores detected in your system. The performance of the program will be optimized by taking into account this number.

We will continue the discussion of how to use multiple cores for matrix transformations in the context of fast Fourier transform (FFT) for image processing in Sect. 15.4.2.

5.1.7 JBlas and Other Matrix Packages

The section dealing with matrices and linear algebra will be incomplete without discussion of the JBlas [7] and JLapack [8] Java libraries. These libraries closely follow the BLas and Lapack packages that originate from the FORTRAN community, and are considered the industry standard for matrix computations. The package includes eigendecomposition, solving linear equations, singular value LU, Cholesky decompositions, and a geometry package. Here is a simple example of initializing a vector and a matrix, performing a multiplication, and calculation of eigenvalues:

```
from org.jblas import DoubleMatrix,Eigen

vector=DoubleMatrix([3,3,3])
matrix=DoubleMatrix([[0,  2,  1],
                     [3,  4,  5],
                     [6,  7,  8]])
m=matrix.mmul(vector) # multiplcation
print "M=",m
print "Eigenvalues=",Eigen.eigenvalues(matrix)
```

Listing 5.15 JBLAS matrix operations

Use this example as an invitation to discover other types of matrices supported by JBLAS, and look at the general Java API of the package `org.jblas`.

The Apache Common Math [9] also includes a number of powerful packages for linear algebra support. Here is a somewhat bulky initialization of a 3×3 matrix using the Apache Common Math interface:

```
from org.apache.commons.math3.linear import *

m=MatrixUtils.createRealMatrix([[0,  2,  1],
                                [3,  4,  5],
                                [6,  7,  8]])
m=m.multiply(m)
inv=LUDecomposition(m).getSolver().getInverse()
print inv
```

Listing 5.16 Apache matrix operations

After matrix creation, we perform the inversion using LU decomposition.

5.1.8 Python Vector and Matrix Operations

In addition to the Java-based packages for matrix operations, DMelt also contains pure Python packages for vector, matrix, and table math operations. These operations are based on the Python package "`statlib`" which is located in the "python/packages" directory. We remind that this directory is imported by the DMelt IDE during startup. More details on how to import Python libraries have been discussed in Sect. 1.4.2.

The `"statlib"` package provides many basic vector operations. Below we show an example of how to call this library and perform a few basic operations with two vectors, `"a"` and `"b"`:

```
from statlib.matfunc import Vec

a = Vec( [ 4,  -2,   5 ] )
b = Vec( [ 3,  10,  -6 ] )
print a.dot(b)        # the dot product of "a" and "b"
print a.norm()        # length of the vector
print a.normalize()   # length 1
print a.cross(b)      # cross product of "a" and "b
```

Listing 5.17 Vector operations

One can learn more about the methods of this package using the code assist or by calling the statement `dir(obj)` (`obj` represents an object of the class `Vec`). Another option is to look at the module directly. Just open the Jython module "matfunc.py" in the directory "python/packages/statlib".

Let us consider several examples of how to work with Python-implemented matrices. A matrix can be defined from a list of lists as in this example:

```
>>> from statlib.matfunc import *
>>> m=Mat( [[1,2,3], [4,5,1,], [7,8,9]] )
>>> print m
1    2    3
4    5    1
7    8    9
>>> print "eigs:", m.eigs()
eigs:  2.692799     -0.846943     13.154144
>>> print "det(m):", m.det()
det(m):  30.0
```

The class `Mat` has many useful methods, such as `tr()` (for transpose), `star()` (for Hermetian adjoints), `diag()`, `trace()`, and `augment()`. All such methods have their usual mathematical meaning. Matrix multiplications are accomplished by the method `mmul()`, and matrix division by `solve(b)`.

Let us consider a more detailed example:

```
from statlib.matfunc import *

C = Mat( [[1,2,3], [4,5,1,], [7,8,9]] )
print C.mmul( C.tr())
print C ** 5
print C + C.tr()
A = C.tr().augment( Mat([[10,11,13]]).tr() ).tr()
q, r = A.qr()
```

```
print "q:\n", q, "\nr:\n",r
q.mmul(r) == A
print "\nQ.tr()&Q:\n", q.tr().mmul(q)
print "\nQ*R\n", q.mmul(r)
b = Vec([50, 100, 220, 321])
x = A.solve(b)
print "x:   ", x
print "b:   ", b
print "Ax: ", A.mmul(x)
inv = C.inverse()
print  "inverse=\n",inv
print "C * inv(C):\n", C.mmul(inv)
```

Listing 5.18 Operations on matrices using the Python approach

Because of the lengthy output, we will not show it in this book. Try to understand all these operations. They are well documented in the file "matfunc.py".

5.1.9   Matrix Operations in SymPy

Another package designed for matrix operations, including operations with symbols, is SymPy [10]. This package has already been discussed in Sect. 6.3. We remind that this library is implemented in Python and included as a third-party package in the directory "python/packages".

Let us build a matrix using the SymPy package. Our matrix can contain numbers as well as symbols that have to be declared with the statement Symbol():

```
>>> from sympy import *
>>> Matrix([[1,2], [3,4]])
[1, 2]
[3, 4]
>>> x = Symbol("x")
>>> Matrix([[1,x], [2,3]])
[1, x]
[2, 3]
```

There are several ways to construct predefined matrices based on the statements eye (the identity matrix), zeros and ones:

```
>>> from sympy import *
>>> eye(2)
[1, 0]
[0, 1]
>>> zeros((2, 3))
[0, 0, 0]
[0, 0, 0]
```

One can use the standard mathematical operations such as *, /, −, + as shown in
the example:

```
>>> from sympy import *
>>> M1=Matrix(([1,2,3],[4,5,6]))
>>> M2=Matrix(([1,1,1],[2,2,2]))
>>> M3=M1-M2
>>> M3
[0,  1,  2]
[2,  3,  4]
```

A few linear algebra operations are shown below:

```
>>> from sympy import *
>>> M1 = eye(3)
>>> M1.det()
1
>>> M1.inv()
[1,  0,  0]
[0,  1,  0]
[0,  0,  1]
```

There are many decomposition methods associated with the class `Matrix` of this
package. Check the SymPy website [10] or look at the package "sympy/matrices"
inside the directory "python/packages".

5.2 Algebraic Manipulations with Tensors

A tensor is an object similar to multidimensional matrix that describes linear relations
or mappings. Tensors are used in a variety of applications ranging from physics to
biology.

Manipulation with tensors in the DMelt can be performed using the Redberry
computer algebra system [11]. The package provides basic tensor operations and
transformations. In addition, one can find more specific libraries dedicated to calculations with Feynman diagrams and solving equations. The original interface is
written in Groovy and is intended to be used together with the Groovy environment.
We will consider how to use Groovy in Sect. 16.4.

Running Redberry in DMelt is straightforward. Let us take a small example put
these lines in a file with the extension ".groovy" and run this file either in the IDE or
in a batch mode:

```
import cc.redberry.groovy.Redberry
import static cc.redberry.core.indices.IndexType.*
import static cc.redberry.groovy.RedberryPhysics.*
import static cc.redberry.groovy.RedberryStatic.*

use(Redberry) {
  def t = 'a * T_mn + g_mn'.t
```

```
    println t.class.simpleName  // print name of Tensor subtype
    println t.indices           // accessing indices of tensor
    println t.size()            // size
}
```

Listing 5.19 Creating a tensor using Redberry

The execution of this script gives several characteristics of the defined tensor. For example, it will say that the tensor name is "Sum", and it will print the indices. Note that our program starts with the `use(Redberry)` statement.

The description of the Redberry environment is beyond the scope of this book. Please read the documentation of this package [11].

5.3 Equations

Reading this section of the book requires only a college algebra background. Here we will illustrate the process of solving equations of various forms using DMelt Java libraries.

5.3.1 Polynomial Equations

A polynomial equation can be written in the form

$$a_n x^n + a_{n-1} x^{n-1} + a_{n-2} x^{n-2} \ldots + a_0 = 0,$$

where a_n are coefficients and x is variable. The simplest way to solve a quadratic equation $ax^2 + bx + c = 0$, where a, b, and c are the coefficients of this equation, is to use the package `jhplot.math` from the core DMelt library. Here is an example that prints the solution:

```
from jhplot.math.Numeric import *
b=solveQuadratic(1, 2, -1) #a=1, b=2, c=-1
print b.tolist()
```

Listing 5.20 Quadratic equation

The equation should have a positive discriminant (imaginary roots are not supported).

A cubic equation of the form $ax^3 + bx^2 + cx + d = 0$ has either one real root, or three real roots. We will solve it as in this example:

```
from jhplot.math.Numeric import *
b=solveCubic(1,2,4,2) # a=1, b=2, c=4, d=2
print b.tolist()
```

Listing 5.21 Solving a cubic equation

There are several other third-party Java packages to solve polynomial equations. For example, Apache Common Math [9] provides the Laguerre's method [12] to solve polynomial equations assuming a given precision. Let us consider an example that solves quartic, cubic, and quartic equations. The comments in this example illustrate the equations.

```
from org.apache.commons.math3 import *
from org.apache.commons.math3.analysis.solvers import *

coeff=[1, -2, 8]    # 1-2*x+8x^2=0
sol=LaguerreSolver(1e-5)
result=sol.solveAllComplex(coeff,0.0)
print result.tolist()

coeff=[1, -2, 8, 1] # 1-2x+8x^2+x^3=0
result=sol.solveAllComplex(coeff,0.0)
print result.tolist()

coeff=[1, -2, 8, -1, 4] # 1-2x+8x^2-x^3+4x^4=0
result=sol.solveAllComplex(coeff,0.0)
print result.tolist()
```

Listing 5.22 Solving a polynomial equation

The output in the form of Python list shows the solutions with real and imaginary terms.

5.3.2   Linear Systems of Equations

Section 5.1.4 discussed the La4J library for the linear algebra. The same library can be used for solving linear equation systems of the form:

$$a_{11}x_1 + a_{12}x_2 + \cdots + a_{1n}x = b_1$$

$$\vdots \qquad\qquad \vdots \qquad\qquad\qquad (5.1)$$

$$a_{m1}x_1 + a_{m2}x_2 + \cdots + a_{mn}x_n = b_m$$

or, in short, $A \cdot x = B$, where A is a matrix and B is a vector. Let us solve a system of linear equations using one of the computing engines of this package:

```
from org.la4j.linear import *
from org.la4j import *

A=Matrix.from2DArray([[9.0, -1.0, -18.0],
                      [-1.0, 6.0,  -3.0],
                      [-18.0, -3.0,33.0]])
B=Vector.fromArray([-45.0, -10.0, 81.0 ])
en=SquareRootSolver(A)
```

```
X=en.solve(B)
print "Solution=",X
print "Check=",B.equals(A.multiply(X), 1e-9)
```

Listing 5.23 Solving a linear system of equations using La4J

The code solves the equation and makes a cross-check of the solution (1, 0, 3). Alternatively, one can use the forward and back substitution method:

```
from org.la4j.linear import *
from org.la4j import *

A=Matrix.from2DArray([[9.0,  -1.0,  -18.0],
                      [-1.0,  6.0,   -3.0],
                      [-18.0, -3.0, 33.0]])
B=Vector.fromArray([-45.0,  -10.0,  81.0 ])
en=ForwardBackSubstitutionSolver(A)
X=en.solve(B)
print "Solution=",X
print "Check=",B.equals(A.multiply(X), 1e-9)
```

Listing 5.24 Solving a linear system of equations using substitutions

There are several other relevant methods in the `org.la4j.linear` Java package.

To solve a linear system in the EJML library discussed in Sect. 5.1.5, look at this example:

```
from org.ejml.simple import *
from java.util import Random

A=SimpleMatrix.random(3,3,0,1,Random(1))
B=SimpleMatrix.random(3,1,0,1,Random(2))
C=A.solve(B)
print C
```

Listing 5.25 Solving a linear system of equations using EJML

This time we created a random matrix and random vector, instead of initializing them by hand.

Typically, all Java packages that deal with the linear algebra include the methods for solving systems of linear equations. For example, JBLAS and Apache Common Math discussed in Sect. 5.1.7 include efficient algorithms for solving systems of linear equations. Here is an example that uses the Apache Common Math library:

```
from org.apache.commons.math3.linear import *

coeff=Array2DRowRealMatrix ([[2,3,-2],[-1,7,6],[4,-3,-5]])
solver=LUDecomposition(coeff).getSolver()
constants=ArrayRealVector([1,-2,1])
print solver.solve(constants);
```

Listing 5.26 Solving a system of linear equations using Apache Math

References

1. Hicklin J et al, Jama, a Java Matrix Package. http://newcenturycomputers.net/projects/dif.html
2. Stewart G, Jampack, a Java package for matrix computations. ftp://math.nist.gov/pub/Jampack/Jampack/
3. Kostyukov V et al, La4J, linear algebra java library. http://la4j.org/
4. Abeles P, EJML, efficient Java matrix library. http://ejml.org/
5. The Colt Development Team, The Colt Project. https://dst.lbl.gov/software/colt/
6. Wendykier P, Nagy J (2010) Parallel colt: a high-performance java library for scientific computing and image processing. ACM Trans. Math. Softw. 1–22. https://sites.google.com/site/piotrwendykier/software/parallelcolt
7. Braun ML et al, JBLAS, linear algebra for Java. http://jblas.org/
8. Dongarra J, Downey A, Seymour K, JLAPACK and the F2J project. http://icl.cs.utk.edu/f2j/
9. The apache common-math library. http://commons.apache.org/proper/commons-math/
10. SymPy development team, SymPy: Python library for symbolic mathematics. http://www.sympy.org
11. Bolotin D, Poslavsky S, Redberry, A computer algebra system for algebraic manipulations with tensors. https://poi.apache.org/
12. Ralston A, Rabinowitz P (2001) A first course in numerical analysis. Dover Publications, New York

Chapter 6
Symbolic Computations

Symbolic calculations, unlike numeric calculations discussed so far, deal with manipulation of mathematical expressions. This book will not be complete without a discussion of this topic.

Symbolic calculations are well supported by a number of commercial programs, such as Wolfram Mathematica, Maple, and by noncommercial, such as GNU Octave. Unlike these programs that use specialized languages for scientific computing, DMelt provides elements of symbolic computations using *general-purpose programming* languages on the Java platform.

It is not immediately clear why you would use Java or Jython for symbolic calculations if there is a well-established tradition to use high-level languages to work with mathematical expressions, such as Matlab, GNU Octave, or Mathematica. The main motivation in using DMelt-supported libraries is to be able to embed symbolic calculations in Java code, or inside codes written in scripting languages (Python, Groovy, Ruby, etc.). This will guarantee a cross-platform robustness, an integration with the reach open-source libraries, and a possibility to create desktop and web applications.

If you want to learn symbolic calculations using the Octave-like language (which is similar to MATLAB), please jump to Chap. 17. Here we will show how to perform analytic calculations using Jython code, or using other scripting language supported by the Java platform. The examples of this chapter can also be converted into pure Java coding.

6.1 Using the Octave Language

To use MATLAB-like (or GNU Octave-like) language for scientific and technical computing, the Java Symbolic class is available from the jhplot package. Let us show how to initialize this class using the Python/Jython language:

© Springer International Publishing Switzerland 2016   207
S.V. Chekanov, *Numeric Computation and Statistical Data Analysis*
on the Java Platform, Advanced Information and Knowledge Processing,
DOI 10.1007/978-3-319-28531-3_6

```
>>> from jhplot.math import *
>>> j = Symbolic ("jasymca") # sets the engine to "
    jasymca"
```

This code creates a computational environment to process symbolic expressions written in Octave-like scripting language. We remind that the Octave language is similar to MATLAB, so many programs are easily portable. The above example is written in Jython, but the same approach can also be used in Java or any other languages supported by the Java platform. This will be discussed in Chap. 16.

To evaluate a mathematical expression, we will need to create a string that contains calculations using the Octave syntax, and pass it to the `Symbolic` class:

```
>>> s = "syms x; trigrat(sin(x)^2+cos(x)^2)"
>>> j.eval(s)
```

As you can see, we use Java to parse strings that contain code written in the Octave-like language. This feature is required, since the Octave-like scripting is not integrated in the standard scripting Java API, compared to Jython.

We will leave the reader here. More details on how to program in an Octave-like language will be described in Sect. 16.6 and in Chap. 17.

6.2 Java Symbolic Computing Library

You can do symbolic differentiation, integration, and simplifications using an alternative engine, called JSCL (Java Symbolic Computing Library [1]). This time we will study this engine in more detail. First, let us show how to initialize this computational engine:

```
>>> from jhplot.math import *
>>> j = Symbolic ("jscl") # set the engine to JSCL
```

As usual, the example is given in Jython, but the same approach can be used for the standard Java.

The JSCL library supports the following computations:

- polynomial system solving;
- vectors and matrices;
- factorization of expressions;
- derivatives;
- integrals (rational functions);
- boolean algebra;
- expression simplification;
- geometric algebra.

Let us consider a simple example of evaluating an expression symbolically. We will use the expand statement as shown below:

```
>>> from jhplot.math import *
>>> j = Symbolic("jscl")  # set the engine to "jscl"
>>> print j.expand("27^(1/3)/3")
```

You will see the answer "1" (string). The symbolic calculations can use the usual operators, such as:

```
a+b,       a-b,        a*b,       a/b,
div(a,b),  mod(a,b),   a*(b+c),   a**b
```

and it has built-in functions shown in Table 6.1.

In the following sections we will turn to the discussion of a few basic symbolic operations using the JSCL library.

Table 6.1 Built-in functions in the JSCL engine

Name(Arguments)	Function
abs(x)	Absolute value
acosh(x)	Hyperbolic area cosine
acos(x)	Arccosine (radian)
acot(x)	Inverse cotangent
asinh(x)	Hyperbolic area sine
asin(x)	Arcsine (radian)
atanh(x)	Hyperbolic arc tangent
atan(x)	Arc tangent (radian)
conjugate(x)	Conjugate
cosh(x)	Hyperbolic cosine
cos(x)	Cosine (radian)
coth(x)	Hyperbolic cotangent
cot(x)	Cotangent
cubic(x)	Cubic root
exp(x)	Exponential
log(x)	Natural logarithm
sgn(x)	Absolute value
sinh(x)	Hyperbolic sine
sin(x)	Sine (radian)
sqrt(x)	Square root
tan(x)	Tangent (radian)

6.2.1 Conversion to Elementary Functions

Here is an example that checks the output in the form of strings. The expressions must return true:

```
from jhplot.math import *
j=Symbolic("jscl")
print "Elementary function convertions"
j.elementary("abs(x)") ==  "sqrt(x^2)"
j.elementary("sgn(x)") ==  "x/sqrt(x^2)"
```

Listing 6.1 Converting to elementary functions

6.2.2 Numeric Calculations

Now we will consider a reduction of symbolic statements to numeric values:

```
from jhplot.math import *
j=Symbolic("jscl")
print "Reducing to numeric values"
j.numeric("pi") ==  "3.141592653589793"
j.numeric("cos(pi)") == "-1.0"
j.numeric("Infinity") == "Infinity"
j.numeric("-1/0") == "Infinity"
j.numeric("0/0")  == "NaN"
j.numeric("-1/0") == "-Infinity"
```

Listing 6.2 Numeric conversions

The example is obvious. The program performs the calculation at a given value and returns the answer.

6.2.3 Simplify

Various types of symbolic expressions can be simplified using the method `simplify`. Often, it is useful to rewrite an expression in terms of elementary functions (log, exp, sqrt, etc.), before simplifying it.

```
from jhplot.math import *
j=Symbolic("jscl")
print j.simplify("exp(sqrt(-1)*pi)")
print j.simplify("cos(x)^2+sin(x)^2")
print j.simplify(j.elementary("cos(x)^2+sin(x)^2"))
```

Listing 6.3 Simplifying expression

The output is shown below:

```
-1
cos(x)^2+sin(x)^2
1
```

As you can see, the expression $\cos^2(x) + \sin^2(x)$ was simplified after calling the command `elementary()`. All trigonometric functions can be reduced to "elementary" functions as shown in this example:

```
from jhplot.math import *
j=Symbolic("jscl")
print j.simplify(j.elementary("atan(tan(x))"))
print j.simplify(j.elementary("cos(x)"))
print j.simplify(j.elementary("sin(x)"))
print j.simplify(j.elementary("acos(x)"))
print j.simplify(j.elementary("asin(x)"))
print j.simplify(j.elementary("atan(x)"))
print j.simplify(j.elementary("acosh(x)"))
print j.simplify(j.elementary("asinh(x)"))
print j.simplify(j.elementary("atanh(x)"))
print j.simplify(j.elementary("tanh(x)"))
print j.simplify("x^(1/-2)")
```

Listing 6.4 Reducing to elementary functions

The output of the above code is:

```
(sqrt(-1)*log(1/exp(sqrt(-1)*x)^2))/2
(1+exp(sqrt(-1)*x)^2)/(2*exp(sqrt(-1)*x))
(sqrt(-1)-sqrt(-1)*exp(sqrt(-1)*x)^2)/(2*exp(sqrt(-1)*x))
sqrt(-1)*log(x+sqrt(-1+x^2))
sqrt(-1)*log(-sqrt(-1)*x+sqrt(1-x^2))
(sqrt(-1)*log((sqrt(-1)+x)/(sqrt(-1)-x)))/2
log(x+sqrt(-1+x^2))
log(x+sqrt(1+x^2))
log((1+x)/(1-x))/2
-(1-exp(x)^2)/(1+exp(x)^2)
sqrt(1/x)
```

The simplification engine uses a substitution approach. It leads to the following trivial statements:

```
from jhplot.math import *
j=Symbolic("jscl")
j.simplify("log(-2)")  == "sqrt(-1)*pi+log(2)"
j.simplify("sqrt(0)")  == "0"
j.simplify("sqrt(1)")  == "1"
j.simplify("sqrt(-2)") == "sqrt(-1)*sqrt(2)"
```

```
j.simplify("sqrt(-4)")  ==  "2*sqrt(-1)"
j.simplify("exp(0)")    ==  "1"
j.simplify("exp(-1)")   ==  "1/exp(1)"
j.simplify("exp(2)")    ==  "exp(1)^2"
j.simplify("exp(-4)")   ==  "1/exp(1)^4"
j.simplify("1/-x")  ==  "-1/x"
j.simplify("conjugate(conjugate(x))")  ==  "x"
```

Listing 6.5 Symbolic substitution rules

6.2.4 Substitutions

While performing expression evaluations, use the command subst() for substitutions. In the next example we substitute "*x*" with "*a*" in the expand command:

```
>>> from jhplot.math import *
>>> j=Symbolic("jscl")
>>> print j.expand("subst(1/x^2, x, a)")
1/a^2
```

One can also substitute a variable with a number, and then apply the method numeric() shown above to perform the actual calculation.

6.2.5 Differentiate

There is a predefined command "*d*" for symbolic differentiation:

```
>>> from jhplot.math import *
>>> j=Symbolic("jscl")
>>> print j.expand("d(cos(f(x)),x)")
-f"(x)*sin(f(x))
```

6.2.6 Integration

Symbolic integration can be done using the integral() operator. Use the expand() function to get the answer.

```
>>> from jhplot.math import *
>>> j=Symbolic("jscl")
>>> j.expand("integral(1/(x+a),x)")  =="log(a+x)"
>>> j.expand("integral(tan(a+b*x),x)")  =="-1/b*log(4*
    cos(a+b*x))"
```

If the expression is too complicated, apply the `simplify()` command.

Let us consider a more complicated example which performs integration. After this task, we apply the simplification method and, at the end, we substitute $x = 2$ and perform the final calculation.

```
from jhplot.math import *
j=Symbolic("jscl")
t1=j.expand("integral((x^4+pi*2*x^2+2*x+1)/x^3,x)")
print "Integral=",t1
t2=j.simplify(t1)
print "Simplified=",t2
t3=j.expand("subst("+t2+",x,2)")
print "substitute x=2:",t3
print "calculate x=2  :",j.numeric(t3)
```

Listing 6.6 Integration and simplification

6.2.7 Factorization

The command `factorize()` factorizes a mathematical formula. Consider this simple example:

```
>>> from jhplot.math import *
>>> j=Symbolic("jscl")
>>> print j.factorize("1-x^16")
(1-x)*(1+x)*(1+x^2)*(1+x^4)*(1+x^8)
```

6.2.8 MathML Output

The symbolic engine provides outputs in the form of Mathematical Markup Language or MathML, a markup language for describing mathematical notation.

```
>>> from jhplot.math import *
>>> j=Symbolic("jscl")
>>> print j.toMathML("x^2*y^2")
```

This code prints the expression $x^2 * y^2$ in the MathML format.

6.2.9 Integration with DMelt Plotting Canvases

To continue with illustrations of symbolic calculations in Java, let us show how to mix symbolic calculations and the standard DMelt Java classes. We will define the

function $10 * \cos(x) + x^2$ and then symbolically integrate and differentiate it. Then
we plot the final result on the canvas using the familiar HPlot and F1D Java classes
considered in the previous chapters. The example also prints the outputs.

```
from jhplot.math import *
from jhplot import *
from java.awt import Color

j=Symbolic("jscl")
func="10*cos(x)+x^2"
print "Function=",func
s1="integral("+func+",x)"  # Integral
ans1=j.eval(s1);  print "Integral=",ans1
s2="d("+func+",x)"              # Differentiate
ans2=j.eval(s2);  print "Differentiation=",ans2

c1  = HPlot()
c1.visible();  c1.setAutoRange()
f0  = F1D( func, 0, 1.0)
c1.draw(f0)
f1  = F1D( ans1, 0, 1.0)
f1.setTitle("Differentiate")
f1.setLineStyle(1);  c1.draw(f1)
f2  = F1D( ans2, 0, 1.0)
f2.setTitle("Integrate")
f2.setLineStyle(3);  c1.draw(f2)
```

Listing 6.7 Integration of symbolic calculations with DMelt Java classes

The resulting Fig. 6.1 shows the original function and the result of symbolic integration and differentiation.

Fig. 6.1 Mixing symbolic
calculations with the Java
plotting libraries. This figure
shows the original function
and the result of symbolic
integration and
differentiation

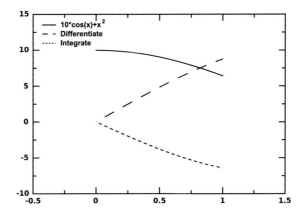

6.3 Using SymPy

Symbolic calculations can be performed using the SymPy library [2]. This library is implemented in Python and included as a third-party package. It is located in the directory "python/packages." This directory is automatically imported by Jython when using the DMelt IDE, so you need not worry about how to install the SymPy library.

The SymPy library can perform differentiation, simplification, Taylor series expansions, and integration. It can be used to solve algebraic and differential equations, as well as systems of equations. The complete list of features can be found on the SymPy web page [2]. Below we show several examples that illustrate integration of this package with the DMelt graphics libraries.

But, first of all, let us show how to use this package for analytic functions. We will start with the example that illustrates how to perform a series expansion of the expression $x^2 / \cos(x)$.

```
from sympy import *

x = Symbol("x")
func=x**2/cos(x)
s=func.series(x, 0, 10)
print   s
print  latex(s)
pprint(s)
```

Listing 6.8 Series expansion

The statement "Symbol()" is rather important: in this package, symbolic variables have to be declared explicitly. The output of this example is printed in the standard ("Python") format when using the print command:

```
x**2 + x**4/2 + 5*x**6/24 + ..
```

Note that we have truncated the output in the above example. In addition, the example shows how to print a nicely formatted output using the pprint() method, close to that used in academic papers and books. In addition, one can transform the output to the LaTeX format which is very popular within the scientific community.

We should note that, similar to DMelt, all mathematical definitions given in Table 3.1 can be used to construct SymPy functions. The functions can be defined in any number of dimensions. For example, if you have a second independent variable, y, in your mathematical formula, specify this variable using the method y=Symbol("y").

Below we show a code snippet that gives you some feeling about what mathematical operations can be performed with the SymPy library. The names of the methods shown below are self-explanatory, so we will be brief here:

```
from sympy import *

x=Symbol("x")
f=x*sin(x)
print "Differential=",diff(f, x)
print "Integral=",integrate(f,x)
print "Definite integral=",integrate(f,(x,-1,1))
print "Simplify=",S(f/x, x)
print "Numerical simplify",nsimplify(pi,tolerance
    =0.01)
print "Limit x->0=",limit(f/(x*x), x, 0)
print "Solve=",solve(x**4 - 1, x)
y=Symbol("y")
z=Symbol("z")
print "Combine together=",together(1/x + 1/y + 1/z)
```

Listing 6.9 SymPy symbolic calculations

The output of this script is given below:

```
Differential= x*cos(x) + sin(x)
Integral= -x*cos(x) + sin(x)
Definite integral= -2*cos(1) + 2*sin(1)
Simplify= sin(x)
Numerical simplify 22/7
Limit x->0= 1
Solve= [-1, I, 1, -I]
Combine together= (x*y + x*z + y*z)/(x*y*z)
```

The DMelt libraries can be used for visualization of the symbolic calculations performed by the SymPy program. Use the DMelt functions, F1D and F2D, and HPlot or HPlot3D (for 2D functions) for graphical canvases. For the conversion between Java and SymPy Python objects, you will need to use the str() method that moves SymPy objects into strings.

Below we show how to differentiate a function and show the result on the HPlot canvas, together with the original function:

```
from jhplot import *

c1 = HPlot("Canvas")
c1.visible(); c1.setAutoRange()

func="2*exp(-x*x/50)+sin(pi*x)/x"
f1 = F1D(func, 1.0, 10.0)

from sympy import * # symbolic calculations
x = Symbol("x")
```

```
d=diff(S(func), x)

f2 = F1D(str(d), 1.0, 10.0)
f2.setTitle("Differential")
c1.draw([f1,f2])
```

Listing 6.10 Differentiation using SymPy

The only nontrivial place of this code is where we moved the result from the differentiation into a string using the `str()` method. For convenience, we also used the simplify method, `"S()"`, which converts the string representing `F1D` function into the corresponding SymPy object. The rest of this code is rather transparent. The result of the differentiation is shown by the green line.

Further discussion of the SymPy package is outside the scope of this book. Please study the original SymPy documentation.

References

1. Jolly R Jscl-meditor, Java symbolic computing library. http://sourceforge.net/projects/jscl-meditor/
2. SymPy development team, SymPy: Python library for symbolic mathematics. http://www.sympy.org

Chapter 7
Histograms

A histogram is a summary graph showing counts of data points falling into various ranges, thus it gives an approximation of the frequency distribution of data. It is an elegant tool to project multidimensional data to lower dimensions and display such projections for visual inspection.

The histogram shows data in the form of a bar graph in which the bar heights display event frequencies. Events are measured on the horizontal axis, X, which has to be binned. The larger the number of bins, the higher the chances that a fine structure of data can be resolved. Obviously, the binning destroys the fine-grain information about original data.

In this respect, the histogram representation is useful when one needs to create a statistical snapshot of a large data sample in compact form. Let us illustrate this: assume we have N numbers, each representing a single measurement. One can store such data in the form of Java or Python lists. Thus, one needs to store $8 \times N$ bytes (assuming 8 bytes to keep one number). In the case of a large number measurements, we need to be ready to store a very big output file, as the size of this file is proportional to the number of events. Instead, one can keep for future use only the most important statistical summary of data, such as the shape of the frequency distribution and the total numbers of events. The information that needs to be stored is proportional to the number of bins, thus the file storage has nothing to do with the size of the original data.

The topic of histograms naturally belongs to Chap. 10 on statistics. Nevertheless, we will examine this topic here since histograms are important tools for data visualization and data mining techniques which will be discussed in several other chapters of this book.

© Springer International Publishing Switzerland 2016
S.V. Chekanov, *Numeric Computation and Statistical Data Analysis
on the Java Platform*, Advanced Information and Knowledge Processing,
DOI 10.1007/978-3-319-28531-3_7

7.1 One-Dimensional Histogram

To create a histogram in one dimension (1D), one needs to define the number of bins,
`Nbins`, and the minimum (`Min`) and maximum (`Max`) values for a certain variable.
The bin width is given by `(Max-Min)/Nbins`. If the bins are too wide (`Nbins` is
small), important information might get lost. On the other hand, if the bins are too
narrow (`Nbins` is large), what may appear to be meaningful information really may
be due to statistical variations of data entries in bins. To determine whether the bin
width is set to an appropriate size, different bin sizes should be tried.

DMelt histograms are designed on the bases of the JAIDA FreeHEP library [1].
For one-dimensional histograms, use the class `H1D`. To initialize an empty histogram,
the following constructor can be used:

```
>>> from jhplot import *
>>> h1=H1D("data", 100, 0, 20)
```

This creates a 1D histogram with the title `"data"`, the number of bins `Nbins=100`,
and the range of axis X to be binned, which is defined by the minimum and the
maximum values, `Min=0` and `Max=20`. Thus, the bin width of this histogram is
fixed to 0.2. The bin size and the number of bins are given by the following methods:

```
>>> d=h1.getBinSize()
>>> i=h1.getBins()
```

We should note that a fixed-size binning is used. In the following sections, we will
consider a more general case when the histogram bin size is not fixed to a single
value.

The method `fill(d)` fills a histogram with a single value, where `"d"` is a double
number. The histograms can be displayed on the `HPlot` canvas using the standard
`draw(h1)` method.

We will discuss the main methods of the histogram class in the following section.
Here, to illustrate the methods discussed before, we give a complete example of how
to fill three histograms with Gaussian random numbers and then display them on
different pads.

```
from java.awt import Color
from jhplot import *

c1 = HPlot("Canvas",600,400,2,1)
c1.visible(); c1.setAutoRange()

h1 = H1D("First",20,-2.0,2.0)
h1.setFill(True)
h1.setFillColor(Color.green)
h2 = H1D("Second",10,-2.5,2.5)
h1.fillGauss(500,0,1)   # mean=0,  sd=1
h2.fillGauss(1000,1,2)  # mean=1,  sd=2
```

```
h3 = H1D("Third",20,0.0,10.0)
h3.setFill(True)
h3.fillGauss(50000,2,1) # mean=2, sd=1

c1.cd(1,1)
c1.setAutoRange()
c1.draw([h1,h2])

c1.cd(2,1)
c1.setAutoRange()
c1.draw(h3)
```

Listing 7.1 Plotting histograms

Here, a convenient method `fillGauss(Max,m,sd)` is used to fill a histogram with Gaussian random numbers. This method is equivalent the Python code:

```
from java.util import Random
r=Random()
for i in range(Max):
    h1.fill(m+sd*r.nextGaussian())
```

that fills `Max` Gaussian random numbers with a mean `m` and a standard deviation `sd`. To do this, we use the package `java.util.Random`. After execution of this script, you should see the plots shown in Fig. 7.1. By default, the lines on the histogram bars are drawn to indicate the size of statistical uncertainty in each bin using the Gaussian

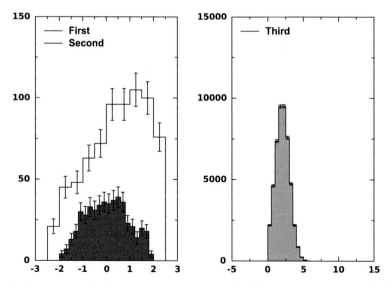

Fig. 7.1 Histograms with Gaussian numbers plotted with different numbers of bins (*left*) and different numbers of events (*right*)

estimation of uncertainties for counting experiments, i.e., Err $= \sqrt{N}$, where N is the number of events in each bin.

We have used two new methods of the `H1D` class in the script shown above. Due to their importance, we will discuss them here:

`setFill(b)`   fill a histogram area when $b = 1$ (Boolean "True"). When $b = 0$ (boolean "False"), the area is not filled;
`setFillColor(c)`   color (Java AWT class) for filling the histogram area.

As before, `"c"` denotes the Java `Color` class discussed in details in Sect. 3.3.1.

The example above illustrates the following important features:

- The height of bins for each histogram depends on the bin size. Even when the number of entries is the same, histograms are difficult to compare in shape when the histogram bins are different, see Fig. 7.1 (left).
- Relative size of errors decreases with increasing the number of entries.

Below we will show a few basic manipulations useful for examining the shapes of histograms, assuming that the underlying mechanism for occurrence of events reveals itself in shapes of event distributions, rather than in the overall statistics or chosen bin size. The shape of the distributions is very important as it conveys information about the probability distribution of event samples.

First of all, let us get rid of the bin dependence. To do this, we will divide each bin height by the bin size. Assuming that `"h1"` represents a histogram, this can be done as

```
>>> width=h1.getBinSize()
>>> h1.scale(1/width)
```

After this operation, all histogram entries (including statistical uncertainties) will be divided by the bin width. You may still want to keep a copy of the original histogram using the method `h2=h1.copy()`, so one can come back to the original histogram if needed.

Different histograms can have different normalizations, thus a visual comparison of histograms might be difficult. If we are interested in histogram shapes, one can divide each bin height by the total number of histogram entries. This involves another scaling:

```
>>> entries=h1.allEntries()
>>> h1.scale(1/entries)
```

Obviously, both operations above can be done using one line:

```
>>> h1.scale(1/(h1.getBinSize()*h1.allEntries()))
```

The second step in comparing our histograms would be to shift the bins of the third histogram. Normally, we do not know the exact shift (what should be don in this case will be considered later). At this moment, for the sake of simplicity, we assume that the shift is known and equals −2. There is a special histogram operation which does such shift:

```
>>> h2.shift(-2)  # shift all bins by -2
```

Now we are ready to modify all the histograms and compare them. Look at the example below:

```
from java.util import Random
from jhplot import *

c1 = HPlot()
c1.visible(); c1.setRange(0,-2,2)

h1 = H1D("First",20,   -2.0,  2.0)
h2 = H1D("Second",50,-2.5,  2.5)
h3 = H1D("Third",20,  0.0,  10.0)
r=Random()
for i in range(50000):
 if(i<5000):
    h1.fill(r.nextGaussian())
    h2.fill(r.nextGaussian())
 h3.fill(2+r.nextGaussian())

h1.scale(1/(h1.getBinSize()*h1.allEntries()))
h2.scale(1/(h2.getBinSize()*h2.allEntries()))
h3.shift(-2)
h3.scale(1/(h3.getBinSize()*h3.allEntries()))

c1.draw([h1,h2,h3])
```

Listing 7.2 Histogram operations

After execution of this script, you will find three overlaid histograms. The shapes of all histograms will be totally consistent with each other, i.e., all bin heights will agree within their statistical uncertainties.

The problem of the histogram comparison discussed above is not a theoretical one. One can find many situations in which you may be interested in how well histogram shapes agree with each other. For example, let us assume that each histogram represents the number of days with rainfall measured during one year for one state. If the distributions are shown as histograms, it is obvious that bigger states have larger number of days with rainfalls compared to small states. This means that all histograms are shifted (roughly by a factor proportional to the area of states, ignoring other geographical differences). The measurements could be done by different weather stations and the bin widths could be rather different, assuming that there is no agreement between the weather stations about how the histograms should be defined. Moreover, the measurements could be done during different time intervals, therefore, the histograms could have different numbers of entries. How can one compare the results from different weather stations, if we are only interested in some regularities in the rainfall distributions? The answer to this question is in the example above: all histograms have to be: (1) normalized; (2) shifted; and (3) bin dependence should be removed.

The only unclear question is how to find the horizontal shifts, since the normalization issue is rather obvious and can be solved with the method discussed above. This problem will be addressed in the following chapters where we discuss a statistical test that evaluates the "fit" of a hypothesis to a sample.

7.1.1 Probability Distribution and Probability Density

The examples above tell that there are several quantities that can be derived from a histogram. One can extract a *probability distribution* by dividing histogram entries by the total number of entries. The second important quantity is a *probability density*, when the probability distribution is divided by the bin width, so that the total area of the rectangles sums to one (which is, in fact, the definition of the probability density).

Both the probability distribution and density distribution can be obtained after dividing histogram entries as discussed above. However, these two characteristics can be obtained easily by calling the following methods:

```
>>> h2=h1.getProbability()
>>> h2=h1.getDensity()
```

which return two new H1D objects: the first represents the probability distribution and the second returns the probability density. In addition to the obvious simplicity, such methods are useful for variable-bin-size histograms, since this case is taken into account automatically during division by bin widths. You can check the density distribution using this statement:

```
>>> print h1.integral()
```

which prints "1" if the histogram is properly normalized.

Note the following: one can save computation time in the case of calculation of the probability distributions if you know the total number of events (or entries) N_{tot} beforehand. In this case, one can obtain the probability distribution using the weight $w1 = 1.0/N_{tot}$ in the method fill(x,w1), without subsequent call to the method getProbability(). After the end of the fill, the histogram will represent the probability distribution normalized to unity by definition. In addition, one can remove the bin dependence by specifying another weight as $w2 = 1.0/bsize$, where *bsize* is the size of the bin. Finally, the density distribution can be obtained using the weight $w3 = w1 * w2$.

7.1.2 Histogram Characteristics

This section continues our discussion of important characteristics of the `H1D` histogram class.

The most popular characteristics of a histogram are the median and the standard deviations (RMS). Assuming that `h1` represents an `H1D` histogram, both (double) values can be obtained as

```
>>> d=h1.mean()
>>> d=h1.rms()
```

We already know that one can obtain the number of entries with the method `allEntries()`. However, some values could fall outside of the selected range during the `fill()` method. Luckily, the histogram class has the following list of methods to access the number of entries:

```
>>> i=h1.allEntries()    # all entries
>>> i=h1.entries()       # number entries in the range
>>> i=h1.extraEntries()  # under and overflow entries
>>> i=h1.getUnderflow()  # underflow entries
>>> i=h1.getOverflow()   # overflow entries
```

All the methods above return integer numbers.

Another useful characteristics is the histogram entropy. It is defined as a negation of the sum of products of the probability associated with each bin with the base-2 log of the probability. One can get the value of the entropy with the method:

```
>>> print "Entropy=",h1.getEntropy()
```

7.1.3 Initialization and Filling Methods

In this subsection we will consider the major histogram methods used for histogram initialization and filling.

Previously, it has been shown how to initialize a histogram with fixed bin sizes. One can also create a histogram using a simpler constructor, followed by a sequence of methods to set histogram characteristics:

```
>>> h1=H1D("Title")
>>> h1.setMin(min)
>>> h1.setMax(max)
>>> h1.setBins(bins)
```

which are used to set the minimum, maximum, and number of bins. These methods can also be useful to redefine these histogram characteristics after the histogram was created using the usual approach.

One can also build a variable-bin-size histogram by passing a list with the bin edges as shown in this example:

```
>>> bins=[0,10,100,1000]
>>> h1=H1D("Title",bins)
```

This creates a histogram with three bins with the bin edges given by the input list. This constructor is handy when a data is represented by a bell-shaped or falling distribution; in this case it is important to increase the bin size in certain regions (tails) in order to reduce statistical fluctuations.

As we already know, to fill a histogram with numbers, use the method `fill(d)`. More generally, one can assign a weight "w" to each value as

```
>>> h1.fill(d, w)
```

where "w" is any arbitrary number representing a weight for a value "d". The original method `fill(d)` assumes that all weights are 1.

But why do we need these weights? We have already discussed in Sect. 7.1.1 that the weights are useful to reduce the computational time when the expected final answer should be either a probability distribution or a density distribution. There are other cases when the weights are useful: We should note again that a histogram object stores the sum of all weights in each bin. This sum runs over the number of entries in a bin only when the weights are set to 1. Events may have smaller weights if they are relatively unimportant compared to other events. It is up to you to make this decision since this depends on a concrete situation.

The method `fill(d)` is slow in Jython when it is used inside loops, therefore, it is more efficient to fill a histogram at once using the method `fill(list)`, where `list` is an array with numbers passed from another program or file. As before, `fill(list, wlist)` can be used to fill a histogram from two lists. Each number in `list` has an appropriate weight given by the second argument.

Instead of Jython (or Java) lists, one can pass a POD array discussed in Sect. 4.1.4:

```
>>> h1.fill(p0d)
```

where p0d represents the POD class.

Analogously, one can fill a histogram by passing a PND multidimensional array discussed in Sect. 4.3.3. This can be done again with the method `fill(pnd)`, where pnd is an array with any size or dimension. One can specify also weights in the form of an additional PND object passed as a second argument to the method `fill(pnd, w)`.

Histograms can be filled with weights which are inversely proportional to the bin size—as it was shown in the previous section, removing the bin size dependence is one the most common operations:

```
>>> h1.fillInvBinSizeWeight(d)
```

It should be noted that this method works even when histograms have irregular binning.

Finally, one can set the bin contents (bin heights and their errors) from an external source as shown below:

```
>>> h1.setContents(values, errors)
>>> h1.setMeanAndRms(mean, rms)
```

where `values` and `errors` are input arrays. Together with the settings for the bin content, the second line in the above example shows how to set the global histogram characteristics, such as the mean and standard deviation. There are more methods dealing with external arrays; advanced users can find appropriate methods in the API documentation of the class `H1D` or using the code assist.

One can create a histogram object from a function. Here is a simple example that shows how to do this:

```
from java.awt import Color
from jhplot  import  *

c1 = HPlot("Canvas")
c1.visible();  c1.setAutoRange()
c1.setNameX("X"); c1.setNameY("Y")

f1 = F1D("2*exp(-x*x/50)+sin(pi*x)/x", -2.0, 5.0)
c1.draw(f1)
h=f1.getH1D("Histogram",500,1,5) # 500 bins between 1
    and 5
h.setFill(1)
h.setFillColor(Color.red); h.setColor(Color.red)
c1.draw(h)
```

Listing 7.3 Histogram created from a function

It should be mentioned that the created histogram does not include the usual event statistics, since it is only a reflection of the input function. Figure 7.2 shows the output of this script.

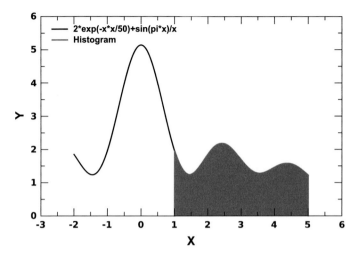

Fig. 7.2 A function and a histogram shown on the same plot

7.1.4   Accessing Histogram Values

One-dimensional histograms based on the `H1D` class can easily be viewed using the following convenient methods designed for visual inspection:

`toString()`   —convert a H1D histogram into a string
`print()`   —print a histogram
`toTable()`   —show a histogram as a table.

Once we know that a histogram is initialized and filled, the next question is to access the histogram values. We will be brief in this section, since most methods are obvious. Table 7.1 shows the most important `H1D` methods.

It should be noted that the bin heights and the numbers of entries are the same when histogram weights used to fill histogram are set to one, i.e., when the method `fill(d)` is called.

Finally, one can view the `H1D` histograms using already known `toTable()` method. This method passes all histogram attributes to a pop-up table for easy visual inspection.

7.1.5   Integration

Histogram integration is similar to the `F1D` functions considered in the previous chapters: We simply sum up all bin heights. This can be done using the method

Table 7.1 Several methods designed to access information about the H1D class

Methods	Returns	Definitions
allEntries()	i	Number of all entries
binCenter(i)	d	Center of ith bin
binCenters()	a	Bin centers
binEntries()	a	Entries in all bins
binEntries(i)	i	Entries in ith bin
binError(i)	d	Errors on entries in ith bin
binErrors()	a	Errors on all entries
binHeight(i)	d	Height of ith bin
binHeights()	a	Heights for all entries
binLowerEdge(i)	d	Low edge of ith bin
binLowerEdges()	a	Low edges for all entries
binMean(i)	d	Mean value in ith bin
binRms(i)	a	RMS value in ith bin
binUpperEdge(i)	d	Upper edge of ith bin
binUpperEdges()	d	Upper edges of all bins
entries()	i	Number of entries in the range
extraEntries()	i	Number of entries outside the range
findBin(d)	i	Find a bin number from a coordinate
getDensity()	H1D	Probability density
getEntropy()	d	Entropy of histogram
getMax()	d	Max value for bins
getMin()	d	Min value for bins
getOverflow()	i	Overflow entries
getOverflowlowHeight()	d	Overflow heights
getProbability()	H1D	Probability distribution
getTitle()	text	Get histogram title
getUnderflowHeight()	d	Underflow heights
getUnderflow()	i	Underflow entries
getValues(i)	a	Arrays with bins, heights and errors $i = 0$ for bin means, $i = 1$ for bin centers
integral(i1, i2)	d	Integrate between $i1$ and $i2$ bins
integralRegion(x1, x2)	d	Integrate region [x1, x2]
maxBinHeight()	d	Maximum bin height
mean()	d	Mean of the histogram
minBinHeight()	d	Minimum bin height
rms()	d	RMS of histogram
sumAllBinHeights()	d	Sum of all bin heights

The following notations are used: "i" denotes an integer, "d" is a double value, "a" corresponds to a 1D array, and "aa" is a 2D array `integral()`. More often, however, it is necessary to sum up heights in a certain bin region, say between a bin `"i1"` and `"i2"`. Then use this method:

```
>>> sum=h1.integral(i1,i2)
```

We should note that the integral is not just the number of events between these two bins: the summation is performed using the bin heights. However, if the weights for the method `fill()` are all set to one, then the integral is equivalent to the summation of numbers of events.

The integration shown above does include multiplication by a bin width. If one needs to calculate an integral over all bins by multiplying the bin content by the bin width (which can be either fixed or variable), use the method:

```
>>> sum=h1.integral(i1,i2,1)
```

where the last parameter should be set to 1 (or to "true" in case of Java codding).

The next question is how to integrate a region in X by translating X-coordinates into the bin indexes. This can be done by calling the method `findBin(x)`, which returns an index of the bin corresponding to a coordinate X. One can call this method every time when you need to identify the bin index before calling the method `integrate()`. Alternatively, this can be done in one line as

```
>>> sum=integralRegion(xmin,xmax,b)
```

The method returns a value of the integral between two coordinates, `xmin` and `xmax`. The bin content will be multiplied by the bin width if the Boolean value `b` is set to 1 (Boolean "true" in Java).

7.1.6 Histogram Operations

Histograms can be added, subtracted, multiplied, and divided. Assuming that we have filled two histograms, `h1` and `h2`, all operations can be done using the following generic method:

```
>>> h1.oper(h2,"NewTitle","operation")
```

where `"operation"` is a string which takes the following values: "+" (add), "−" (subtract), "*" (multiply) and "/" (divide). The operations are applied to the histogram `h1` using the histogram `h2` as an input. One can skip the string with a new title if one has to keep the same title as for the original histogram. In this case, the additive operation will look as `h1.oper(h1,"+")`.

To create an exact copy of a histogram, use the method `copy()`. Previously, we have already discussed the `scale(d)` and `shift(d)` operations.

A histogram can be smoothed. This topic will be described in Sect. 14.3, since smoothing and interpolation are widely used data analysis techniques. Here we just mention that a smoothing of histogram can be done using this method:

```
>>> h1=h1.operSmooth(b,k)
```

This is done by averaging over a moving window. If `"b=1"` then the bins will be weighted using a triangular weighting scheme favoring bins near the central bin (`"b=0"` for the ordinary smoothing) One should specify the integer parameter `"k"` which defines the window as "2*k + 1". Smoothing may be weighted to favor the central value using a "triangular" weighting. For instance, for "k = 2", the central bin would have weight 1/3, the adjacent bins 2/9, and the next adjacent bins 1/9. For all these operations, errors are kept to be the same as for the original (non-smoothed) histogram.

One can also create a Gaussian smoothed version of an H1D histogram. Each band of the histogram is smoothed by a discrete convolution with a kernel approximating a Gaussian impulse response with the specified standard deviation.

```
>>> h2=h1.operSmoothGauss(rms)
```

where `rms` is a double value representing a standard deviation of the Gaussian smoothing kernel (must be nonnegative).

One useful technique is histogram re-binning, i.e., when groups of bins are joined together. This approach could be used if statistics in bins is low; in this case, it makes sense to make bins larger in order to reduce relative statistical uncertainty for entries inside bins (we remind that in case of counting experiments, such uncertainty is $\sqrt{N}$, where N is a number of entries). The method that implements this operation is called `rebin(group)`, where `group` defines how many bins should be merged together. This method returns a new histogram with a smaller number of bins. However, there is one restriction: the method `rebin` cannot be used for histograms with nonconstant bin sizes.

7.1.7 Accessing Low-Level Jaida Classes

The H1D class is based on the two classes, `IAxis` and `Histogram1D` of the Jaida FreeHep library. Assuming `h1` represents a H1D object, these two Jaida classes can be obtained as

```
>>> a=h1.getAxis()  # get IAxis object
>>> h=h1.get()      # get Histogram1D class
```

Both objects are useful. Although they do not contain graphical attributes, they have many methods for histogram manipulations, which are not present for the higher level H1D class. The description of these Jaida classes is beyond the scope of this book. Please look at the Java documentation of these classes or use the code assist.

7.1.8   Graphical Attributes

Sometimes, one has to spend a lot of typing and playing with various graphical options to present analysis results in an attractive and clear form. This is especially important when one needs to show several histograms inside a single canvas. DMelt provides many methods designed to draw histograms using different styles.

First of all, histograms can be shown either by lines (default) or by using symbols. For the default option (lines), one can consider either to fill histogram area or keep this area empty. The following methods below can be useful:

```
>>> h1.setFill(b)
>>> h1.setFillColor(c)
```

For the first method, Jython Boolean "b=1" means to fill the histogram, while "b=0" (false) keeps the histogram empty. If the histogram area has to be filled, you may consider to select an appropriate color by specifying Java AWT Color object "c". How to find the appropriate color has been discussed in Sect. 3.3.1.

Histograms can be shown using symbols as

```
>>> h1.setStyle("p")
```

The style can be set back to the default (histogram bars). This can be done by passing the string "h" instead of "p". One can also use symbols connected by lines; in this case, use the character "l" (draw lines) or the string "lp" (draw lines and symbols).

Table 7.2 lists the most important graphical attributes of the H1D class. The graphical attributes can be retrieved back using similar methods after substituting the string "set" by "get" in the method names.

7.2   Histogram in 2D

A histogram in two dimensions (2D) is a direct extension of the 1D histogram class discussed in the previous section. To initialize such histograms, one should define ranges and bins in X and Y. A 2D histogram can be visualized using the 3D canvas discussed before.

The 2D histograms are implemented using the H2D class, which is based on the JAIDA FreeHEP Histogram2D class [1]. The H2D class "decorates" the Histogram2D class with additional graphical attributes and methods that are not available in the original JAIDA FreeHEP library. To create a 2D histogram, one needs

Table 7.2 Methods for graphical representation of the H1D histograms

Methods	Definitions
setStyle("text")	"p"—show symbols, "l"—show lines
	"lp"—lines and symbols, "h"—as histogram bars
setPenWidthErr(i)	Line width
setPenDash(i)	Dashed line style,
	"i" is the length of dashed line
setColor(c)	Set color for drawing
setFill(b)	b = 1—fill histogram area (b = 0 not fill)
setFillColor(c)	Set AWT Java color for fill area
setFillColorTransparency(d)	Set the transparency ($0 \leq d \leq 1$)
setErrX(b)	Show or not errors on X
setErrY(b)	Show or not errors on Y
setErrColorX(c)	Set Java color for errors on X
setErrColorY(c)	Set Java color for errors on Y
setSymbol(i)	Symbol type:
	0: circle
	1: square
	2: diamond
	3: triangle
	4: filled circle
	5: filled square
	6: filled diamond
	7: filled triangle
	8: plus (+)
	9: cross
	10: star
	11: small dot
	12: bold plus
setSymbolSize(i)	Set symbol size "i"

The following notation are used: "i" means an integer, "d" means a double value, "b" corresponds to a Boolean ("b = 1" means Java true and "b = 0" means false), "c" is Java AWT Color class

to define the number of bins for X and Y axes, and the minimum and the maximum values for each axis. Then the histogram can be initialized as

```
>>> from jhplot import *
>>> h2 = H2D("Title", binsX, minX, maxX, binsY, minY, maxY)
```

where `binsX(binsY)` is the number of bins for X (Y), `minX(minY)` and `maxX(maxY)` are the minimum and the maximum values for the X (Y) axis, respectively.

In addition to the fixed-bin-size case, one can create histograms with variable bin sizes. One should call the constructor shown below that requires a list with the bin edges:

```
>>> from jhplot import *
>>> h2= H2D("Title",[1,2,3],[1,2,4,8])
```

This constructor shows how to define the bin edges in X (the first input list) and Y (the second input list).

The method `fill(x,y)` fills the histograms, where "x" and "y" values for the X and Y axes. It should be noted that the bin heights and the numbers of entries are the same as when weights used to fill the histogram are set to one. Non-unity weights "w" can be specified in the method `fill(x,y,w)`.

Table 7.3 shows the main methods of the H2D histogram class. Unlike the H1D histogram, the H2D class has the "setter" and "getter" methods for each axis. For example, `getMeanX()` returns the mean value in X, while `getMeanY()` returns the mean value in Y.

You may find that the methods given above are not enough for complicated operations. We should remind that the H2D class is based on the Jaida classes, `IAxis` and `Histogram2D`, which are located in the Jaida package `hep.aida.ref.histogram.*`. As a consequence, one can build a H2D histogram by creating the object of the class `IAxis`, which represents the axis in one dimension, and then pass it to the H2D constructor.

```
>>> from hep.aida.ref.histogram import *
>>> from jhplot import *
>>> xAx=FixedAxis(10,0.0,1.0)
>>> yAy=FixedAxis(20,0.0,1.0)
>>> h2=H2D("Title",xAx,yAy)
```

Again, as for the one-dimensional case, both Jaida classes can be obtained as

```
>>> aX=h2.getAxisX()  # get IAxis object for axis X
>>> aY=h2.getAxisY()  # get IAxis object for axis Y
>>> h=h2.get()        # get Histogram2D class
```

assuming that h2 represents a H2D object. We will return to the Jaida histogram package in Sect. 7.3.

Table 7.3 Some methods of the H2D class

Methods	Returns	Definitions
Fill methods		
fill(x,y)	–	Fill x and y
fill(x,y,w)	–	Fill x and y with weight w
clear()	–	Clean from all entries
getter methods		
allEntries()	i	Number of all entries
binEntries(ix,iy)	i	Entries in bins (ix,iy) bin
binError(ix,iy)	d	Errors on a (ix,iy) bin
binHeight(ix,iy)	d	Height of a (ix,iy) bin
copy()	H2D	Exact copy
entries()	i	Number of entries in the range
extraEntries()	i	Number of entries outside the range
getBinsX()	i	Number of bins in X
getBinsY()	i	Number of bins in Y
getDensity()	H2D	Density distribution
getMaxX()	d	Get max value in Y
getMaxY()	d	Get max value in Y
getMeanX()	d	Get the mean value in X
getMeanY()	d	Get the mean value in Y
getMinX()	d	Get min value in X
getMinY()	d	Get min value in Y
getOverflowlowHeightX()	d	Overflow heights in x
getOverflowlowHeightY()	d	Overflow heights in y
getOverflowX()	i	Overflow entries in x
getOverflowY()	i	Overflow entries in y
getProbability()	H2D	Probability distribution
getRmsX()	d	Get RMS in X
getRmsY()	d	Get RMS in Y
getTitle()	text	Get histogram title
getUnderflowHeightX()	d	Underflow heights in x
getUnderflowHeightY()	d	Underflow heights in y
getUnderflowX()	i	Underflow entries in x
getUnderflowY()	i	Underflow entries in y
integral(i1, i2, j1, j2)	d	Integral in the range
integralRegion(d1, d2, d1, d2)	d	Integral in the range (coordinates)
sumAllBinHeights()	d	Sum of all bin heights
setter methods		
setContents(d[][], e[][])		Set heights and errors (double arrays)
setBinError(ix, iy, d)		Set the bin error for (ix,iy) bin
setMeanX(d)		Set the mean value for X
setMeanY(d)		Set the mean value for Y
setRmsX(d)		Set RMS for X
setRmsY(d)		Set RMS for Y

The table uses the following notation: "i" indicates an integer value, "d" means a double value, "a" corresponds to a 1D array, "aa" denotes a 2D array

7.2.1   Histogram Operations

All histogram operations are exactly the same as for the H1D class discussed in
Sect. 7.1.6. The H2D histograms can be added, subtracted, multiplied, and divided
using the generic method:

```
>>>  h1=h1.oper(h2,"NewTitle","operation")
```

where "operation" is a string which can have the following values: "+" (add),
"−" (subtract), "*" (multiply) and "/" (divide) histograms, and h1 and h2 are objects
of the class H2D. The histograms can be scaled using the method scale(d).

The probability and density distributions can be obtained as discussed in
Sect. 7.1.1:

```
>>>  h2=h1.getProbability()
>>>  h2=h1.getDensity()
```

One can obtain the sum of the bin contents in the range defined in the terms of
the bin indexes, or coordinate values using the methods shown in Sect. 7.1.5.

Finally, one can calculate an integral of all histogram entries using the method
integral(). A histogram region can be integrated by passing bin indices.

7.2.2   Graphical Representation

The 2D histograms should be visualized using the three-dimensional canvas HPlot3D
which has been used before plotting F2D functions. One can find details in Sect. 3.4.
The example below shows how to fill and display an H2D histogram:

```
from jhplot import *
from java.util import Random
c1 = HPlot3D()
c1.visible()
c1.setNameX("X")
c1.setNameY("Y")
h1=H2D("2D Test",30,-3.0, 3.0, 30, -3.0, 3.0)
r=Random()
for i in range(1000):
    h1.fill(r.nextGaussian(),r.nextGaussian())
c1.draw(h1)
```

Listing 7.4 Filling a 2D histogram

Here we fill a 2D histogram with Gaussian random numbers and then plot it using exactly the same method `draw(obj)` (with `obj` being an object for drawing) as that used for the F2D functions.

Below we show a more complicated example. We will fill several histograms using Gaussian numbers after shifting the means for the sake of better illustration. In addition, we show how to: (1) Plot two histograms on the same plot; (2) Plot a histogram and a 2D function on the same plot. In both cases, we use the method `draw(obj1,obj2)`, where `obj1` and `obj2` could be either H2D or F2D object.

```
from java.awt import Color
from java.util import Random
from jhplot import *

c1  = HPlot3D("Canvas",600,700, 2,2)
c1.visible(); c1.setGTitle("H2D drawing options")

h1=H2D("H2D Test1",30,-4.5,4.5,30,-4.0, 4.0)
h2=H2D("H2D Test 2",30,-3.0, 3.0, 30, -3.0, 3.0)
f1=F2D("8*(x*x+y*y)", -3.0, 3.0, -3.0, 5.0)

r=Random()
for i in range(10000):
  h1.fill(r.nextGaussian(),0.5*r.nextGaussian())
  h2.fill(1+0.5*r.nextGaussian(),-2+0.5*r.
   nextGaussian())

c1.cd(1,1)
c1.setScaling(8)
c1.setRotationAngle(30)
c1.draw(h1)

c1.cd(1,2)
c1.setScaling(8)
c1.setColorMode(2)
c1.setRotationAngle(30)
c1.draw(h1,h2)

c1.cd(2,1)
c1.setColorMode(4)
c1.setLabelFontColor(Color.red)
c1.setScaling(8)
c1.setRotationAngle(40)
c1.draw(f1,h2)

c1.cd(2,2)
c1.setColorMode(1)
c1.setScaling(8)
c1.setElevationAngle(30)
c1.setRotationAngle(35)
c1.draw(h1)
```

Listing 7.5 Showing several 2D histograms in 3D

H2D drawing options

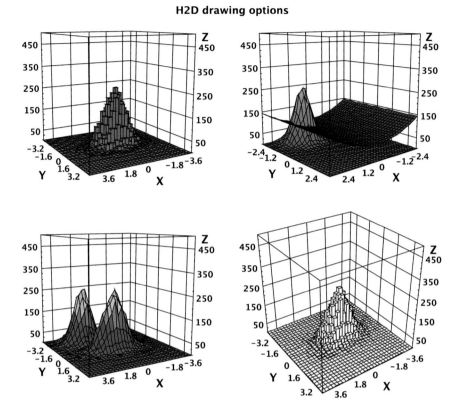

Fig. 7.3 2D histogram and a 2D function shown in an interactive 3D canvas

The resulting plots are shown in Fig. 7.3. One can see that the default drawing option for the H2D class are histogram bars. The 2D functions are usually shown using a surface-type representation.

We should note that, at the time this book is written, there was no support for drawing 2D histograms with variable bin sizes.

The 2D histograms (as well as 2D functions) can be shown as a contour (or density) plot, an alternative graphical method when each region in X and Y is represented by different color, depending on the density population of each area with data points. To show such plot, one can use the method setContour(). The small code snippet below illustrates this:

```
from jhplot   import *
from java.util import Random

c1  = HPlot3D ("Canvas" ,600 ,600)
c1.visible ()
c1.setNameX ("X")
```

```
c1.setNameY("Y")
c1.setRangeZ(0,20)
c1.setContour()
c1.setContourLines(10)

h1 = H2D("Contour",30,-2.0,2.0,30,-2.0,2.0)
h1.fillGauss(1000,0,1.0,0,0.2)        # X=0,Y=0 (sX=1,sY
    =0.2)
h1.fillGauss(800,1.0,0.5,-1.0,0.2) # X=1,Y=1 (sX=0.5,
    sY=0.2)
c1.draw(h1)
```

Listing 7.6 2D contour histogram

The execution of this script brings up a window with the contour plot shown in Fig. 7.4. One can also show the 2D histogram as a "density" plot (or intensity map). Such plots create a grid and fill each cell with colors that correspond to the height (or intensity) of histogram bins. The grid for such plots is created using bin edges. Try to change setContour() by setDensity() in the above example to see how it works.

As for the F2D and P1D objects, one can use the canvas HPlot2D for showing H2D histograms. This canvas class was specifically designed to show the density and contour plots as will be discussed in Sect. 8.8.

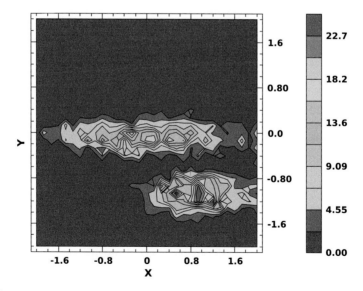

Fig. 7.4 A H2D histogram shown as a contour plot in 2D

7.3 Histograms in Jaida

We have already mentioned that there is a way to access the so-called Jaida histograms from the FreeHep scientific library. Essentially, every histogram class in DMelt originates from the corresponding Jaida classes that do not contain graphical attributes.

The Jaida histograms can be created after importing the Java package hep.aida. ref.histogram. Generally, one can use the so-called Java factories to create the Jaida histograms. In this section, however, we will concentrate on a more basic histogram construction using Jaida.

Before building a Jaida histogram, first you have to define the "axis" object, which can contain fixed or variable size bins. The example below shows how to create one-dimensional Jaida histograms from the class Histogram1D:

```
>>> from hep.aida.ref.histogram import *
>>> ax=FixedAxis(bins,min,max)
>>> h1=Histogram1D("name","title",ax)
```

where "bins" represents the number of bins, "min" and "max" is the minimum and the maximum value for the X range. The code builds a 1D Jaida histogram using fixed-size bins. One should also specify the histogram name and the title (they may not need to be the same). The passed name is used internally by the Jaida class, so one can just set it to the histogram title.

Using the code assist, try to check the available methods of the object h1. You may notice that they are rather similar to those of the H1D class. The only difference is that the Histogram1D does not have any graphical attributes.

Analogously, one can build a Jaida histogram using a variable bin size. The code is essentially the same, with the only one difference: the axis should be replaced by the line with axis=VariableAxis(edges) with edges being an array with the bin edges.

A two-dimensional Jaida histogram, Histogram2D, can be constructed similarly:

```
>>> from hep.aida.ref.histogram import *
>>> ax=FixedAxis(binsX,minX,maxX)
>>> ay=FixedAxis(binsY,minY,maxY)
>>> h1=Histogram2D("name","title",ax,ay)
```

Again, to make a 2D histogram with variable bin sizes, use the VariableAxis() instead of FixedAxis.

To add graphical attributes to these Jaida classes, we have to move them to the full-featured DMelt histograms, H1D or H3D. One can do this easily using the following constructors:

```
>>> from hep.aida.ref.histogram import *
>>> from jhplot import *
>>> h1d=H1D(h1)  # build H1D from Jaida histogram
>>> h2d=H2D(h2)  # build H2D from Jaida histogram
```

One can retrieve the Jaida histograms back as

```
>>> h1=h1d.get()  # get Histogram1D object
>>> h2=h2d.get()  # get Histogram2D object
```

But why do we need to use the Jaida histograms, if the DMelt histograms are direct derivatives of the corresponding Jaida classes? The answer is simple: DMelt does not map every single method of the Jaida histogram classes. In most cases DMelt histograms inherent the most common methods, and add extra methods not present in Jaida. Therefore, if you find that DMelt histograms do not have necessary methods, try to access the IAxes, Histogram1D and Histogram2D objects that may have the methods you need for your work.

7.4 Histogram in 3D

Histograms in three dimensions are slightly tricky. We cannot use them for graphical representation, therefore, DMelt does not add extra features compared to those present in the Jaida Histogram3D class.

In DMelt, 3D histograms are implemented using the H3D class. In the case of fixed-size binning, these histograms can be defined by building three axes in X, Y and Z, and passing them to the histogram constructor:

```
>>> from jhplot import *
>>> from hep.aida.ref.histogram import *
>>> ax=FixedAxis(binsX,minX,maxX)
>>> ay=FixedAxis(binsY,minY,maxY)
>>> az=FixedAxis(binsZ,minZ,maxZ)
>>> h3=H3D("title", ax, ay, az)
```

The methods of the H3D class are similar to those discussed for H2D. The only difference is this: now we should take care of the additional Z axis. For example, to fill this histogram with weights, one needs to use the method fill(x,y,z). We remind again that this type of histograms does not have any graphical attributes—essentially, it is the exact mapping of the corresponding Histogram3D Jaida class.

7.5 Profile Histograms

A profile histogram is used to show the mean value in each bin of a second variable. Errors on the bin heights usually represent statistical uncertainties on the mean values or data spreads (i.e., standard deviations) of the event distributions inside bins.

The profile histograms in one dimension are implemented in the `HProf1D` class. Such histograms are filled using the method `fill(x,y)`, similar to two-dimensional histograms. The first variable x is used for binning, while the second argument represents the variable for which the mean is calculated.

To show a profile histogram, use the method `getH1D()`. This method converts the profile histogram into the usual `H1D` discussed in Sect. 7.1, which then can be used for graphical representation. It can also accept a string to define a new title after the conversion. By default, errors on the mean values in each bin are shown by vertical lines. One can also display the mean values of y and their root-mean-square (RMS) deviations from the mean for each bin. The RMS values are shown as errors on the histogram heights when using the option `"s"` during the conversion with the method `getH1D("title","s")`.

Below we calculate the mean values of a Gaussian distribution as a function of the second variable with uniform random numbers between zero and ten.

```
from jhplot import *
from java.util import Random

c1 = HPlot()
c1.setGTitle("Profile histogram")
c1.setRange(0,11,0.0,5)
c1.setNameX("X")
c1.setNameY("Gaussian mean")
c1.visible()
h2=HProf1D("Profile1D",10,0.0, 11.0)
r=Random() # fill X and Y with random numbers
for i in range(2000):
  h2.fill(10*r.nextDouble(),r.nextGaussian()+2)
h1=h2.getH1D("Mean and RMS","s") # errors show RMS
h1.setStyle("p")
c1.draw(h1)
```

Listing 7.7 Profile histogram

The result of this script is shown in Fig. 7.5. Since we used the option "s", the error bars show the RMS of the distributions. If you would remove this option in the method `getH1D`, you will see that the statistical uncertainties on the mean values are quite small.

In contrast to the `HProf1D` histogram, the class `HProf2D` is designed to construct a profile histogram in two dimensions. As before, such histograms represent the mean of some distribution in each bin of two additional variables in X and Y.

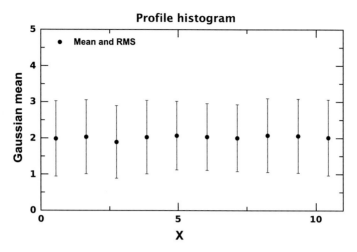

Fig. 7.5 A profile histogram which shows the mean values as a function of a second variable. Error *bars* show the RMS of the distribution in each bin

An HProf2D histogram can be created by specifying the number of bins in X and Y, as well as the minimum and maximum values for each axis. Alternatively, one can pass arrays with the bin edges, if the bin sizes are not fixed to a constant value. The HProf2D histograms should be converted to H2D histograms (see Sect. 7.1) for graphical representation. The conversion can be done by either calling the method getH2D() (error on the mean in each bin) or getH2D("title", "s") (errors correspond to RMS in each bin). Obviously, the HPlot3D canvas should be used for plotting.

Below we show a simple example of how to display the mean of a Gaussian distribution in two dimensions. Note that "Z" variable is given by a Gaussian distribution with a constant term that depends on the two other variables, X and Y.

```
from jhplot import *
from java.util import Random
from math import *

c1=HPlot3D()
c1.visible()
c1.setRange(0,10,0.0,10,0,200)
h2=HProf2D("Profile2D",10,0.0,10.0,10,0.0,10.0)
r=Random()
for i in range(5000):
    x=10*r.nextDouble()
    y=10*r.nextDouble()
    z=abs(x)*abs(y)
    h2.fill(x,y, z+5*r.nextGaussian())
c1.draw( h2.getH2D() )
```

Listing 7.8 Working with a 2D profile histogram

Fig. 7.6 A profile
histogram in 3D as a function
of two variables, *X* and *Y*

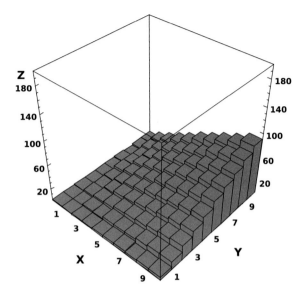

Figure 7.6 shows the image of the 2D profile histogram.

7.6   Histogram Input and Output

All the histograms discussed above can be written in files and restored later. For
example, one can use the method `toFile()` to write a histogram into an ASCII
file. As for any DMelt object, one can save collections of histograms using Jython
lists, tuples, or dictionaries. Alternatively, one can use the Java containers, such as
arrays, sets, or maps.

In the case of Python lists, one should always remember the order that was used
to store histograms inside the lists. In the case of dictionaries or Java maps, one can
use a human-readable description as the key for each histogram entry. Below we
illustrate how to use a Python dictionary to store several histograms (including their
graphical attributes) in a serialized Java file: The code below writes a collection of
histograms filled with random numbers into the file `"file.jser"`:

```
from jhplot   import  *
from java.awt import  Color
from java.util import  Random

h1=H1D("Simple1",20,-2.0,2.0)
h1.setFill(1)
h1.setFillColor(Color.green)
h2=H2D("Simple2",20,-3.0,  3.0,  20,  -3.0,  3.0)
h3=HProf1D("Profile1D",10,0.0,  11.0)
```

```
r=Random()
for i in range(1000):
  h1.fill(r.nextGaussian())
  h2.fill(r.nextDouble(),r.nextGaussian())
  h3.fill(10*r.nextDouble(),r.nextGaussian()+50)

hold = {}                        # define a dictionary
hold["describe"]="Collection" # use dictionary with
    keys
hold["h1"]=h1
hold["h2"]=h2
hold["h3"]=h3
IO.write(hold,"file.jser")       # write the collection
```

Listing 7.9 Saving multiple histograms in a file

Next, we will restore the histograms using this file. First, we will read the dictionary and then will fetch all the histograms from the file using appropriate keys:

```
from jhplot  import *

c1 = HPlot()
c1.visible(); c1.setAutoRange()

hold=IO.read("file.jser") # read dictionary
print hold.keys()            # print all keys
print "Description:"+hold["describe"]
c1.draw(hold["h1"])          # plot histograms
c1.draw(hold["h3"].getH1D())
```

Listing 7.10 Reading histograms from a file

In this example, we draw two histograms retrieved from the serialized file, H1D and HProf1D. The latter histogram has to be converted to an H1D object for visualization. We will remind that, in order to draw the H2D histogram, we will need a 3D drawing canvas, HPlot3D.

Object serializations into XML files can be done using the writeXML() and readXML() methods from the same IO class. In addition, one can convert a histogram into an XML string using the toXML() method.

Finally, we recall that one can store and retrieve multiple objects, including DMelt histograms, using the class HFile. It is designed to work with a large sequence of any Java objects (see Chap. 9 for detail). In addition, one can use a GUI browser to look at all stored objects inside the files created with the class HFile and then plot them.

When we save histograms of the class H1D into files, we write the histogram data and graphical attributes for visualization. To reduce disk usage, one can only save the histogram data. To do this, use the method get() that returns the Jaida histogram class without graphical attributes. This approach is similar to that used for the F1D functions discussed before. The next example shows this.

```
from jhplot  import *

h1=H1D("Simple1",20,-2.0,2.0)
IO.write(h1.get(),"file.jser") # write data of H1D
h1=H1D(IO.read("file.jser"))   # restore the
   histogram
```

Listing 7.11 Writing and reading histograms without graphical attributes

Here we write only data of the H1D class. Then we restore this histogram by creating a H1D object using the default graphical settings. Thus the graphical attributes are lost in this approach.

7.6.1   External Programs for Histograms

7.6.1.1   CFBook Package

Histograms can be filled by an external C++ or FORTRAN program. For this, use the CFBook or CBook packages [2]. The first package can be used to fill histograms from a C++ or FORTRAN program, and write them into specially designed XML files (which are optimized for storing numerical data). The second package is based on compressed records constructed using the Google's Protocol Buffers. This approach will be discussed in Sect. 9.4.

The package CFBook package generates two static libraries: libcbook.a (to be linked with C++) or libfbook.a (to be linked with FORTRAN). Histograms filled by the CFBook library can be retrieved and visualized using the HBook class. Let us give one example. We assume that histograms are filled by a C++ external program and are kept in the "cpp.xml". One can read and retrieve histograms from this file as

```
>>> from jhplot import *
>>>
>>> hb = HBook()          # create HBook instance
>>> hb.read("cpp.xml")    # read input XML file
>>> print hb.listH1D()    # list all histograms
>>> h1 = hb.getH1D(10)    # get H1D histogram with ID
   =10
>>> c1.draw(h1)           # draw it
```

In the code above, "c1" represents an instance of the HPlot class.

The HBook class can also be used to save histograms or other objects into XML files. In this case, one should use its method write():

```
>>> from jhplot import *
>>>
```

```
>>> hb = HBook ("hbook")
>>> h3=H1D("test",2,0.0,1.)
>>> h4=H2D("test",5,0.0,1.,4,0.,1.)
>>> hb.add(30,h3)        # add to HBook with ID=30
>>> hb.add(40,h3)        # add to HBook with ID=40
>>> p1=P1D("test")       # create P1D objects
>>> hb.add(10,p1)        # add to HBook with ID=10
>>> print hb.listH1D()   # list all histograms
>>> h1 = hb.getH1D(30)   # get H1D histogram with ID
    =30
>>> ...
>>> hb.write("out.xml")  # write to an XML file
```

This example illustrates that one can insert the histogram H1D, H2D or even containers like P1D into the HBook holder using some identification numbers (10, 20, 30). These numbers can be used later for object retrieval. Also, we show how to write all such objects into an external XML file.

7.6.1.2   ROOT Package

DMelt also supports a library to read histograms saved into ROOT files [3, 4]. One can read ROOT histograms using Jython scripts or using a GUI browser in order to navigate to a certain histogram object inside the ROOT files. These topics will be discussed in Sect. 9.7.

7.7   Analyzing Histograms from Multiple Files

This section will discuss a practical example based on an analysis of multiple histograms. It is a direct extension of the example discussed in Sect. 2.16. What we what to do now is more complicated than what was shown in that example: (1) We will need to identify all files with the extension ".dat" and read back. The files contain numbers (one per line) and can be compressed. (2) Then we will analyze all numbers in these files by putting them into a histogram. This histogram can be used to build statistical summaries of data, such as the number of entries, the mean value, and the standard deviation.

While the described tasks are notably more complicated, our code will be at least a factor 2 smaller than that discussed[1] in Sect. 2.16. The files with the extension ".dat" will be located in the directory "/tmp/" (of course, you will need to change this directory).

First, let us generate 10 files. Each file will contain 100 random numbers distributed using a normal distribution. The mean of the distribution is 0 and the standard deviation is 2:

---

[1]Below we will skip the part which was necessary to remove duplicate files, since this task described in Sect. 2.16 was used for an illustration only.

```
from jhplot import *

for i in range(10):               # write 10 files
  p=P0D("data"+str(i))            # data array
  p.randomNormal(100,0,2.0)       # mean=0, SD=2
  p.toFile("/tmp/file"+str(i)+".dat")
```

Listing 7.12 Writing 10 arrays to files

Below we show a Jython code which does the complete data analysis:

```
from jhplot import *
from jhplot.utils import *

c1 = HPlot("Analysis")
c1.visible(); c1.setRangeX(-30,30)

# list of files with extension .d
list=FileList.get("/tmp/",".dat$")
sum=P0D("sum")          # sum all numbers

for f in list:
   p0=P0D("data",f)    # read data from a file
   sum=sum.merge(p0)   # merge with the rest
h1=H1D("test",40,-20,20)
h1.fill(sum)            # fill histogram
c1.draw(h1)
c1.drawStatBox(h1)      # show statistics
```

Listing 7.13 Reading histograms

The execution of this script brings up a frame with the filled histogram. A box with calculated statistics, displayed after calling the method drawStatBox(), tells about all major statistical features of the data sample. Note that the method FileList was used to scan all files which have the extension ".dat". Obviously, one can use rather sophisticated Java regular expressions to find a necessary patten in the file names. Also, it should be noted that we first created a P0D object from ASCII files. If the input files are zipped (gzipped), use the methods readZip("name") or readGZip("name") instead. If you know that files are serialized using Java, replace the read methods with the method readSerialized("name").

In exactly the same fashion one can use various DMelt containers, such as P1D, PND, etc. To plot a particular slice of data (row or column), one can convert these objects into Java arrays, Jython lists or P0D objects, which then can be passed to either H1D or H2D histograms.

References

1. FreeHEP Java Libraries. http://java.freehep.org/
2. Chekanov S, CFBook histogram library. http://jwork.org/dmelt/cbook/
3. Brun R, Rademakers F, Canal P, Goto M (2003) Root status and future developments, ECONF C0303241, MOJT001
4. Brun R, Rademakers F (1997) ROOT: An object oriented data analysis framework. Nucl Instrum Meth A 389:81. http://root.cern.ch/

Chapter 8
Scientific visualization

Scientific visualization deals with various methods of presenting data and results of numerical calculations in graphical form. This plays a crucial role in numerical computations and statistical data analysis, where charts, tables, histograms, plots of data, and mathematical expressions are rather common for knowledge discovery and for gaining insight into various mathematical concepts.

DMelt is a complete suite for visual exploration of data and mathematical functions. It includes dozens of Java libraries for interactive charts in 2D and 3D. Some of them will be discussed in this chapter.

8.1 Graphical Canvases

In this section, we will step back from the numerical and statistical computations and discuss somewhat technical issues related to the question of presenting numerical results in graphical form. We will discuss what graphical canvases should be used to plot data points, histograms, functions, geometrical shapes, and mathematical objects.

The task of choosing the right graphical canvas may look daunting at first. It is further complicated by a large choice of canvases and objects which can be visualized for better understanding of data and mathematical concepts. Below we summarize several most important graphical canvases included in DMelt:

HPlot   2D canvas and contour plots for `P1D`, `F1D`, `H1D` and other graphical primitives;

HPlotJa   2D canvas with interactive editor for drawing diagrams. Support for the `H1D`, `P1D`, `F1D`, graphical primitives and diagrams;

SPlot   a lightweight 2D canvas, supports `H1D`, `P1D`, and arrays;

HPlot2D   contour (or density) plots in 2D for classes `P1D`, `H2D`, and `F2D`;

HPlotXY   2D canvas for X–Y data and `F1D` functions;

© Springer International Publishing Switzerland 2016

S.V. Chekanov, *Numeric Computation and Statistical Data Analysis on the Java Platform*, Advanced Information and Knowledge Processing, DOI 10.1007/978-3-319-28531-3_8

`HPlot3D`	used for interactive 3D plots, `P2D`, `P3D`, `H2D`, and `F2D`;
`HPlot3DP`	used for interactive 3D surfaces for parametric and nonparametric functions (`FPR`);
`HChart`	used to show 2D charts. It supports plotting `P1D`, X–Y charts, area and bar charts, histogram and pie charts;
`HGraph`	used for interactive interconnected graphs;
`HPlotRT`	used to show data in real time;
`WPlot`	for 2D canvas for X–Y data and `F1D` functions using custom components.

Usually, all such canvases originate from different Java base classes, and are created for different visualization tasks. Some of these canvases are also implemented as Java singleton classes (see our discussion later). In the following sections, we will describe these canvases and help to identify the most appropriate canvas for representation of your results.

The drawing canvases discussed above implement the method `draw(obj)`, where `obj` is data or mathematical object to be shown. Therefore, the best method to figure out what object can be shown is to look for this method. In addition, all the canvases have the common method `export(file)`, which is used to export images to high-quality vector-graphics images (SVG, EPS, PDF, etc.), so they can easily be included in text-processing systems.

We should remind that DMelt is a graphics-intensive program, and the burden of plotting graphs on various canvas on the CPU could be immense. The advice is to build graphical canvas after performing all CPU-consuming numerical calculations. This book does not always use this rule since our examples are not too CPU demanding. But in real situations, one should consider to reorganize analysis codes such that creation of graphical canvas goes after numerical calculations.

In the next section, we will discuss the most popular general-purpose DMelt approach to plot data and functions in 2D. Then we will discuss other Java classes that can be used to show data and other objects in 2D and 3D.

8.2 HPlot Canvas

In Sect. 3.3.1, we have already discussed the `HPlot` class. We will remind that it can be used to show `F1D`, `P1D` and `H1D` objects on X–Y plane. In this section, we will discuss this canvas in more detail.

First, we will remind how to build this canvas and make it visible:

```
>>> from jhplot import HPlot
>>> c1=HPlot("Canvas")
>>> c1.visible(1)
```

The title "Canvas" used in the initialization can be ignored. If the canvas should not be shown, use the method `visible(0)` for Jython—in this case it will be created in the computer memory. We remind that, Jython "1" means Java "true" and "0" corresponds to Java "false". One can also use the usual Boolean "True" and "False" types. As a shortcut, the method `visible()` can be used instead of `visible(True)`.

The size of the canvas on the screen can be customized as

```
>>> c1=HPlot("Canvas",600,400)
```

This initializes a plotting canvas of the size 600 by 400 pixels (explicitly defined). One can resize the canvas frame later by dragging the edges of the canvas frame with the mouse. Most objects, such as titles, labels, symbols, lines, etc., should be resized proportional to the canvas size.

The constructor:

```
>>> c1=HPlot("Canvas",600,400,1,2)
```

also creates a canvas of size 600 by 400 pixels. In addition, the last two numbers show that two plot regions (or pads) will be created inside the same canvas. If the last two numbers are, say 3×2, then 6 pads are created, (3 pads are in X direction and 2 pads are in Y direction). One can navigate to the current pad using the `cd(i1,i2)` method, where `i1` is the location of the pad in X, and `i2` is the location in Y. For example, if one needs to plot a DMelt object `obj` on the first pad, use:

```
>>> c1.cd(1,1)
>>> c1.draw(obj)
```

where `obj` could either be F1D, P1D, or H1D. It should also be noted that one can plot a list of objects at once:

```
>>> c1.draw([f1,f2])    # draw f1 and f2 functions
```

One can navigate to the second pad using the method `cd()`. For example, if an object `obj` should be shown on the second pad, use

```
>>> c1.cd(1,2)
>>> c1.draw(obj)
```

By default, the HPlot canvas has a range between 0 and 1 for the X or Y axis. One should specify a necessary range using the method

```
>>> c1.setRange(Xmin,Xmax,Ymin,Ymax),
```

where Xmin (Xmax) and Ymin (Ymax) are the minimum (maximum) ranges for *X* and *Y* axes. Alternatively, one can set auto-range mode using the method setAutoRange(). After calling this method, the minimum and maximum values for the *X*-axis range will be determined automatically.

One can change color for many attributes of the canvas, as well as to annotate the canvas. First, one should import the classes Color or Font from the Java AWT library. How to use the class Color and Font has been discussed in Sect. 3.3.1. How to set user-defined annotations for the canvas c1 is shown in this example:

```
>>> from java.awt import Color
>>> c1.setGTitle("Global title",Color.red)
>>> c1.setNameX("X axis title")
>>> c1.setNameY("Y axis title")
>>> c1.setName("Current pad title")
```

All the statements above are self-explanatory. One may add a background color to the canvas as

```
>>> c1.setBackgroundColor(Color.yellow)
```

or one can specify custom fonts for the legends as

```
>>> from java.awt import Font
>>> font=Font("Lucida Sans",Font.BOLD, 12)
>>> c1.setLegendFont(font)
```

Read the Java documentation of the HPlot class [1].

Finally, you can edit pad titles and titles of the axes using the mouse. Simply double-click on the area with the text and drag the mouse. We will discuss this in detail in the following section.

During the debugging stage, it is often necessary to execute a script (using the key [F8]) and then manually close the HPlot canvas. If you do not close the HPlot canvas, a new instance of the canvas will be created next time you execute the same (or different) script. It will overlie on the existing canvas, but the worst thing is that it will consume the computer memory.

Several HPlot frames shown on the same desktop is useful in certain situations, for example, when it is necessary to compare plots generated by different Jython scripts. But in many cases, the last thing you want is to create a new HPlot object and manually dispose it in order to prepare for the next script execution.

One can avoid the creation of new canvas frame using the SHPlot class:

```
>>> from jhplot import *
>>> c=SHPlot.getCanvas()
>>> c.visible()
>>> c.setAutoRange()
```

Here "c" represents an SHPlot object, which is a Java singleton extension of the HPlot class. You can work with this class exactly as with the HPlot canvas. For example, in order to set the overall size of the canvas to 600×400 pixels and to create 2×2 plotting pads, just type:

```
>>> c=SHPlot.getCanvas("Canvas",600,400,2,2)
```

When you create the canvas using SHPlot.getCanvas method, the HPlot frame is instantiated only once, and only once. If the canvas frame already exists from the previous run, all graphics will be disposed, and then a new graph will be drawn on the same canvas. In this case, you do not close the canvas frame manually.

8.2.1 Working with the HPlot Canvas

There are several important operations you should know while working with the HPlot canvas frame:

8.2.1.1 Find USER or NDC Coordinators

To determine the coordinate of the mouse in the USER coordinate system (i.e., determined by the range of X or Y axis) or the NDC (a user-independent, given by the actual size of the canvas on the screen) coordinate system, click the middle mouse button. The mouse pointer should be located *inside* the area with the drawing pad. The located coordinates will be displayed at the bottom of the canvas frame.

8.2.1.2 Zoom in to a Certain Region

To zoom in, use the middle mouse button. The mouse location should be *below* the X-axis or on the *left* side of the Y-axis. Drag the mouse holding the middle button. A red line indicating a zoom region in X (or Y) will be shown. After releasing the middle mouse button, you will see an updated HPlot canvas with the new axis range. To set the axis range to the default, use the right mouse button and select the pop-up menu called "Default axis range".

One can also click-and-drag the mouse pointer to create a zoom rectangle. To perform a zoom to rectangle, press the middle mouse button, hold, and drag the mouse to draw a rectangle. Release the middle mouse button to complete the rectangle.

8.2.1.3   How to Change Titles, Legends, and Labels

If one needs to change titles, legends, and labels, select the appropriate object and double click on the right mouse button. This brings up a window with all graph settings. One can change the location of a selected object by dragging it while holding the mouse button.

8.2.1.4   Edit Style of Data Presentation

Click on the right mouse button. You should see a pop-up menu with several submenu options. Click on "Edit" and this will bring up a new window that allows the necessary changes. In particular, one can change:

- axis ranges, set auto-range, logarithmic or linear scales;
- ticks size, numbers of ticks, colors, axis colors, etc.;
- labels, legends, title, the names of X and Y ranges;
- the presentation style of data points or histograms. One can select the fill style, points, lines, colors for points. One can also remove (or add) statistical or systematical error bars. Refresh the canvas to update drawn graphics.

Of course, all these attributes can be changed using Jython macros. Read the API documentation of the `HPlot` class. We remind that one can access API documentation by calling the method `doc()` of the `HPlot` class (or any class).

8.2.1.5   Clear the Plotting Area

All graphs on the same `HPlot` canvas can be removed using several methods. If one needs to clean canvas from plotted objects (histogram, function, etc.), use the method `cleanData()`. Note: in this case, only the current plot defined by the `cd(i1,i2)` method will be updated. If one needs to remove all objects from all plots on the same `HPlot` canvas, use `cleanAllData()`.

It is also useful in many cases to remove all user settings from a certain graph, as well as to remove input objects. In this case, use the method `clear()`. One can also use the method `clear(i1,i2)` to remove graphics on any arbitrary pad, since `i1` and `i2` specify the pad location. The method `clearAll()` removes drawings on all pads, but keeps the main canvas frame untouched. The method `close()` removes the canvas frame and disposes the frame object.

8.2.1.6   Modifying Global Margins

To change the global title, which can be set using the `setGTitle()` method, navigate the mouse to the title location at the very top of the frame, and click the right mouse button. A new pop-up menu will appear. Using this menu, one can increase or decrease the divider location, make the divider invisible, change fonts, and colors.

One can edit left, right, and bottom margins of the main frame using exactly the same approach: navigate to the frame boarder and use the mouse pop-up menu. One can access all attributes of the margins using this method:

```
>>> c1.panel()
```

This method returns the GHPanel class (an extension of the swing JPanel) which keeps attributes of all four margins. Here are several operations associated with the positions on margins:

```
>>> c1.setMarginSizeLeft(n)
>>> c1.setMarginSizeRight(n)
>>> c1.setMarginSizeBottom(n)
>>> c1.setMarginSizeTop(n)
>>> c1.setTextBottom("X")             # text of   margins
>>> c1.setTextLeft("Y")
>>> c1.setTextTop("X")                # same  as setGTitle()
>>> c1.setTextBottom("Y")
>>> c1.setMarginBackground(color) # margin background
```

("n" is the number of pixels). If you need to make a space between the X and Y axis and the title labels, use the methods setMarginBottom(n) and setMarginLeft(n), where n is again the number of pixels. Look at the API documentation for more details.

8.2.2 Saving Plots

One can save all drawings shown on the HPlot canvas to external files using the File menu. The plots can be restored later using the same menu. This works for most of the plotted objects, like HLabel, Primitives, and other graphics attributes to be discussed below.

The output files have the extension "*.jhp" and contain the XML file which keeps all attributes of the HPlot canvas and the data files necessary to recreate figures without running Jython macros. Look inside this file after unzipping it (below we show how to do this using a Linux/UNIX prompt):

```
>>> unzip [file].jhp
```

This creates the directory with the name [file] with [file].xml, where
[file] is the specified file name and several data files. The general form of the file
names is "plotXY-dataZ.d", where X and Y are the positions of the pads inside the
HPlot canvas and Z indicates the data set number.

It should be noted that all data files are just the outputs from P1D objects, see
Sect. 4.2. Therefore, one can easily read such files using the methods of the P1D
class. This is useful if the automatic procedure from the File menu fails, and the
user wants to replot the data using a different macro.

It should be noted that not all objects can be saved in this approach. Therefore,
the best method to save a plot in a file is to save the macro that makes it, and rerun
it again when needed.

8.2.3 Reading Data

One can open a serialized (the extension "jser") or PFile (the extension "jpbu")
file for browsing Java objects stored inside this file using the menu [File] and [Read
data] of the HPlot canvas. This file should be created using the class HFile as
discussed in Sect. 9.3 (using the default compression). Select a file with the extension
"jser" or "jpbu" and open it. You will see a dialog with all objects inside the file.
Select an object and click on the "Plot" button. If the object can be plotted, you will
see it inside the canvas. Most objects can be visualized on the HPlot canvas, such
as POD, P1D , H1D histograms, functions, and Java strings. The latter are converted
into an HLabel object on the fly for drawing inside the HPlot canvas. In addition,
one can store GUI dialogs based on the JFrame class of the Java swing library.

We remind that the browser is based on the BrowserData class to be called
from analysis scripts. See Sects. 9.3.2 and 9.4 for details.

8.2.4 Axes

The axis range can be set automatically by calling

```
>>> c1.setAutoRange()
```

This method tells that the canvas determines the X and Y ranges automatically from
the objects passed to the method draw(obj).

A user can specify the X–Y range manually by calling the method

```
>>> c1.setRange(xMin,xMax,yMin,yMax)
```

If you need to set a fixed range only for *X* axis, use the methods

```
>>> c1.setAutoRange()
>>> c1.setRangeX(xMin,xMax) # X is fixed, Y has autorange
```

Analogously, one can use the method `setRangeY(min,max)` for the *Y* axis.
One can remove all drawn axes using:

```
>>> c1.removeAxes()
```

If only one axis should be drawn instead of all four, first remove all axes and then call the method:

```
>>> c1.setAxisY() # show only Y axis
>>> c1.setAxisX() # show only X axis
```

One can draw a small arrow at the end of axes as

```
>>> c1.removeAxes()
>>> c1.setAxisX()
>>> c1.setAxisArrow(1) #  arrow type 1
>>> c1.setAxisArrow(2) #  another arrow type 2
```

If no mirror axes should be drawn, use these methods

```
>>> c1.setAxisMirror(0,0) # no mirror axis on X
>>> c1.setAxisMirror(1,0) # no mirror axis on Y
```

If no ticks should be drawn, use

```
>>> c1.setTickLabels(0,0) # no mirror axis on X
>>> c1.setTickLabels(1,0) # no mirror axis on Y
```

Finally, call the method `update()` to redraw the canvas.

8.2.5 Summary of the HPlot Methods

Table 8.1 shows the major methods of the `HPlot` class. This list is incomplete, therefore, use the code assist or Java API documentation to find more methods.

Table 8.1 The main methods of the `HPlot` class

Methods	Definitions
add(obj)	add a graphical object
cd(i1, i2)	navigates to a specific region (i1 x i2)
draw(obj)	draw an object
draw([obj,...])	draw a list of objects
drawStatBox(H1D)	draws statistical box for a histogram
setAutoRange(axis, b)	sets auto-range for axis
setAutoRange(b)	sets auto-range for X and Y
setAutoRange()	sets auto-range
setBackgroundColor(c)	background color of the graph
setBox(b)	bounding box around the graph (b = 1) or not (b = 0)
setBoxColor(c)	color of the bounding box
setBoxFillColor(c)	fill color of the bounding box
setBoxOffset(d)	offset of the bounding box
setGrid(axis, b)	shows grid (b = 1) or not (b = 0) for axis
setGridColor(c)	sets grid color
setGridToFront(b)	grid in front of drawing (b = 1) or not (b = 0)
setGTitle(string, f, c)	sets the global title with Font and Color
setLabel(string, f, c)	sets a label at random position
setLegend(b)	draws the legend when b = 1 (if b = 0, do not draw it)
setLegendFont(f)	sets the legend font
setLegendPosition(axis, d)	sets legend for axis
setLogScale(axis, b)	sets log scale (b = 1) or not (b = 0) for axis
setMarginBottom(n)	set n-pixels between X title and X axis
setMargineTitle(i)	defines the region size for the global title
setMarginLeft(n)	set n-pixels between Y title and X axis
setRange(axis, min, max)	set the range for axis (axis = 0,1)
setRange(minX, maxX, minY, maxY)	ranges for X and Y
setTicsMirror(axis, b)	sets ticks (b = 1) or not (b = 0) for axis
showMargineTitle(b)	do not show the global title
updateAll()	updates all regions
update()	updates a region defined by cd(i1,i2)
viewHisto(b)	shows Y starting from 0 (for histograms)
visible(b)	makes it visible (b = 1) or not (b = 0)
visible()	sets to visible

The following notations are used: "i" denotes an integer value, "d" means a double value, "b" corresponds to a boolean ("b = 1" is True and "b = 0" is False), "c" is Java AWT color, "f" is the Java AWT class. Finally, "axis" is an integer that defines axis number: "axis = 0" for X and "axis = 1" for Y

8.2.6   Exporting to Image Files

One can export a graph shown on the HPlot canvas (including all its pads) into an image using the method export(file), where "file" is a string representing the name of the image file. With this method, the graphs can be saved to a variety of vector-graphic formats as well as bitmap image formats.

The export(file) statement should always be at the end of your script, when all objects have been drawn with the draw(obj) method or after the update() statement.

In the example below

```
>>> c1.export("file.ps")
```

we export the drawing of the canvas HPlot (c1 represents its instance) into a PostScript image file. One can also export it to SVG, SVGZ, PNG, JPEG, EPS, PDF, etc., formats using the appropriate extension for the output file name. Here are more examples:

```
>>> c1.export("file.eps")   # create EPS file
>>> c1.export("file.png")   # create PNG file
>>> c1.export("file.jpg")   # create JPG file
>>> c1.export("file.pdf")   # create PDF file
>>> c1.export("file.svg")   # create SVG file
>>> c1.export("file.svgz")  # create compressed SVG file
```

If you are not sure about what extension to use, look at the [File]-[Export] menu which gives you some ideas about the supported graphics formats. One can use this menu for exporting graphs into images without calling the method export() inside your scripts.

It is also useful to create an image file using the same name as the name of your script.

```
>>> c1.export(Editor.DocMasterName()+".ps")
```

where the method Editor.DocMasterName() accesses the file name of the currently opened script.

It is also possible to save the HPlot canvas to an image file with a pop-up dialog. Use the method exportDialog(file) for this task.

8.2.7 *Labels and Keys*

8.2.7.1 Simple Text Labels

Labels can be shown on the HPlot canvas using the Text class. It is impossible to interact with such simple labels using the mouse since this class is based on the standard Java 2D graphics. However, due to a low memory footprint, such labels can be rather useful. The Text class is located in the Java package jhplot.shapes. The example below shows how to access such labels:

```
from jhplot.shapes import Text
from jhplot import *
c1=HPlot("Canvas with a text")
c1.visible()
lab=Text("Label in USER system", 0.5, 0.2)
c1.add(lab)
c1.update()
```

Listing 8.1 Text label example

You may notice that, instead of the draw() method, we use the add() and update() methods. This could be rather handy since now we can add many objects to the same canvas and then trigger update of the canvas to display all added objects at once. The text label will be drawn in the USER coordinate system at $X = 0.5$ and $Y = 0.2$. For the NDC system, use the method setPosCoord("NDC").

```
>>> lab=Text("Text in the NDC system", 0.5, 0.2)
>>> lab.setPosCoord("NDC")
>>> lab=Text("Text in USER system", 0.5, 0.2)
>>> lab.setPosCoord("USER")
>>> c1.add(lab)  # add to HPlot canvas
>>> c1.update()  # update the canvas
```

As before, one can set the text fonts, color, and transparency level using setFont(f) and setColor(c) methods, where "f" and "c" are Java AWT Font and Color classes, respectively. The transparency level can be set using the setTransparency(d) method ($0 < d < 1$, with $d = 1$ for completely transparent objects).

8.2.7.2 Interactive Labels

After initialization of an HPlot object, one can insert an interactive text label and show it on this canvas. Such labels are significantly more memory consuming objects than those created with the help of the Text class. The interactive labels are created using the HLabel class. They can be dragged to adjust their positions. One can also edit the text of the labels using a GUI dialog (after double clicking on the label text). As before, use the method `add(obj)`, and then make it visible by calling the `update()` method. Here is a typical example:

```
>>> lab=HLabel("HLabel in NDC", 0.5, 0.2)
>>> c1.add(lab)  # add it to HPlot object
>>> c1.update()  # trigger update
```

In the above code, the `HLabel` object is inserted at positions 0.5 and 0.2 in the USER coordinate system of the `HPlot` canvas. One can see this by clicking the middle mouse button and by looking at text message at the bottom of the `HPlot` frame. The status panel at the bottom of the frame indicates the mouse position in the USER system.

Alternatively, one can set the label location in the user-independent coordinate system ("NDC"). We remind that this coordinate system is independent of the window size and is defined with two numbers in the range from 0 to 1. Again, one can study this system by clicking the middle mouse button.

The same label in the NDC system should be created as

```
>>> lab=HLabel("HLabel in NDC",0.5,0.2,"NDC")
```

One can modify the label attributes using the usual `setFont(f)`, as well as `setColor(c)` method. The position of the label can be adjusted using the mouse. A double-click on the label brings up a label property window.

One can also show a multiline interactive label on the `HPlot` canvas using the `HMLabel` class. It is similar to `HLabel`, however, instead of a string, it takes a list of strings. Each element of such list will be shown on a new line. We will show a relevant example in Sect. 8.2.9.

It should be noted that in order to show a legend, global title, or title for an axis, the `HLabel` class is not necessary; one should use the special methods of the `HPlot` canvas, such as `setGTitle()`, `setNameX()`, and `setNameY()` instead.

8.2.7.3 Interactive Text Labels with Keys

Unlike the `HLabel` class, the `HKey` class creates an interactive label with a text and a symbol describing the shown data. It should be noted that it behaves differently than the legend which is automatically shown with the corresponding data set. The `HKey` object is not related to any data set and can be shown even if no data are plotted. This class is rather similar to `HLabel` and has all the methods that the `HLabel` class has. To make it visible, call the `update()` method. Here is a typical example using the Jython shell:

```
>>> #... assume c1 was created before
>>> h1=HKey("key type=32",55,62)   # key at x=55 and y=62
>>> h1.setKey(32,7.0,Color.blue)   # key of type 32, size=7
>>> h1.setKeySpace(4.0)   # space between a key and text
```

```
>>> c1.add(h1)
>>> c1.update()
```

Various key types are shown in Fig. 8.1. This figure was generated by the script shown below:

```
from java.awt import Font,Color
from jhplot   import  *

c1=HPlot("Canvas",600,450)
c1.visible(); c1.setRange(0,100,0,100)
c1.setGridAll(0,0)
c1.setGridAll(1,0)
c1.setGTitle("HKey types")
c1.removeAxes()

for j in range(1,13):
     title="key type="+str(j)
     hh=HKey(title,15,97-8*j)
     c=Color(0,65 + j*10,0)
     hh.setKey(j,2.0,c)
     hh.setKeySpace(4.0)
     c1.add(hh)
h=HKey("key type=20",55,90)
h.setKey(20,7.0,Color.blue)
h.setKeySpace(4.0); c1.add(h)

h1=HKey("key type=21",55,83)
h1.setKey(21,7.0,Color.blue)
h1.setKeySpace(4.0); c1.add(h1)

h1 =HKey("key type=30",55,76)
h1.setKey(30,7.0,Color.blue)
h1.setKeySpace(4.0); c1.add(h1)

h1 =HKey("key type=31",55,69)
h1.setKey(31,7.0,Color.green)
h1.setKeySpace(4.0); c1.add(h1)

h1 =HKey("key type=32",55,62)
h1.setKey(32,7.0,Color.red)
h1.setKeySpace(4.0); c1.add(h1)

c1.update()
```

Listing 8.2 Keys and their descriptions

Fig. 8.1 Various types of the keys used by the `setKey()` method

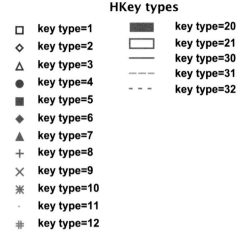

8.2.8 Geometrical Primitives

The package `jhplot.shapes` can be used to display several (noninteractive) geometrical primitives, including the text label discussed before:

`Arrow(x1,y1,x2,y2)`	shows an arrow from (x1,y1) to (x2,y2)
`Circle(x1,y1,R)`	inserts a circle with the radius R
`Ellipse(x1,y1,rX,rY)`	inserts an ellipse with the radius rX (rY for Y axis)
`Text("text",x1,y1)`	inserts a text label
`Line(x1,y1,x2,y2)`	inserts a line from (x1,y1) to (x2,y2)
`Picture(x1,y2,file)`	inserts a PNG or JPG figure
`Rectan(x1,y1,w,h)`	rectangle with the width "w" and height "h"

In all cases, the objects will be drawn in the USER coordinate system. But one can also insert the primitives in the NDC system using the usual method `setPosCoord` (`"NDC"`) to be applied to the objects above. As before, the method `setPosCoord` (`"USER"`) sets the user coordinates.

To show all such graphical primitives on a canvas, use the `add()` method and execute `update()` when you want all objects to be shown. To add a different color or a line width, use additional arguments for the constructor. For example,

```
>>> from java.awt import Font,BasicStroke
>>> from jhplot.shapes import *
>>> stroke= BasicStroke(1.0)
>>> c=Circle(x1,y1,R,stroke,Color.red)
```

or, alternatively, one can use several "setter" methods:

```
>>> c=Circle(x1,y1,R)
>>> c.setFill(1)
>>> c.setColor(Color.red)
>>> c.setStroke( BasicStroke(1.0) )
>>> c.setDashed(3.0)
>>> c.setTransparency(0.5)
```

The last line in the above code makes a circle filled with red color and sets its transparency to 50 %.

The arrow lines can be of two different types, depending on the arrow style. The style can be set using the setType(i) method, where $i = 1, 2, 3$, while the length and the width of the arrows can be set as setEndLength(d) and setEndWidth(d), where "d" is a double value.

To fill the primitives with a certain color, use the method setFill(1). Color and a transparency level is set by the setColor(c) and setFont(f) methods, respectively.

8.2.9   Text Strings and Symbols

In the previous section, we have discussed several important classes to add annotations to the HPlot plots. In many occasions, you would need to show special symbols inside HKey, HLabel, Text labels, or inside HPlot methods designed to show global titles and axis names. The text attributes can be set manually or via the property dialog. Let us discuss several rules:

- Subscripts and superscripts in the text should be included as for the standard LaTeX text files, i.e., use the "underscore" symbol (_) to show subscripts and the "hat" symbol to show a superscript.
- To indicate overline, use the reserved word #bar{symbol}. To display a square root, use #sqrt{symbol}.
- To make a space between characters, use the usual space.
- For a backspace, the predefined keyword #bs2{} should be used. It creates a backspace with the width of one character. In order to make a smaller backspace, equal to 1/2 of the character width, use #bs1{}.
- Symbols for the DMelt labels must be encoded in HTML using the entity-reference notations of the Unicode Consortium [2]. For instance, use the symbol ω to show the Greek letter ω.

Below we give a small code, which makes a label with several special symbols:

jHPlot symbols I
Add '&' in front and terminate by ';'

d_3 subs	± plusmn	Å Aring	Ù Ugrave	í iacute	A Alpha	Φ Phi
d^3 super	² sup2	Æ AElig	Ú Uacute	î icirc	B Beta	X Chi
đ s̄ ā	³ sup3	Ç Ccedil	Û Ucirc	ï iuml	Γ Gamma	Ψ Psi
nbsp	´ acute	È Egrave	Ü Uuml	ð eth	Δ Delta	Ω Omega
¡ iexcl	µ micro	É Eacute	Ý Yacute	ñ ntilde	E Epsilon	α alpha
¢ cent	¶ para	Ê Ecirc	Þ THORN	ò ograve	Z Zeta	β beta
£ pound	· middot	Ë Euml	ß szlig	ó oacute	H Eta	γ gamma
¤ curren	¸ cedil	Ì Igrave	à agrave	ô ocirc	Θ Theta	δ delta
¥ yen	¹ sup1	Í Iacute	á aacute	õ otilde	I Iota	ε epsilon
¦ brvbar	º ordm	Î Icirc	â acirc	ö ouml	K Kappa	ζ zeta
§ sect	» raquo	Ï Iuml	ã atilde	÷ divide	Λ Lambda	η eta
¨ uml	¼ frac14	Ð ETH	ä auml	ø oslash	M Mu	θ theta
© copy	½ frac12	Ñ Ntilde	å aring	ù ugrave	N Nu	ι iota
ª ordf	¾ frac34	Ò Ograve	æ aelig	ú uacute	Ξ Xi	κ kappa
« laquo	¿ iquest	Ó Oacute	ç ccedil	û ucirc	O Omicron	λ lambda
¬ not	À Agrave	Ô Ocirc	è egrave	ü uuml	Π Pi	µ mu
shy	Á Aacute	Õ Otilde	é eacute	ý yacute	P Rho	ν nu
® reg	Â Acirc	Ö Ouml	ê ecirc	þ thorn	Σ Sigma	ξ xi
¯ macr	Ã Atilde	× times	ë euml	ÿ yuml	T Tau	o omicron
° deg	Ä Auml	Ø Oslash	ì igrave	ƒ fnof	Υ Upsilon	π pi

Fig. 8.2 Symbols for DMelt strings (Set I)

```
from jhplot import *
c1=HPlot()
c1.visible()
s1="&omega;,F^{2},F_{2},&gamma;&rarr;e^{+} e^{-}"
s2="g &rarr; q#bar{q}"
s=[s1,s2]
lab=HMLabel(s,0.3,0.7)
c1.add(lab)
c1.update()
```

Listing 8.3 Showing special symbols

This creates a multiline label with the text:

$$\omega, F^2, F_2, \gamma \rightarrow e^+e^-$$

$$g \rightarrow q\bar{q}$$

Now let us consider how to draw special characters using Unicode. Here is a small example:

```
>>> text="W^{ &#8723;}"
```

which shows $W^{\mp}$. You can find this using Unicode lookup tables.

Figures 8.2 and 8.3 show lists of symbols supported by DMelt. These figures are created by the example code that comes with the DMelt program.

Let us give a complete example that uses mathematical symbols. Our next code shows data and several labels with mathematical annotations and equations. The output of this code is shown in Fig. 8.4.

```
from jhplot import *
from java.awt import *

c1=HPlot("plot")
c1.visible(); c1.setRange(1,100, 0, 7)
c1.setMarginLeft(80); c1.setMarginBottom(50)
c1.setNameX("L [ab^{-1}]")
c1.setNameY("M_{&phi;} [GeV^{-1}]")
c1.setLogScale(0,1)

p1=P1D('#sqrt{s}#bs2{}=33 TeV')
p1.setStyle("l"); p1.setPenWidth(4)
p1.add(1,1.2);    p1.add(3,1.8)
p1.add(10,2.0);   p1.add(80,5.0)

s="#sqrt{&beta;+j}&ge;&sum;&alpha;_{i}"
lab1=HLabel(s,0.5,0.75,"NDC")
lab1.setFont(Font("Serif", Font.ITALIC, 24))
lab1.setColor(Color.blue)
c1.add(lab1)

s="#bar{q}#bs2{} q &rarr; &delta; &rarr; &mu;"
lab2=HLabel(s,0.50,0.65,"NDC")
lab2.setFont(Font("Arial", Font.BOLD, 24))
lab2.setColor(Color.red)
c1.add(lab2)

c1.draw(p1)
```

Listing 8.4 Using mathematical symbols

jHPlot symbols II
Add '&' in front and terminate by ';'

τ tau	← larr	Σ sum	□ sub	" quot
υ upsilon	↑ uarr	− minus	□ sup	& amp
φ phi	→ rarr	□ lowast	□ nsub	< lt
χ chi	↓ darr	√ radic	□ sube	> gt
ψ psi	↔ harr	□ prop	□ supe	Œ OElig
ω omega	□ crarr	∞ infin	□ oplus	œ oelig
ϑ thetasym	□ lArr	□ ang	□ otimes	Š Scaron
Υ upsih	□ uArr	□ and	□ perp	š scaron
ϖ piv	□ rArr	□ or	□ sdot	Ÿ Yuml
• bull	□ dArr	∩ cap	□ lceil	ˆ circ
... hellip	□ hArr	□ cup	□ rceil	˜ tilde
′ prime	□ forall	∫ int	□ lfloor	ensp
″ Prime	∂ part	□ there4	□ rfloor	emsp
‾ oline	□ exist	□ sim	□ lang	thinsp
/frasl	□ empty	□ cong	□ rang	zwnj
℘ weierp	□ nabla	≈ asymp	◊ loz	zwj
ℑ image	□ isin	≠ ne	♠ spades	lrm
ℜ real	□ notin	≡ equiv	♣ clubs	rlm
™ trade	□ ni	≤ le	♥ hearts	- ndash
ℵ alefsym	Π prod	≥ ge	♦ diams	— mdash

Fig. 8.3 Symbols for DMelt strings (Set II). Some symbols could not be available for some platforms, in which cases they are shown as open squares

8.3 Interconnected Objects

To visualize interconnected graphical objects, one should use the HGraph class. It is similar to HPlot, since it also extends the GHFrame class used as a basis for the HPlot Swing frame. This means that one can set titles, margins, and plotting pads in exactly the same way as for the HPlot canvas. The HGraph canvas is fully interactive: one can drag the connected objects using the mouse to modify their

Fig. 8.4 Plotting data with special symbols and equations

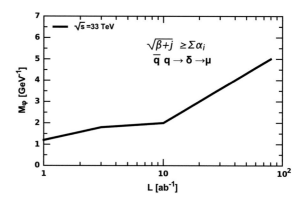

locations and even edit their attributes. The example below shows how to build a graph with interconnected objects:

```
from jhplot   import  *

c1  =  HGraph()
c1.setGTitle("Bayesian  network")
c1.visible()
v1,v2,v3="rain","sprinkler","grass"
c1.addVertex( v1 )
c1.addVertex( v2 )
c1.addVertex( v3 )

c1.setPos( v1,  400,  50 )
c1.setPos( v2,  100,  50 )
c1.setPos( v3,  240,  200 )

c1.addEdge( v1,  v2 )
c1.addEdge( v2,  v3 )
c1.addEdge( v1,  v3 )
```

Listing 8.5 A graph with interactive objects

Figure 8.5 shows the result of the above script. Note that this graph will be used for more practical application, namely the Bayesian network discussed in Sect. 13.5.

Fig. 8.5 Using the HGraph canvas to show interconnected objects

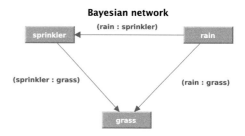

8.4 Showing Charts

The HChart class is also similar to the HPlot class. The HChart canvas allows plotting various charts, such as pie, bar, histogram, line, and area charts. All charts are based on the jFreeChart package library [3]. This is a simple example of how to create two pads with a pie-like and a bar-like charts:

```
from jhplot import *
c1 = HChart("Canvas",700,250,2,1)
c1.setGTitle("Chart examples")
c1.visible()

c1.setChartPie()
c1.setName("Pie chart")
c1.valuePie("Hamburg",1.0)
c1.valuePie("London",2.0)
c1.valuePie("Paris",1.0)
c1.valuePie("Bern",1.0)

c1.cd(2,1)
c1.setChartBar()
c1.setName("Bar charts")
c1.valueBar(1.0, "First", "category1")
c1.valueBar(4.0, "Second", "category2")
c1.valueBar(3.0, "Third",  "category3")

c1.update()
```

Listing 8.6 Chart examples

The result of the execution of this script is shown in Fig. 8.6.

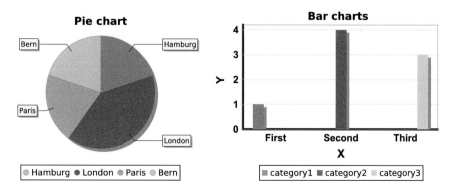

Fig. 8.6 A pie and a bar charts using the HChart canvas

When the `HChart` canvas is created, one can set the following types of charts:

```
>>> c1.setChartXY()          # create a X-Y chart
>>> c1.setChartPie()         # create a Pie chart
>>> c1.setChartPie3D()       # create a 3D Pie chart
>>> c1.setChartLine()        # create a line chart
>>> c1.setChartAre()         # create an area chart
>>> c1.setChartBar()         # create a bar chart
>>> c1.setChartBar3D()       # create a 3D bar chart
>>> c1.setChartHistogram()   # create a histogram
```

Then one can add values using the `value+ChartName()` method. For example, to add a value to the bar chart, use `valueBar()` method. Check the `HChart` API documentation for details. Finally, to display a chart, execute the usual `update()` method.

One can access many jFreeChart components via several "getter" methods. For example, `c1.getChar()` returns the `JFreeChart` class for further manipulations. Use `help.doc(c1)` to view the complete Java documentation of this class.

Note that one can also use JFreeChart directly. For this purpose, call the `HPlotChart` canvas that helps to create JFreeChart plots using its native methods. This gives quite significant flexibility. Here is an example that creates a chart shown in Fig. 8.7:

```
from org.jfree.chart import ChartFactory
from org.jfree.data.general import DefaultPieDataset
from jhplot import HPlotChart

data =DefaultPieDataset()
data.setValue("Apr",   10)
data.setValue("May",   30)
data.setValue("June",  40)
chart = ChartFactory.createPieChart("Pie Chart",data,1,1,1)
c1 = HPlotChart( chart )
c1.visible()
```

Listing 8.7 A pie chart using JFreeChart

`HPlotChart` is based on the `JFreeChart` class, so now one can use the usual method `c1.export(f)` to create an EPS, SVG, EPS, and PDF images ("*f*" is the file name with the appropriate extension). To learn about the methods of the `HPlotChart` class, use the standard DMelt help system.

8.5 Lightweight Canvases

There are several graphic canvases in DMelt that are well-suited for simple tasks of drawings of *X–Y* data. For example, if there is no need to show many graphical objects, or interactions with a canvas using the mouse are not required, use a lightweight canvas implemented using the `SPlot` Java class ("*S*" originates from the

Fig. 8.7 A pie chart using the native JFreeChart Java class

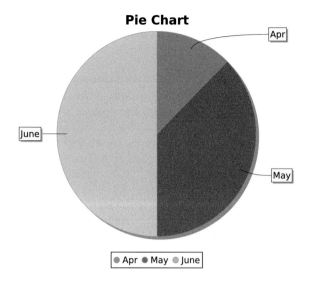

word "simple"). The class SPlot is based on the PTPlot package [4] which was originally designed to make scatter plot planes and to draw functions and simple histograms.

The methods of the SPlot canvas are similar to those of the HPlot class, but the number of such methods is not too large. However, this canvas requires much less computer memory and thus is better suited for applets or to show streams of data at runtime.

One can initialize the SPlot canvas as

```
>>> from jhplot import *
>>> c1 = SPlot ( "Canvas" )
>>> c1 = visible ()
```

This creates a default plotting canvas of the size 600×400 pixels. Now one can show the names for X and Y axes as

```
>>> c1 . setNameX ( "X" )
>>> c1 . setNameY ( "Y" )
>>> c1 . setAutoRange ()
```

Both axes can be set to an auto-range mode, so you need not to worry about setting proper values for the axis ranges:

```
>>> c1 . setAutoRange ()
```

To set the ranges, use the method setRange(xmi,xma,ymi,yma), where the arguments define the ranges for the X and Y axes.

To add a single point at a location (x, y), use the method:

```
c1.addPoint (0,x,y,  b)
```

Usually, "b = 0" (Java false) if points are not connected by lines. When "b = 1" (Java value `true`), then the point should be connected to the next one plotted by calling the `addPoint()` method again. The data set is characterized with an integer number (0 in the above example). Finally, the method `setMarksStyle("various")` tells that each new dataset will be shown with different symbols.

To add an error bar for a point at the location (x, y), use the method:

```
>>> addPointErr (dataset ,x,y,yLow ,yHigh ,con)
```

where `dataset` is an integer number, `yLow` and `yHigh` are the lower and upper errors for the Y axis, and `con` is a Boolean value defining whether the points are connected (=1) or not (=0). When all data points are added, one should call `update()` to make data visible on the canvas.

The method `addPoint()` adds a single point to the canvas. As we already know, to call methods inside each loop iteration in Jython is CPU and memory consuming. However, as for the `HPlot` canvas, one can also use the method `draw(obj)` to draw Java high-level objects, such as histograms (`H1D`), data containers (`P1D`), or simply arrays of numbers with x and y values. Data can be shown with symbols connected by the lines if `setConnected(1,set)` is called with the first argument 1 (or Boolean `true` for Java), where `set` specifies the data set identification number (integer). There is no need to call the update method after calling the `draw(obj)` method, since the plot will be updated automatically.

The plot can be zoomed into a specific rectangular area by clicking and dragging, to draw a rectangle where desired; this feature is also different from the `HPlot` canvas where you can rescale axis ranges one at the time. Finally, one can save the plot into an image using the usual `export(file)` method. Look at all other methods associated with this class using the code assist.

Below we give a small example of how to plot two different sets of data points:

```
from java.util import Random
from jhplot  import *

c1 = SPlot ()
c1.visible (); c1.setAutoRange ()
c1.setNameX ("Time")
c1.setNameY ("Data")
p1,p2=P1D ("data1"),P1D ("data2")
r = Random ()
for i in range (20):
    x=100*r.nextGaussian ()
    y=200*r.nextDouble ()
    p1.add (x,y)
    p2.add (x+x,y+y)
```

```
c1.draw(p1); c1.draw(p2)
c1.addLegend(0,"Data1"); c1.addLegend(1,"Data2")
c1.update()
```

Listing 8.8 SPlot example

The execution of this script brings up a window with two sets of data points and the inserted legends indicating each data set.

In Sect. 8.10, we will discuss how to use the class SPlot to draw streams of data in real time, without using the draw(obj) method.

8.5.1 Henon Attractor Again

Let us give another example illustrating the usage of the SPlot canvas. In Sect. 4.2.9, we have shown how to build the Henon strange attractor using the P1D class. One feature of that script was that we could not see (x, y) points populating the canvas during script execution, i.e., we have to wait until the P1D container is filled and only then the image of the attractor is displayed. Being inconvenient, this also leads to large memory usage.

Let us rewrite the same code using the addPoint() method of the SPlot class. We will update the graph axes after generating 100 events. Execution of the script shown below illustrates how points populate the attractor at run time, i.e., while the program is running.

```
from jhplot  import *

c1 = SPlot()
c1.setGTitle("Henon attractor")
c1.setMarksStyle("pixels")
c1.setConnected(0, 0)
c1.setNameX("x")
c1.setNameY("y")
c1.visible()
a = 1.4;  b = 0.3
x=0; y=0
for i in range(100000):
   x1=x
   x=1+y-a*x*x
   y=b*x1
   c1.addPoint(0,x,y,1)
   if i%1000==0: c1.update()
```

Listing 8.9 Creating a Henon attractor

We will leave the reader here for more experimentation.

8.6   Canvas for Interactive Drawing

For complicated tasks which involve drawing diagrams together with the usual data-driven plots, one should use the HPlotJa canvas. From the point of view of displaying functions, arrays and histograms, this canvas is similar to the HPlot class, i.e., one can use the same method draw(obj) to display instances of lists, F1D, H1D and P1D objects. Yet, it has many advanced features such as

- plots are more interactive. One can easily manipulate with different pads, make inserts, overlays etc.;
- one can draw diagrams using scripts or using the object editor;
- it has a built-in editor to move, cut and remove plotted objects.

One should emphasize that this canvas is easy to use for making inset plots, i.e., showing one pad inside the other. This is technically impossible for the HPlot canvas, in which each pad is based on the JPanel Java class. In the case of HPlotJa, the pads are located inside a single JPanel instance.

The HPlotJa canvas is based on the JaxoDraw package [5] designed to draw Feynman diagrams used in high-energy physics. The original package was significantly modified by adding the possibility to plot the standard DMelt objects and after enhancing vector-graphics capabilities.

To build an instance of the HPlotJa canvas, use the following code snippet:

```
>>> from jhplot import *
>>> c1=HPlotJa()
>>> c1=visible()
```

The methods associated with the HPlotJa canvas are similar to those of the HPlot class and we will not repeat them here. The largest difference with the HPlot is that there are many methods related to drawing axes which cannot be called directly, but only via the method getPad() as shown in this example:

```
>>> from jhplot import *
>>> c1=HPlotJa("Canvas",600,400,1,2)
>>> c1.visible()
>>> c1.cd(1,2)
>>> pad=c1.getPad()          # get current pad
>>> pad.setRange(0,0,1)      # set axis range [0,1] on X
```

As for the original JaxoDraw package, the HPlotJa canvas has a complete graphical user interface that can be used to carry out all actions in a mouse click-and-drag fashion. To bring up the graphical editor, go to the "[Option]" menu and select "[Show editor]". Now you can create and edit the graph objects using the mouse clicks. One can draw many graphical objects, lines, circles, etc. One can remove, drag, and resize all plotted objects, including the pad regions. Finally, one can edit properties of all plotted objects.

8.6.1 Drawing Diagrams

Once an object of the HPlotJa canvas is created, one can draw diagrams using the JaxoDraw mouse click-and-drag fashion. To do this, you have to select the [Editor] option from the [Option] menu.

The flexibility of DMelt scripting allows drawing diagrams interactively or by executing macro files. This will require importing the static methods of the Diagram class from the package jhplot.jadraw. Below we show a typical example of how to draw "gluon" and "fermion" lines typically used for representation of Feynman diagrams:

```
from jhplot import *
from jhplot.jadraw import *

c1=HPlotJa("Canvas",500,400,1,1,0)
c1.visible(); c1.showEditor(1)

gl=Diagram.GlLine(0.3,0.2)  # gluon line
gl.setRelWH(0.0,0.5,"NDC")
c1.add(gl)

gl=Diagram.GlLoop(0.56,0.4)  # gluon loop
gl.setRelWH(0.0,0.1,"NDC")
c1.add(gl)

gl=Diagram.FLine(0.7,0.2)  # fermion line
gl.setRelWH(0.0,0.5,"NDC")
c1.add(gl)

c1.update()
```

Listing 8.10 Drawing "gluon" and "fermion" lines

It should be noted the way the HPlotJa canvas is created; in this example, we do not show axes since the last argument in the constructor HPlotJa is set to zero (Java false). Also, we use the add() and update() methods, as we usually do when showing labels and graphical primitives.

Figure 8.8 shows the output of this example. One can further edit the diagram using the editor panel. Please refer to the Java API documentation of the package jhplot.jadraw to learn more about the classes and methods of this package.

You can also create graphs showing some building blocks of software programs, flowcharts, etc. Figure 1.1 of the introductory chapter of this book illustrates the structure of the DMelt program. You can build this image as shown here

Fig. 8.8 Drawing diagrams
on the HPlotJa canvas
using Jython

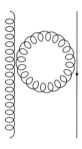

```
from java.awt import *
from jhplot import *
from jhplot.jadraw import Diagram

X,Y=0.35,0.4 # relative X and Y positions
c1=HPlotJa("Canvas",400,300,1,1,0)
c1.setGTitle("DMelt structure", Color.blue)
c1.visible()

g1=Diagram.TextBox("DMelt IDE", X+0.26,Y-0.15)
g1.setColor(Color.gray)
g1.setRelWH(0.30,0.1,"NDC") # width and hight
c1.add(g1)

g1=Diagram.FLine(X+0.1,Y-0.1)
g1.setRelWH(-0.15,0.0,"NDC")
g1.setStroke(2)
c1.add(g1)

g1=Diagram.Blob(X-0.05, Y-0.1) # show a blob
g1.setRelWH(0.18,0.05,"NDC")
c1.add(g1)

g1=Diagram.Text("Web services",X-0.17,Y-0.08)
c1.add(g1)

c1.update()
```

Listing 8.11 Building blocks of the DMelt program

Note that we have shown only a few building blocks, but you can easily extend
this code to create the complete image shown in Fig. 1.1.

8.6.2   SHPlotJa Class

Similar to the SHPlot class, one can create a singleton representing the HPlotJa canvas object using the static class SHPlotJa. In this case, every execution of a script does not create a new object of the canvas frame, but it just redraws the existing one. The example below shows how to create such singleton:

```
>>> from jhplot import SHPlotJa
>>> c1=SHPlotJa.getCanvas()
>>> c1.setGTitle("Global title");
>>> c1.setNameX("X")
>>> c1.setNameY("Y")
>>> c1.visible(1)
```

Of course, all methods of the HPlotJa canvas are also applicable to the SHPlotJa Java class.

8.7   Custom Plotting in XY

8.7.1   HPlotXY Canvas

Another class that builds flexible mathematical canvases is called HPlotXY. It can be used to draw data similar to the HPlot canvas. But you can also draw each graphical component separately. Let us consider an example:

```
from org.jplot2d.data import *
from org.jplot2d.element import *
from jhplot import *
from array import *

c1=HPlotXY(False)
c1.visible()
fac,plot = c1.getFactory(),c1.getPlot()
title=fac.createTitle("Axis Demo")
title.setFontScale(2)
plot.addTitle(title)

xaxis = fac.getInstance().createAxis();
xaxis.getTitle().setText("x axis")
plot.addXAxis(xaxis)
yaxis = fac.createAxis();
xaxis.getTitle().setText("y axis")
plot.addYAxis(yaxis)

x=array('d',[0, 2, 4, 6, 8, 10,50])
y=array('d',[ 0, 0.6, 1, 0.4, 0.5, 0.8, 0.4])
```

```
graphData = XYGraphData(ArrayPair(x, y))
graph = fac.createXYGraph(graphData)
layer0 = fac.createLayer()
layer0.addGraph(graph)
plot.addLayer(layer0, xaxis, yaxis)
```

Listing 8.12 Drawing X-Y data using HPlotXY

that shows a simple *X–Y* drawing of a line using arrays. This time we build the
plot using the concept of "layers". The good thing about such a concept is that the
layers are sufficiently flexible, so one can create subplots or plots with multiple axes.
DMelt provides a number of examples implemented in Jython and Java. Some of
such examples are based on the JPlot2D [6] package which works behind the scenes
of the HPlotXY class. The package can create scatter plots, line charts, linear, and
logarithmic axis transformations. The vector-graphic export is supported via the usual
export(file) method. Please read the API of this class.

8.7.2   WPlot Canvas

The Java class WPlot is another alternative canvas for plotting data. It also supports
plotting lines, scatter plots, and histograms. We will give a simple example that
illustrates the basics of showing a scatter plot in polar coordinates. The data for such
plots should be transformed to the R-θ space, where R is the distance from the center
of the plot (origin) and θ is the angle from a reference angle. This example shows
two data sets in the polar coordinates:

```
from jhplot   import   *
from java.awt import  Color

p1=P1D("data Nr1")
p2=P1D("data Nr2")
for i in range(10):
    p1.add(0.3*i,i)       # theta=0.3*i, R=i
    p2.add(-0.2*i,2*i)    # theta=-0.2*i, R=2*i
p1.setColor(Color.blue)
p2.setColor(Color.red)
c1 = WPlot()
c1.visible()
c1.draw(p1,"PolarScatter")
c1.draw(p2,"PolarScatter")
```

Listing 8.13 Showing X-Y data using the WPlot canvas

The output image of the above code is shown in Fig. 8.9.

Fig. 8.9 Showing two data sets in the polar coordinate system using the `WPlot` class

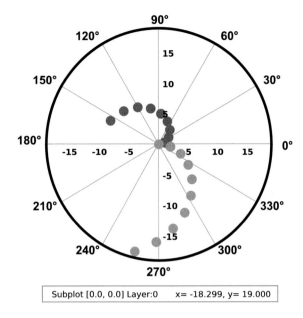

8.7.3 HPlotJas Canvas

The `HPlotJas` canvas uses a different approach to display data on interactive plots. This canvas is based on the Jas2 package (a more recent version can be found in [7]) of the FreeHep [8] Java library and, as a consequence, it has a different look and feel for showing data and histograms compared to the previous plots. Let us create a short code snippet that fills 1D arrays and two histograms.

```
from java.util import *
from jhplot  import *

h1 = H1D("1D histogram",100, -2, 2.0)
h1.fillGauss(1000,0,1)    # fill with 1000 random numbers

h2 = H2D("2D histogram",50, -2, 2.0,50,-2,2)
h2.fillGauss(1000,0,1,0,1) # fill with 1000 random numbers

p0=P0D("Normal distribution")
p0.randomNormal(1000,0,10) # fill with 1000 random numbers
c=HPlotJas("JAS",ArrayList([h1,h2,p0])) # pops-up a menu
```

Listing 8.14 Different types of data showed using `HPlotJas`

Note that, instead of plotting graphical objects on the canvas, we collect all objects in an array. Running this code brings up a menu with the data and the canvas. With the mouse, select "Data" and click the object representing a histogram or data array. You should see the data plotted on the canvas. One distinctive feature of this canvas is that it has several useful options for plotting and data overlaying. In addition, one can interactively fit data using built-in analytic functions.

8.8 HPlot2D Canvas

Although one can use the HPlot canvas to show 2D histograms, functions, or arrays as contour or density plots, this is not the best way to this. This is because this canvas was not designed from the ground to support such types of plots. Instead, use the HPlot2D canvas to display data as contour / density plots. This canvas is partially based on the SGT project [9].

The HPlot2D canvas is very similar to HPlot, and it shares many common methods. The initialization of this canvas also looks similar:

```
>>> from jhplot import *
>>> c1=HPlot2D("Canvas")
>>> c1.visible()
```

This brings up a frame with the HPlot2D canvas.

Now let us walk through several examples which show how to use this canvas. First, let us draw a F2D function as a contour plot:

```
from jhplot import *
from java.awt import *

f1=F2D("x^2+sin(x)*y^2",-2,2,-2,2)
c1=HPlot2D("Canvas",600,700)
c1.visible()
c1.setName("2D function")
c1.setNameX("X")
c1.setNameY("Y")
c1.setStyle(2) # set style of the plot
c1.draw(f1)
lab1=HLabel("&omega; test",0.66,0.25, "NDC")
lab1.setColor(Color.white)
c1.add(lab1,0.1)
c1.update()
```

Listing 8.15 Visualizing 2D functions

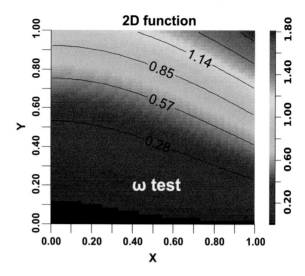

Fig. 8.10 A 2D function shown as a density plot using the HPlot2D canvas

The resulting plot is shown in Fig. 8.10. The methods used for the canvas object c1 are similar to those of the HPlot canvas. The only new feature is that when we add a text label, we use the method add(obj,d), where "d" specifies the label height.

We should mention that the HPlot2D canvas is fully interactive. One can move and edit labels, click-and-drag the mouse pointer to create a zoom rectangle. One can edit axis attributes by clicking on an axis (this will bring up a dialog where all modifications can be made).

There are several options for showing an object inside the canvas. They are controlled with the method seStyle(int), where

style=0   draws data using a raster style;
style=1   draws using a contour style;
style=2   combined style (raster and contour);
style=3   draws data using filled areas.

To modify the plots, you should access the methods of the HPlot2D Java class inside your script. First, one can obtain the axis object as getAxis(axis), where axis is either 0 (for X) or 1 (for Y). Using the methods of this axis object, one can make necessary modifications, like setting new fonts, redefine colors, etc. One can also access the color bar-key as getColorBar(), which also has several useful methods.

One important method you should keep in mind is

```
>>>   setRange(axis,min,max,bins)
```

which sets the range for X (axis = 0), Y (axis = 1) and Z (axis = 2). The variable bins specifies how many divisions between min and max values should be used.

In the case of the Z-axis, the variable `bins` specifies how many contour levels to draw.

Let us give another example. This time we will plot data using different styles. We display a histogram and 2D array. We change the range for X and Y axes for the histogram using the `setRange()` method. We note that the last argument for this method, which usually defines the number of bins between the minimum and maximum values, does not do anything. This is because we plot the histogram which has its own binning, and this cannot be changed when the histogram is shown.

```
from jhplot import *
from java.util import *

h1=H2D("Data",30,-3.0, 3.0, 30,-3.0, 3.0)
p1=P1D("Data")
r=Random()
for i in range(1000):
  x=r.nextGaussian()
  y=r.nextGaussian()
  h1.fill(0.6*x-0.5, y)
  p1.add(0.4*x+1,   y+0.5)

c1=HPlot2D("Canvas",600,400,3,1)
c1.visible()

c1.cd(1,1)
c1.setName("H2 histogram");
c1.setStyle(0);  c1.draw(h1)

c1.cd(2,1)
c1.setName("H2 range")
c1.setRange(0,-2.0,2.0,50)
c1.setRange(1,0.0,1.0,50)
c1.setStyle(1);  c1.draw(h1)

c1.cd(3,1)
c1.setName("2D array")
c1.setStyle(1);  c1.draw(p1)
```

Listing 8.16 Showing data on the HPlot2D canvas

The output of this script is shown in Fig. 8.11.

8.9 Visualization in 3D

Sections 3.4.2, 4.3.1.1, and 7.2 have already discussed plotting canvases for scientific visualization in 3D. Here we remind again that one should use the classes HPlot3D and HPlot3DP (from the package jhplot) for 3D representations of mathematical functions and data. The first class is used for plotting H2D, P2D, and F2D objects,

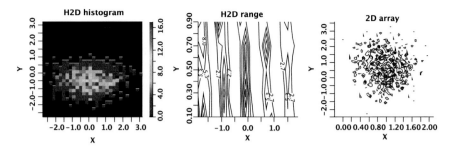

Fig. 8.11 A histogram and *X–Y* array shown using different styles and ranges of the `HPlot2D` canvas

while the second one is used for parametric equations based on the `FPR` class, see Sect. 3.8. This section discusses these 3D plotting canvases in detail.

8.9.1 HPlot3D Canvas

This canvas is used for visualization of `H2D`, `P2D`, and `F2D` objects. They can be shown with the same `draw(obj)` method as for the 2D case. If two objects, say `obj1` and `obj2`, should be shown on the same canvas, on the same canvas, one should call the method `draw(obj1,obj2)`. Table 8.2 lists the most important methods of the `HPlot3D` Java class.

We should remind that if you are debugging a script, and do not want to create many pop-up windows with canvas frames after each script execution, you may want to instantiate a singleton using the class `SHPlot3D`. In this case, every new macro execution redraws the existing canvas, instead of creating a new canvas object. One can build a singleton as usual:

```
>>> c1=SHPlot3D.getCanvas("3D",600,400,2,2)
```

This creates four plotting pads inside a 3D canvas of the size 600×400.

8.9.2 HPlot3DP Canvas

The second 3D canvas, `HPlot3DP`, is used to draw surfaces defined by an analytic parametric or a nonparametric function of the type `FPR`, as shown in Sect. 3.8.

The `HPlot3DP` canvas can be constructed exactly in the same way as the `HPlot3D` Java class. Specifically, one can define the frame size during object instantiation and set any number of drawing pads. The major difference with `HPlot3D` is

Table 8.2 The main methods of the `HPlot3D` class

Methods	Definitions
cd(iX, iY)	go to the pad (iX, iY)
clearAll()	clean all pads
clear()	clean the current region
clear(iX, iY)	clean the pad (iX, iY)
draw(obj)	draw an object (P2D, F2D,…)
getDistance()	get distance from object
getElevationAngle()	get elevation angle
getLabelOffset(a)	get label offset for axis
getPenWidthAxes()	get pen with for axes
getRotationAngle()	get rotation angle
getScaling()	get scaling
quite()	remove frame
setAxesFontColor(c)	set fonts for axes labels
setColorMode(i)	set color mode (from 1 to 4) 0: wire-frame, 1: hidden 2: color spectrum 3: gray scale 4: dual shades
setDistance(d)	set distance to objects
setElevationAngle(i)	set elevation angle to "i" degrees
setLabelColor(c)	set label color
setLabelFontColor(c)	set color for labels
setLabelFont(f)	set fonts for axes numbers
setLabelFont(f)	set fonts for labels
setNameX("text")	set name for axis X
setNameY("text")	set name for axis Y
setNameZ("text")	set name for axis Z
setPenWidthAxes(i)	set line width for axes
setRange(Xmin, Xmax, Ymin, Ymax)	set plot ranges
setRangeZ(Zmin, Zmax)	set range for Z
setRotationAngle(i)	set rotation angle to "i" degrees
setScaling(i)	set scaling factor to "i" (default is 12)
setTicFont(f)	set fonts for ticks
setTicOffset(d)	set ticks offset
updateAll()	update all pads
updateData(iX, iY)	update data shown in the pad (iX, iY)
updateData()	update data on the current pad
update(iX, iY)	update the pad (iX, iY)

"b" indicates a boolean value (1 for true and 0 for false), while "i" is an integer value. The notation "d" denotes a float value. The attributes "c" and "f" correspond to the `Color` and `Font` AWT Java classes. "text" represents a string value. The character "a" is a shortcut to "axis" (a = 0 for X, a = 1 for Y, a = 2 for Z)

Table 8.3 The main methods of the `HPlot3DP` class

Methods	Definitions
cd(i1, i2)	navigates to a i1 x i2 pad
clear(object)	clear the frame
draw(obj)	draws an object (FPR)
setFog(b)	sets the fog style for 3D
setAxes(b)	shows or not axes
setAxes(b1, b2, b3)	shows axes for X, Y, Z
setAxesColor(c)	axes color
setBackgColor(c)	background color
setNameX(text)	text for X axis
setNameY(text)	text for Y axis
setNameZ(text)	text for Z axis
setCameraPosition(d)	set camera position zoom in for positive "d" zoom out for negative "d"
setEyePosition(x, y, z)	set eye positions
update()	updates the canvas

"b" denotes a Boolean value (1 for true and 0 for false), while "i" is an integer value. The notation "d" indicates a float value. The attributes "c" and "f" correspond to the `Color` and `Font` AWT Java classes

as follows: if several objects should be shown on the same pad, then one can use the `draw(obj)` methods several times, one after the other. The main methods of this canvas are given in Table 8.3.

One can zoom in the pad area using the right mouse button. In source codes, one can zoom in and zoom out using the following method:

```
>>> c1.setCameraPosition(d)
```

where d is a double value, which should be positive for zooming in, and negative for zooming out. One can change the location of axes and the object position using the method `setEyePosition(x, y, z)`. The graphs can be edited using the GUI dialog and the [Edit] menu by clicking the mouse button.

Below we show how to use several methods of the `HPlot3DP` Java class. Our next code draws two parametric functions using two independent variables, *u* and *v*:

```
from java.awt import Color
from jhplot import *

c1 = HPlot3DP("Canvas",600,600)
c1.setGTitle("HPlot3DP examples")
c1.visible()
f1=FPR("u=2*Pi*u; x=cos(u); y=sin(u); z=v")
c=Color(0.5,0.2,0.5,0.5) # color+transparency
f1.setFillColor(c)
```

```
f1.setLineColor(Color.green)
f2=FPR("u=2 Pi u; v=2 Pi v; r=0.6+.2cos(u); \
        z=.8 sin(u); x=r cos(v); y=r sin(v)")
f2.setFillColor(Color.blue)
f2.setFilled(1)
c1.setFog(0)
c1.setAxes(1)
c1.setNameX("X axis")
c1.setNameY("Y axis")
c1.setAxesColor(Color.gray)
c1.setAxesArrows(0)
print c1.getEyePosition()
print c1.getCameraPosition()
c1.setCameraPosition(-1.2) #zoom out
c1.draw(f2); c1.draw(f1)
```

Listing 8.17 Showing a parametric function in 3D

The above code brings up a frame with the image of a cylinder and a torus as shown in
Fig. 8.12. One can rotate objects and zoom into certain area using the mouse button.
In addition, one can further edit the figure using the [Edit] menu.

Fig. 8.12 Showing a
cylinder and a torus in 3D
using the HPlot3DP canvas

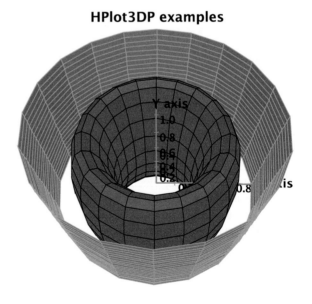

8.9.3 Mathematical Objects in 3D

Another useful class to show various mathematical objects is HPlotMX from the same jhplot package. This class is based on the package called 3D-XplorMathJ[10]. It should be noted that this 3D canvas was not designed to show data. Instead, it can be used to show surfaces and other mathematical objects. The objects can be rotated to any desired orientation by dragging them with the mouse.

Let us show an example of creating a curvature surface called "Kuen Surface". It is given by the following parametric equation [11]:

$$x(u, v) = 2 \cosh(v)(\cos(u) + u \sin(u))/(\cosh(v)^2 + u^2),$$
$$y(u, v) = 2 \cosh(v)(-u \cos(u) + \sin(u))/(\cosh(v)^2 + u^2),$$
$$z(u, v) = v - (2 \sinh(v) \cosh(v))/(\cosh(v)^2 + u^2)$$

Readers interested in the mathematical definition should refer to the appropriate sources. To plot this parametric equation, one can use the corresponding predefined class as shown in this example:

```
from jhplot import *
from vmm3d.surface.parametric import KuenSurface
c1=HPlotMX()          # build interactive 3D canvas
ks=KuenSurface()      # create Kuen Surface
c1.draw(ks)
c1.visible()
```

Listing 8.18 Drawing a Kuen Surface in 3D

That is all it takes. Run this example, and you will see an image shown in Fig. 8.13. One can edit this image using the built-in editor of this canvas.

In order to show other mathematical surfaces given by parametric equations, look at the predefined Java classes in the package vmm3d.surface.

Our next goal is to show how to create a custom parametric surface using the HPlotMX class and Python codding for a parametric equation. We will consider the torus equation:

$$x(u, v) = (a + b \cos(v)) \cos(u),$$
$$y(u, v) = (a + b \cos(v)) \sin(u),$$
$$z(u, v) = b \sin(v)$$

where u and v run from 0 to 2π for a complete torus. In the example shown below, we will create a half-torus using the abstract Java SurfaceParametric class:

Fig. 8.13 Shows the Kuen
Surface created by the code
example given in
Listing 8.18

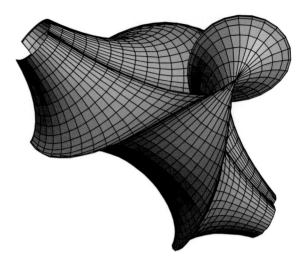

```
from jhplot import *
from vmm3d.surface.parametric import
    SurfaceParametric
from vmm3d.core import RealParam
from vmm3d.core3D import Vector3D,View3DLit
from java.lang import Math

class torus(SurfaceParametric):
    global aa,bb,cc
    aa=RealParam("genericParam.aa",1.75)
    bb=RealParam("genericParam.bb",0.4)
    cc=RealParam("genericParam.cc",0.4)
    def __init__(self):
        self.setU(0,2*Math.PI)  # min and max for U
        self.setV(0,Math.PI)    # min and max for V
        self.setPatchCount(8,8) # divisions in X nad Y
        self.setViewpoint(10,-10,10)
        self.setDefaultWindow(-2.5,2.5,-2.5,2.5)
        self.addParameter(aa)
        self.addParameter(bb)
        self.addParameter(cc)
    def surfacePoint(self, u, v):
        A=aa.getValue()
        B=bb.getValue()
        C=cc.getValue()
        x = (A + B * Math.cos(u))* Math.cos(v)
        y = (A + B * Math.cos(u))* Math.sin(v)
        z = C * Math.sin(u)
        return Vector3D(x,y,z)
c1=HPlotMX()
c1.draw(torus())
c1.visible()
```

Listing 8.19 Drawing a half-torus in 3D

Fig. 8.14 Showing a half-torus defined by Listing 8.19

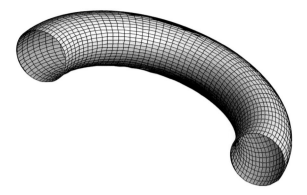

Figure 8.14 shows the expected output. You can complete the torus after replacing `setV(0, Math.PI)` by `setV(0,2*Math.PI)`. You can further explore such an approach by looking at the API of this package.

8.10 Plotting Real-Time Data

Showing streams of data in real time is common for many scientific and engineering applications. Instead of collecting data in data containers, such as Java or Python arrays, followed by plotting data at once using the `draw(obj)` method, one can plot a fraction of data without waiting until the complete data set becomes available for drawing.

In this section, we will show how to fill histograms or data arrays, and update the canvas after each operation. Let us illustrate this using the Python `sleep()` function: we will fill a histogram and update the plot after each random Gaussian number. Thus we will see how the histogram is filled at runtime. At the end of the loop, we will clean up the canvas and then replot the final histogram. How to clear the canvas from plotted data has been discussed in Sect. 8.2

```
from java.util import Random
from jhplot  import *
import time

c1 = HPlot()
c1.setGTitle("data stream")
c1.visible(); c1.setAutoRange()
c1.setLegend(False)
h1 = H1D("Updated histogram",20,-2.0,2.0)
h1.setFill(1); h1.setErrX(0); h1.setErrY(1)

r = Random()
for i in range(100):
    h1.fill(r.nextGaussian())
```

```
    time.sleep(0.1)  # 0.1 sec delay
    c1.clearData()   # the next fill
    c1.draw(h1)
c1.drawStatBox(h1)
c1.draw(h1)
```

Listing 8.20 Showing real-time data using HPlot

One may notice that before each draw(obj) statement, we remove data from the canvas using the method clearData(). This is necessary in order to avoid generating many objects during object drawings since each call of the method draw(obj) creates a new object and this leads to an extensive memory usage.

8.10.1 Real-Time Data Using SPlot

For showing steams of data, it is good idea to use a lightweight canvas, such as SPlot discussed in Sect. 8.7. The reason is that, typically, you do not need a GUI-type approach to interact with canvases while they collect data in real time. The SPlot uses significantly less memory since it does not have interactive Java components.

In this example we fill a canvas with data points and rescale the plot to fit all data:

```
from java.util import Random
from jhplot   import *
import time

c1 = SPlot()
c1.visible(); c1.setAutoRange()
c1.setMarksStyle("various")
c1.setConnected(1, 0)
c1.setNameX("Time")
c1.setNameY("Data")

r = Random()
for i in range(20):
    x=r.nextGaussian()
    y=r.nextGaussian()
    c1.addPoint(0,x,y,1)
    c1.update()
    time.sleep(1)
```

Listing 8.21 Real-time data showed in SPlot canvas

The output of this example is rather fancy: you will see data points connected with various lines. When one uses the addPoint() method, there is no need in removing data from the canvas after each call to this method, since the method addPoint() does not create a new object after each call.

8.10.2   Real-Time Data Using HPlotRT

This canvas was developed keeping in mind real-time data. It is based on the JChart2D [12] plotting package. The `HPlotRT` canvas is fast and has low memory footprint. However, the backside of this feature is that it has a smaller number of drawing options and less interactivity. The canvas comes in handy when showing frequently changing data, such as CPU loading or stock quotes. Let us make two line traces using a small Jython code:

```
from jhplot import *
from java.awt import Color
from java.util import Random
from info.monitorenter.gui.chart.traces import *
import time

trace = Trace2DSimple()
c1=HPlotRT()
trace1=Trace2DSimple("Trace1") # trace 1
trace1.setColor(Color.red)
c1.add(trace1)

trace2=Trace2DSimple("Trace2") # trace 2
c1.add(trace2)

rand = Random()
for i in range(100):
    time.sleep(0.05)
    trace1.addPoint(i,rand.nextGaussian())
    trace2.addPoint(i, 5+rand.nextGaussian())
```

Listing 8.22   Showing real-time line traces

The code updates the line traces in real time and then creates the final plot shown in Fig. 8.15.

8.11   Graphs and Java GUI Components

Section 2.17 shows how easy it is to use the Java Swing components in Jython. Jython scripting allows creations of graphical user interfaces (GUI) using the Java GUI widgets, such as windows, menus, buttons, etc. One can make them respond to physical events (keyboard, mouse, etc.) with a few lines of Jython code. This is important when it is necessary to make "control panels" or "graphical user interfaces" for applications in order to manipulate with Java programs at runtime. Such applications can be deployed as self-contained programs in jar files after compiling Jython codes into the Java bytecodes, the machine language of the Java virtual machine. But can

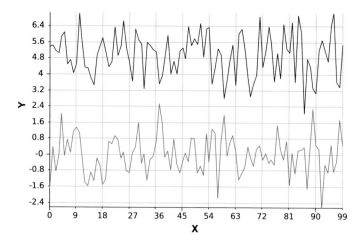

Fig. 8.15 Showing real-time data using the `HPlotRT` canvas

we use the DMelt canvas together with such components? Yes, you can. We will briefly discuss this topic below.

Numerous methods of the canvases described in the previous sections are so rich that one can always find a necessary method suited for your GUI application. The most important method for the GUI development is that which returns the Swing `JFrame` object holding the entire canvas panel. In the example of Sect. 2.17, an instance of the Java class `JFrame` was created using the standard constructor. When you are working with the DMelt canvases, there is no need to do this. What you will need is to use the `getFrame()` method which returns an appropriate `JFrame` instance:

```
>>> c1 = HPlot("GUI")
>>> jframe=c1.getFrame()
```

Once you have retrieved the frame object, you can add the necessary Swing GUI components (menu, buttons, sliders, etc.) in order to build a "control panel" the displayed data.

Let us show an extensive example that illustrates how to attach a panel with two Swing `JButton` buttons and a `TextArea` component directly to the `HPlot` canvas object. One button will be used to generate a histogram populated with Gaussian numbers. The second will be used to erase graphs from the canvas. The text area displays a short message about what happens when you press these buttons.

```
from java.awt import *
from javax.swing import *
from jhplot   import *

c1 = HPlot()
c1.setGTitle("Plot area with GUI")
c1.visible(); c1.setAutoRange()
```

```
fr=c1.getFrame()

h1 = H1D("Histogram",20, -2.0, 2.0)
h1.setFill(1)
h1.setFillColor(Color.blue)

pa0 = JPanel(); pa1 = JPanel()
pa2 = JTextArea("GUI test",6,20)
pa2.setBackground(Color.yellow)

def act1(event):  # fill random numbers
    h1.fillGauss(1000,0,1) # 1000 events
    c1.draw(h1)
    pa2.setText("Generated 100 random numbers")
def act2(event): # clear canvas
    c1.clearData()
    pa2.setText("Clear plot")

pa0.setLayout(BorderLayout());
bu1=JButton("Gaussian", actionPerformed=act1)
pa0.add(bu1,BorderLayout.NORTH)
bu2=JButton("Clear", actionPerformed=act2)
pa0.add(bu2,BorderLayout.SOUTH)
pa0.add(pa2,BorderLayout.WEST)
fr.add(pa0,BorderLayout.EAST)
fr.pack()
```

Listing 8.23 Combining plotting area with Swing components

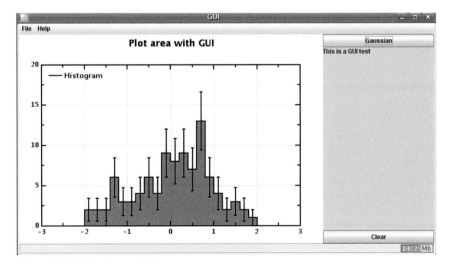

Fig. 8.16 HPlot canvas together with a Swing control panel

After execution of this script, you will see the HPlot canvas inside an attached panel with the Swing components, see Fig. 8.16. Press the buttons to see the output of this program.

After examining the script, you may wonder why we need the method pack() at the very end of this script. We call this method to avoid changes in the size of the HPlot canvas, so the extra panel just extends the canvas without distorting its default size (which is 600×400 pixels). If you want to append extra canvas by keeping the same size of the entire frame, remove the method pack() and call any method that updates the frame (calling setVisible(True) can do this).

References

1. Chekanov S The jhplot package https://github.com/chekanov/dmelt_jhplot
2. The Unicode Consortium http://www.unicode.org/
3. jFreeChart package http://www.jfree.org/
4. Ptplot 5.6. Ptolomy project http://ptolemy.berkeley.edu/java/ptplot/
5. Binosi D, Theussl L (2004) Jaxodraw: A graphical user interface for drawing feynman diagrams. Comput Phys Commun 161: 76–86 http://jaxodraw.sourceforge.net/
6. Li J JPLOT2D. a high-performace multi-threaded 2d plot library in Java https://github.com/jplot2d/
7. FreeHep, Jas, Java analysis studio. http://jas.freehep.org/jas3/
8. FreeHEP Java Libraries http://java.freehep.org/
9. Denbo DW Scientific Graphics Toolkit http://www.epic.noaa.gov/java/sgt/
10. T. D.-X. Consortium, The 3d-xplormathj project. http://3d-xplormath.org/
11. Kuen T (1884) Ueber flchen von constantem krmmungsmass. Akad Wiss Math-Phys Classe, Heft II, 193–206
12. Westermann A, JChart2D. a 2d plot library http://jchart2d.sourceforge.net/

Chapter 9
File Input and Output

DMelt library provides many classes and methods for file input and output (I/O) beyond those discussed in Sect. 2.15. Many of them are implemented using external Java libraries that contain nearly every data format you might ever need to use for efficient data storage and processing.

Let us summarize the I/O approaches used by DMelt:

- I/O streams used to write or read data using the native Java platform. These Java libraries are located in the `java.io` and `java.nio` packages that can be imported using the usual `"import"` statement. We will not discuss the Java I/O streams in this book as they are described in detail in any Java book [1–5]. We have already discussed the class `DataOutputStream` in Sect. 2.15 in the context of writing and reading data using Java classes;
- native Python I/O discussed in Sect. 2.15;
- libraries included into the DMelt Java package `jhplot.io`. In this chapter, we will discuss these libraries in more details;
- third-party Python libraries. We will discuss the package `DIF` from these libraries in Sect. 9.6.4;
- third-party Java libraries included into the DMelt. For example, we will discuss the Apache Derby relational database in Sect. 9.11.

File I/O, data streaming and format conversions are essential for any stage of data analysis. Unfortunately, in the standard manuals and tutorials, details of file I/O often are buried at the back. In this book, we will discuss the I/O topic in great detail in the context of data analysis and numeric computation.

9.1 Nonpersistent Data: Memory-Based Data

In simple words, nonpersistent objects contain data that are kept in the computer memory and cannot be restored after the end of the program execution. In contrast, persistent objects are stored in files and can be restored at any moment.

© Springer International Publishing Switzerland 2016
S.V. Chekanov, *Numeric Computation and Statistical Data Analysis
on the Java Platform*, Advanced Information and Knowledge Processing,
DOI 10.1007/978-3-319-28531-3_9

Most scripts discussed in this book hold objects in a nonpersistent way. For example, when you create a histogram or an array with random numbers, data are kept in the computer memory. This is exactly what happens when creating an array with random numbers:

```
>>> from jhplot import *
>>> p=P0D("data")
>>> p.randomUniform(100000,0,1)
```

where the object "p" is created in the computer memory. Watch out for the memory status monitor at the bottom right corner of the DMelt editor: you will see a notable increase in the used memory. By increasing the number of random numbers in the above example, one can eventually run out of memory, and an OutOfMemoryError message will be thrown.

Thus, you are limited by the available memory of your computer (and the available memory assigned to the virtual machine). The memory assigned to store objects can be released manually: in Python/Jython, to remove an object, say "p", call the statement "del p". In addition, the Jython garbage collector takes care of reclamation of memory space automatically. One can also invoke the Jython garbage collector manually as:

```
>>> del p
>>> import gc
>>> gc.collect()
```

After calling the statement above, you will see a decrease in the used memory (again, look at the bottom right corner of the DMelt IDE).

The computer memory is like the work table. It is fine to work on it using small portions of data because it is fast and effective. But in order to handle large pieces of data, you have to put data on "a shelf", i.e., copy them to a file storage. Below we will discuss how to save numerical data in files, and how to bring them back into the computer memory for fast operation.

9.2   Object Serialization

Previously, we have already discussed how to save separate objects into files. Most objects, like P1D or H1D, contain methods for writing them into files, which can be either text files or binary files. This means we persist objects in a serialized file using the standard Java serialization.

These objects also have methods to instantiate themselves from files. Look at the previous sections: Sect. 3.10 discusses this topic for functions, while Sects. 4.1.5, 4.2.8 and 4.3.4 for P0D, P1D and PND arrays, respectively. Section 7.6 describes the histogram I/O operations. In these sections we have illustrated how to save and restore all such objects using the standard Java serialization, including the serialization into a human-readable XML file format.

We also should remind that DMelt objects can be saved into Python/Jython lists, dictionaries, or Java collection classes using the IO Java class. They can be used in all scripting languages on the Java platforms, and in Java programs as well.

This is a generic example of how to save Python lists:

```
>>> from jhplot import *
>>> list=[]
>>> list.append(object1)
>>> list.append(object2)
>>> IO.write(list,"file.jser") # write to a file
>>> list=IO.read("file.jser")  # restore the list
```

By convention used in this book, the file extension ".jser" is reserved for binary serialized files. You can read the data in the form of the Python list, or as ArrayList if you are working with Java.

Similarly, one can use Python dictionaries with the keys. They are rather handy for labeling and extracting objects:

```
>>> from jhplot import *
>>> map={}
>>> map["object1"]=object1
>>> map["object2"]=object2
>>> IO.write(map,"file.jser")  # write to file.jser
>>> map=IO.read("file.jser")   # restore "map"
```

The object map is either Python dictionary, or Java Map if you are working with Java.

The class IO compresses data using the GZip format. One can also save and read objects without compression using the following methods:

```
>>> IO.write(obj,"file.jser",comp)
>>> obj=IO.read("file.jser",comp)
```

where a Boolean comp should be set to zero (or "False") if no compression is applied.

One can also use the XML format for output files. This format is implemented in the methods writeXML(obj) and readXML() from the same IO Java class. However, this is not the most economic approach since string parsing is CPU intensive and the XML files are significantly larger than files with binary data.

The methods discussed above have some significant drawbacks: we write data at the end of numerical computations, so first we put the entire collection in a nonpersistent state, and then move it into a persistent state. This can only work if the data are not too large and one can fit them in the computer memory.

Below we will discuss how to avoid such problems by writing objects with data directly into files sequentially, without building collections of objects first and then writing them using the IO class.

9.3   Persistent Event Records

9.3.1   Sequential Input and Output

To load all data into a Jython or Java container and write such a container at once into a file using the IO class is not the only approach. One can also write or read selected portions of data, i.e., one can sequence through data by writing or reading small chunks of data, or "event records" that are easier to handle in the computer memory.

This time we are interested in truly persistent way of saving objects: instead collecting them in the collections before saving them into a file, we will write each object separately into a file. In this case one can write and read back huge data sets to/from the disk.

The Java serialization mechanism for objects could be slower than the standard Java I/O for streaming primitive data types. As usual, there is a trade-off between convenience and performance and, in this book, we usually prefer convenience. In the case of Java serialization, restored objects carry their types, and such "self-description" mechanism is very powerful and handy feature as there is no need to worry about what exactly data have been retrieved from a file.

If the serialization is slow for your particular tasks, one can always use the I/O methods of DMelt objects which are not based on the Java serialization. See, for example, Sects. 4.1.5, and 4.2.8.

To write objects sequentially into a file and read them back, one can use the HFile class. A general call to create an instance of this class is

```
>>> f=HFile(file,option,comp,buff)
```

where file is a string with the file name, "option" is a string which can be either "w" (write into file) or "r" (read from a file). If the Boolean variable "comp" is set to zero ("False" in Python), objects inside the file will not be compressed. An integer value, "buff", specifies the buffer size. The class HFile is constructed such that it writes data chunks to the memory buffer first, before recording them on the disk. This dramatically increases the speed of I/O operations compared to the situations when one needs to write one byte at a time. Selecting a correct buffer size is important. Typically, this number should be multiples of 1024 bytes. If the data chunks to be written are large, consider to increase the buffer size. This point will be illustrated in detail later.

The following brief methods for initialization of HFile can also be used:

```
>>> f=HFile(file,option,comp)
>>> f=HFile(file,option)
>>> f=HFile(file)
```

where `file` is a file name. These initializations assume the buffer size of 12 kilo-bytes (KB). The second constructor assumes the gzip compression, while the third constructor assumes that the file is opened in the read-only mode.

Let us give an example of how to store several objects from the DMelt library in a file. The example below shows how to create an output file with several DMelt objects. Some of them (like P1D) can contain data:

```
from jhplot.io import *
from jhplot import *
from javax.swing import JFrame

li=[] # create a list with objects
li.append(P0D("Data array"))
li.append(P1D("2D data array"))
li.append(H1D("Histogram",100,-5,5))
li.append(F1D("x*sin(x)",1,5))
li.append(F2D("y*sin(x)"))
li.append("My String")
li.append(JFrame("Swing Frame"))
f=HFile("output.jser","w")
for i in li:
   err=f.write(i)
f.close()
```

Listing 9.1   Writing objects to a file

As mentioned earlier, data from the serialization contain a self-description. This is a great feature since now we can easily restore the contents of the objects. Let us read the file from the above example and check the types of the retrieved objects. In one case, we will use the Jython `type(obj)` method, in the second case, we will use the Java-oriented type of check:

```
from jhplot.io import *
from jhplot import *

f=HFile("output.jser")
while 1:
  obj=f.read()
  if obj == None:
     break
  print "Type=",type(obj)
f.close()
```

Listing 9.2   Reading objects from a file

The output of this script is

```
Type= <type 'jhplot.P0D'>
Type= <type 'jhplot.P1D'>
Type= <type 'jhplot.H1D'>
Type= <type 'jhplot.F1D'>
```

```
Type= <type 'jhplot.F2D'>
Type= <type 'unicode'>
Type= <type 'javax.swing.JFrame'>
```

In the case if an object belongs to a Java class, one can also access the information about this object using the getClass() method from the native Java API. This has already been discussed in Sect. 3.2.1, but here we will remind how to do this again:

```
>>> try:
>>>     cla=obj.getClass()
>>>     print "Java=",cla.getCanonicalName()
>>> except:
>>>     print "Non-Java class";
```

9.3.2 Opening Data in a Browser

If a serialized file has been created with the help of the HFile class using its default attribute (i.e., GZip compression), one can browser objects inside such a file using a specially designed browser. Start the DMelt IDE, select the menu [Plots] and then [HPlot] menu, which brings up an empty HPlot canvas. Then select [Read data file] from the menu [File] on the toolbar. This opens an object browser which will be discussed in detail in Sect. 8.2.3.

In this section, instead of the GUI-driven approach, we show how to look at the serialized objects using Java scripting. In our next example we create a canvas and open a HFile file in the BrowserData dialog. You will see a list of objects stored in the file (together with their titles). One can select an object and plot it on the HPlot canvas. The selected object can be either a function (F1D), H1D histogram, P0D or P1D array. This code snippet opens the file "output.jser" created before in the object browser:

```
from jhplot.io import *
from jhplot import *

c1=HPlot("canvas")
c1.visible()
BrowserData("output.jser",c1)
```

Listing 9.3 Using Java object browser for files

Select an entry, say "String" (which, obviously, means "Java String" object) and push the button "Plot". You will see the plotted text on the canvas. In exactly the same way, one can bring up a serialized JFrame (note that we did not set the size of this frame, so you will see only a small box). One can also plot other objects (histograms, DMelt arrays) if they have been filled before saving them into the file

"output.jser". If the selected object cannot be plotted on the given canvas, you will
see an error message.

9.3.3 Saving Event Records Persistently

Now let us come to the question of how to store large data sets in sequential form.
This can be done as well, since the argument `obj` in the method `write(obj)` is
designed to keep data records with various types of containers and their graphical
attributes.

Below we will show a simple example of how to write 10,000 events into a
serialized file. Each event is a list with the event identification (a string), `POD` array
with random numbers and a histogram `H1D`. We print a debugging message after
every 1000 event, showing how the file is growing after inserting the data.

```
from jhplot.io import *
from jhplot import *
import os

def makeEvent(entry):
    "event with 3 objects"
    label="Event="+str(entry)
    p=P0D(label)
    p.randomUniform(100,0,1) # uniform distribution
    h=H1D(label,10,-1,1)
    h.fillGauss(100,0,1)      # Gauss distribution
    return [label,p,h]

name="events.jser"
f=HFile(name,"w")
for i in range(10000):
  ev=makeEvent(i)
  if (i%1000 == 0):
     mb=os.path.getsize(name)/(1024.0*1024)
     print ev[0]+" size (MB)=",mb," MB"
  f.write(ev)
f.close()
```

Listing 9.4 Writing event records with objects

One should note that the data stream is flushed[1] after each 100 event by default. One
can increase or decrease this number by using the method `setFlush(i)` of the
`HFile` class, where "i" is the number of entries after which data should be flashed to
the disk. This is necessary since a serialization of multiple objects without resetting
the stream causes an increase of the memory used by the Java virtual machine.

---

[1]For Java experts: flushing here means resetting `ObjectOutputStream`.

We can call getEntries() to obtain the actual number of written entries and getBufferSize() to return the default size of the buffer used to write data.

Now, let us read the saved file back and restore all the objects from the list. We will read the events from the file in an infinite loop. If method read() returns None (no remaining entries), then this indicates that the end of file is reached and the loop should be terminated:

```
from jhplot  import *
from jhplot.io import *

f=HFile("events.jser")
while 1:
   event=f.read()
   if event == None:
      print "End of events"; break
   print event[0]
   p=event[1]
   h=event[2]
   # print p.toString()
   # print h.toString()
print "No of processed events=",f.getEntries()
f.close()
```

Listing 9.5   Reading random events from a file

We have commented out some debugging print statements intentionally: for a very large loop, the Jython console will be very quickly filled with large output.

We remind again that when using the Java serialization, the retrieved objects from the HFile files are self-described. This can be checked in the above example by either uncommenting the debugging statements or checking the object types with the method type(obj).

9.3.4   Buffer Sizes

We have mentioned before that in order to optimize the execution speed of an application with I/O intensive operations, it is important to find a proper buffer size for the HFile class. The buffer size depends on several factors, but one important factor is the size of the objects used for persistent storage on the disk. To determine the most optimal buffer size, write a code to benchmark the read/write operations. In the next example we will create a relatively large array with random numbers and then write this array multiple number of times into a file. The objects will be written using the buffer size of 2048 bytes.

```
from jhplot.io import *
from jhplot import *
import time
```

```
p1=P0D("data")
p1.randomUniform(1000000,0,1)

buffer=2048
start = time.clock()
f=HFile("tmp.jser","w",0,buffer)
for i in range(10000):
  if (i%1000 == 0):
     print "pocessed=",i
  f.write(p1)
f.close()
print " time (s)=",time.clock()-start
```

Listing 9.6 Benchmarking I/O operations

Note the required time to complete the execution of this code. Then change the buffer size. If you set it to a small number (say, `buffer=2`) you will see a slow-down in the execution of this script. Increasing the buffer size usually speeds up the I/O operations for large data sizes.

Of course, instead of the `P0D` array, one can use any DMelt or Java object that implements the `Serializable` interface (almost all DMelt objects do). After running the script above, do not forget to remove the temporary file "tmp.jser" (which is about 130 MB) as:

```
>>> import os
>>> os.remove("tmp.jser")
```

9.3.5 XML File Format

Analogously, one can write and read data using the XML format. This can be achieved using the class `HFileXML`, i.e., replacing `HFile` by the `HFileXML` statement in all examples shown above. You will see that the output files are significantly larger than in the case of `HFile`. This is not too surprising since XML tags use some disk space. However, this time the data will be written to a human-readable text files.

9.4 PFile Data Format

Despite the convenience of Java serialization, it is important to note several features of the serialization mechanism when making decisions about input and output of numerical data:

- Java serialization can be slower in comparison with the usual I/O approach. It is a good approach for storing complex objects, but it has unnecessary overhead for numerical data streams;
- It is not intended to be platform neutral. One cannot write or read data using other programming languages (such as C++). In fact, it is not even guaranteed that a serialized object can be restored when reading it with the library that has a modified class;
- The size of the files created by the Java serialization can be larger than expected (even after compression). In fact, what we really need in many data analysis applications is simply to write or read numerical values, rather than objects with all associated graphical attributes.

This poses a certain trade-off between the power of being able to write essentially any Java object into a Java serialized file and the shortcomings listed above. If Java serialization is too inconvenient for numerical data, consider an alternative I/O approach using the Java class called PFile. The methods of this class allow writing and reading compressed data records constructed using the Google's Protocol Buffers package [6] (which gives the origin of the first letter "P" in the class name PFile). The Protocol Buffers program encodes structured data in a platform-neutral format, similar to XML. But unlike the XML, the Protocol Buffers approach is faster, simpler, and the output file is smaller. This will be discussed in detail in Sect. 9.8.

There are several advantages in using the class PFile: it is faster for writing compared to HFile and the output sizes of the files created by HFile are smaller. But, more importantly, one can use external programs to write and read files as long as a Protocol Buffers file which specifies data format is provided (again, see Sect. 9.8 for details). The class PFile is designed to keep data without their graphical attributes, which leads to smaller sizes of output files. However, a limited set of objects is supported by this class. At present, only objects from the package jhplot can be dealt by PFile.

In the previous section we have considered several constructors of the HFile Java class. Similarly, an object of the class PFile can be created. Below we list the major constructors to open a file "data.jpbu":

```
>>> f=PFile("data.jpbu","w")       # open for writing
>>> f=PFile("data.jpbu","r",map)   # open for reading and map
>>> f=PFile("data.jpbu","r")       # as above with map=1
>>> f=PFile("data.jpbu")           # as above with map=1
```

The file extension of the file is "jpbu", which originates from the underlying Protocol Buffers format. The only difference from the HFile class is additional Boolean attribute "map". If this attribute is set to 1 ("True" in Python, or "true" in Java), we will create an association between a numerical record position and an object name which can usually be set using the setTitle() method. This is handy since one can retrieve an object by using its name (which can be obtained using the method `getTitle()`), rather than using its position in the `PFile` file. The price to pay is that it takes more resources in the case of large files since the constructor must preprocess records in order to make a map between object names and positions.

Let us give a detailed example. Below we generate two one-dimensional arrays with 100 numbers and then we create a two-dimensional array. We write all arrays in a compressed file:

```
from jhplot.io import *
from jhplot import *

f=PFile("data.jpbu","w")
x=P0D("X")
x.fill(100,0,1);   f.write(x)
y=P0D("Y")
y.fill(100,0,1);   f.write(y)
xy=P1D("XY",x,y); f.write(xy)
f.close()
```

Listing 9.7   Writing DMelt objects in PFile files

Now let us read the objects from this file. As usual, we will open the file and check what is inside:

```
from jhplot.io import *
from jhplot import *

f=PFile("data.jpbu")
print f.listEntries()
print "No of entries=",f.getNEntries()
for i in range(f.getNEntries()):
    a=f.read(i+1)
    print "entry=",i+1,type(a)
f.close()
```

Listing 9.8   Reading DMelt objects from PFile files

We open the file in the read-only mode using the "name-mapping" option which will allow us to retrieve objects by calling their names. We also print all entries together with the record sizes using the method `listEntries()`. Finally, we read all records in a sequential order and verify their types:

```
1   ->  X   -->   809
2   ->  Y   -->   809
3   ->  XY --> 1613
No of entries= 3
entry= 1 <type "jhplot.P0D">
entry= 2 <type "jhplot.P0D">
entry= 3 <type "jhplot.P1D">
```

Since we have opened the file using the "mapped-name" option, one can take advantage of this by retrieving objects using their names, instead of record positions:

```
from jhplot.io import *
from jhplot import *

f=PFile("data.jpbu")
x=f.read("X");    print type(x)
y=f.read("Y");    print type(y)
xy=f.read("XY"); print type(xy)
f.close()
```

Listing 9.9 Reading objects using their names

Now we have a convenient (and fast) access to the records using their names. We should note that when the name-mapping option is enabled, objects inside PFile cannot contain duplicate names; check this by calling the method getTitle(). The name returned by this method will be used as a key to access the object.

At present, all "named" objects can be processed by the PFile class, such as P0D-PND, F1D-F3D, FND, FPR, H1D and H2D histograms. One can also write strings.

One should point out again that the file format with the extension "jpbu" is platform neutral. It has been mentioned before that the record structure of the class PFile is constructed using a Protocol Buffers template file. This file, called HRecord.proto and located in the directory "macros/system", can be used to build a C++ application which can read the files created by the PFile class (and vice versa). How to write more general data structures using the Google's Protocol Buffers will be discussed in Sect. 9.8.

If one needs to write DMelt-specific objects from a C++ program, one can use the CBook C++ programming library [7] which is designed to write several major objects, such as H1D, H2D, P0D, and P1D, into the "jpbu" files. Such files can later be opened by the Java PFile class, or can be analyzed and plotted using the generic BrowserData browser to be discussed below.

9.4.1 Browser for PFile File Data

The files created by the PFile class can be studied using the BrowserData generic browser. This functionality is very similar to that discussed in the section with the HFile class description. Start the DMelt IDE, select [Plots] and then [HPlot] canvas. This brings up an empty HPlot canvas. Then select the menu [Read data file] using [File] on the toolbar menu. This pops up a file dialog for file loading. Select a file with the extension "jpbu". This automatically opens an object browser which is discussed in detail in Sect. 8.2.3.

One can also call the browser from a code. In our next example we create a canvas and open a PFile file in the BrowserData dialog. This script shows a typical example:

```
from jhplot.io  import *
from jhplot  import *

fi="test.jpbu"
f=PFile(fi,"w")
for i in range(10):
  p0= POD("Random="+str(i))
  p0.randomNormal(1000,0.0,1.0)
  f.write(p0)
f.close()

c1=HPlot()            # plotting canvas
c1.visible()
BrowserData(fi,c1) # browser the file
```

Listing 9.10   Browsing data in PFile files

Select one entry, and push the button "Plot". You will see a plotted histogram with the entries from the saved POD object. Similarly, one can plot H1D histograms, functions, and multidimensional arrays.

9.5 HBook XML Data Output

Essentially any object (data arrays, histograms, functions) of DMelt can be saved into an XML file without using the standard Java serialization. The Java class HBook can help with this. This class is rather lightweight, and can create significantly smaller files than those created by the HFileXML class considered before. In particular:

- it keeps only numeric content of objects without graphical attributes (color, fonts);
- it does not use the XML tags to keep separate values of arrays. This means that this class generates substantially smaller outputs than for the standard XML approach. This is especially important for large data volumes, 2D matrices, and histograms;
- it is better suited for cross-platform data exchange. In particular, C++ and FORTRAN packages are available to read and write data in this format.

Let us give a small example that creates a file with 1D array, histogram, and a function:

```
from jhplot  import *
from jhplot.io import *

hb = HBook("data.jdat","w") # initialize
```

```
p1=P1D("array")
p1.add(1.0,2.0); p1.add(2.0,3.0)
hb.write(10,p1) # add 1D array with key 10

h1=H1D("My histogram",100,0,50)
h1.fillGauss(100,0,1)
hb.write(20,h1) # add histogram with key=20

f1=F1D("x*cos(x*10)")
hb.write(30,f1) # add a function with key=30

hb.close()
```

Listing 9.11 Writing an XML file with numeric data

It is easy to understand this code: We use numeric keys (10, 20, 30) to identify saved objects. Likewise, one can use strings as the keys. Open the file "data.jdat" in a text editor and examine it.

Now we will restore the saved objects as

```
from jhplot  import *
from jhplot.io import *

hb = HBook("data.jdat") # read file
print hb.getKeys()       # list keys of data
p=hb.get("10")           # extract object using the key 10
h=hb.get("20")
f=hb.get("30")
print p.toString()
print h.toString()
print f.toString()
hb.close()
```

Listing 9.12 Reading data from an XML HBook file

First, we list all keys stored in the file and print them. Then we restore all objects and close the file. One can also use `getAll()` method that extracts all the objects and put them into a map.

Finally, it is good idea to check the version of `HBook` file using `getVersion()`. This is needed to make sure that one can restore the saved objects using the correct version of software.

As before, one can open the file in the DMelt data browser.

```
from jhplot  import *
from jhplot.io import *

c1=HPlot()
c1.visible()
BrowserData("data.jdat",c1)
```

Listing 9.13 Reading XML HBook using a data browser

You will see a plotting canvas and a list of stored objects. Use the "Plot" button to visualize the data.

9.6   Text File Formats

Numeric data can be stored as plain text files, in which case they can visually be examined in text editors. There is a number of data formats that organize data as ASCII characters such that one can keep relatively complex data structures. Such formats are useful for data exchange between different software programs. To save large volumes of data in these formats is not recommended since the files should be compressed to reduce file sizes.

9.6.1   Working with ASCII Files

In order to read and write ASCII files with numeric data, one can use the standard Java API, or one can mix Java classes with the methods from scripting languages supported by Java.

In many cases, we will need to read numeric data organized in columns and rows. For example, let us consider the following data:

```
1 2 3
4 5 6
7 8 9
```

We will save these numbers into a file "data.txt". Note that this file can be made from the double list using this small code snippet using Python:

```
f=open("data.txt","w")
data=[[1,2,3], [4,5,6], [7,8,9]]
for row in data:
  for l in row:
     f.write(str(l)+" ")
  f.write("\n")
```

Listing 9.14   Writing ASCII data

Now let us read this file by mixing Java and the Python syntax:

```
from java.util import *
from java.io import *
from java.lang import *

sc=Scanner(File("data.txt"))
```

```
while(sc.hasNext()):
  tok = sc.nextLine().split() # split line into list
  line=[]                     # list of floats
  for s  in tok:
    line.append(Float.valueOf(s))
  print line
```

Listing 9.15 Reading ASCII data using Java API

Here we apply the `split()` method to split each line using the white space. If the numbers are separated by commas, use `split(",")`. The example prints each row as a Python list, and uses the `print` statement from the Python language, while the classes `Scanner`, `File`, and `Float` come from the Java API. To convert the numbers to integers or double values, use the Java classes `Integer` or `Double`.

The same calls to Java classes in the above example can be used for any other scripting language supported by the Java platform, such as BeanShell, Groovy, or JRuby considered in the next chapters. You only need to use the appropriate syntax, and to remove the Python list and the `print` statement in the above example.

The above example can be rewritten using 100 % Python:

```
with open("data.txt") as xfile:
  for l in xfile:
    l = l.strip()
    line=[]  # list of floats for each row
    for number in l.split():
      line.append(float(number))
    print line
```

Listing 9.16 Reading ASCII data in Python

9.6.2 CSV File Format

The comma separated value files, CSV, are used to store information in the form of tables. This format is especially popular for import and export in spreadsheets. Data in such files are either separated by a comma or a tab, or by any other custom delimiter.

We have already considered how to read and write the CSV files using the Python module `csv` (see Sect. 2.15.4). Below we will consider high-performance Java libraries to work with such files.

9.6.2.1 Reading CSV Files

DMelt supports reading the CSV files as well as writing data into such files. Let us create a typical CSV file using any text editor:

```
Sales,  Europe,  Russia,  USA
January,   10,   20,      20
February, 30,    20,      50
March,     10,   40,     100
April,     80,    7,      30
May,      300,  400,      90
June,      50,   10,      70
```

Listing 9.17 File "table.csv"

Dealing with the CSV format is rather easy when using the DMelt Java classes—
just open a CSV file using the class CSVReader and iterate over all its entries as
shown in the example below:

```
from jhplot.io.csv import *

r=CSVReader("table.csv",",")
while 1:
  line= r.next()
  if line == None: break
  print line
r.close()
```

Listing 9.18 Reading the file "table.csv"

As one can see, the method next() returns a list of strings for each line of the CSV
file. File reading stops when the end of the file is detected. If you want to read arrays
of lists, use the method next() instead. One can easily convert strings inside the
list representing each row to floats and integers, as float(str) (returns a float
value) or int(str) (returns an integer).

One can read all entries at once, instead of looping over all rows. This can be
done using the methods:

```
l=r.readList()    read all elements into a 2D list;
l=r.readAll()     read all elements into into a list where each row is an array.
```

Let us take a closer look into the CSVReader class. It has several constructors:

```
>>> CSVReader(file, sep, quote, line)
>>> CSVReader(file, sep, quote)
>>> CSVReader(file, sep)
>>> CSVReader(file)
```

where file is the string with a file name, sep is the character used for separating
entries in each row, quote is the character to use for quoted elements. Finally,
line is an integer number representing the line number to skip before start reading.
When no sep is given, then the default separation is done using a comma. If no
quote is given, the default escaping quotes is double quotes ("), and if no line is
given, the default end of line is the new line symbol.

You can display a CSV file in a spreadsheet using the SPsheet class discussed in Sect. 15.1.3. Once CSVReader object has been created, just pass it to the SPsheet object as:

```
from jhplot import  *
from jhplot.io.csv import *

r=CSVReader("table.csv",",")
SPsheet(r)
```

Listing 9.19 Showing "table.csv" file in a spreadsheet

Now let us move to the next subject where we will discuss how to write data into the CSV files.

9.6.2.2   Writing CSV File

To write data to CSV files is as easy as reading them. Instead of the CSVReader class, one should use the CSVWriter class. It has the following constructors:

```
>>> CSVWriter(file, sep, quote, line)
>>> CSVWriter(file, sep, quote)
>>> CSVWriter(file, sep)
>>> CSVWriter(file)
```

Here, file is an output file name. Other arguments are exactly as for the CSV Reader class.

Let us show an example illustrating how to write lists of objects line by line:

```
from jhplot.io.csv import *

w=CSVWriter("out.csv",",")
w.write( ["Test1","20","30"] )
w.write( ["Test2","100","50"] )
w.write( ["Test3","200","100"] )
w.close()
```

Listing 9.20 Writing "out.csv" file

Check the output file. It has the following structure:

```
"Test1","20", "30"
"Test2","100","50"
"Test3","200","100"
```

One can write the entire file at once using the writeAll(list) method.

9.6.3 EDN File Format

DMelt includes support for the extensible data notation format, abbreviated as EDN
[8]. The EDN files can hold any Java and Python objects in as ASCII text files.
Therefore, EDN is simple, since files can be examined, yet are powerful enough to
keep complex data structures. The EDN files are typically smaller than similar XML
files, and this fact makes the EDN format attractive to keep numeric data. The EDN
files have the extension ".edn".

Let us show how to store complex data structures in the EDN format. As usual,
we will use the Python language, but the same login can be implemented in other
scripting language supported by Java. We will create Java and Python objects, fill
them with values, and then write them to the EDN file format.

```
from us.bpsm.edn.printer import Printers
from java.util import *
from java.io import *

sw=StringWriter()
p=Printers.newPrinter(sw)

al = ArrayList()
al.add(1); al.add(2)
p.printValue(al)  # Write a Java list
s=[[22],[33,33]]
p.printValue(s)   # Write Python double list
map = HashMap()
map.put("name", 100)
p.printValue(map) # Write a map
map={}
map["key1"]="val1";  map["key2"]="val2"
p.printValue(map) # Write a Python dictionary
data=sw.toString()
fw = FileWriter("file.edn")
fw.write(data)     # write all objects
fw.close()
print "Data in file = ",data
```

Listing 9.21 Writing lists to EDN files

The comments in this code show each step for object creation. In addition, we write
the data to the computer screen. This helps you see the internal structure of the EDN
format:

```
[1 2]((22)(33 33)){"name"100}{"key1""val1""key2""val2"}
```

Now we will attempt to restore the saved objects.

```
from us.bpsm.edn.parser import IOUtil
from us.bpsm.edn.parser import Parsers
```

```
from us.bpsm.edn.parser.Parsers import defaultConfiguration

edn=IOUtil.stringFromResource("file.edn")
pbr = Parsers.newParseable(edn)
p = Parsers.newParser(defaultConfiguration())

a=p.nextValue(pbr); print a
a=p.nextValue(pbr); print a
a=p.nextValue(pbr); print a
a=p.nextValue(pbr); print a
```

Listing 9.22 Reading data from an EDN file

The print command shows that all objects have been restored as

```
[1, 2]
[[22], [33, 33]]
{name: 100}
{key1: val1, key2: val2}
```

9.6.4 DIF File Format

The DIF ("Data Interchange Format") format has a limited support in DMelt. Files
in this format have the extension ".dif". This ASCII format can be used to import or
export spreadsheets between various applications.

The current implementation of this format [9] is only for read-only mode. There
is no possibility for saving data into a DIF file. To read such files, one should import
the module dif located in the "macros/system" directory. In the example below,
we read the file "nature.dif", extract all information into Jython tuples and print the
entries:

```
import dif

f=open("nature.dif","r")
d = dif.DIF(f)
print d.header
print d.vectors
print d.data
```

Listing 9.23 Reading files in the DIF format

Below we give some explanation:

d.header   dictionary of header entities, with names shifted to lowercase for
               ease of handling;
d.data   list of tuples with data;
d.vectors   list of vector names (or, better, column names).

Read more about Python implementation of this module in Ref. [9].

9.7 Reading ROOT and AIDA Files

DMelt has several Java classes to read files created by the ROOT data analysis framework [10, 11] and AIDA (Abstract Interfaces for Data Analysis). Both data formats are used in high-energy physics applications.

9.7.1 ROOT Files

At this moment, one cannot write ROOT files, but to read and extract histograms or graphs written in the ROOT format should not be a problem. DMelt uses the FreeHEP library to read ROOT files.

First, open a ROOT histogram viewer as

```
>>> from jhplot import *
>>> BRoot()                  # open a viewer, or
>>> BRoot("Example.root") # open a ROOT file
```

Obviously, to run this example, you will need the file "Example.root" shipped with DMelt (see the directory "macros/examples/data/"). The latter version of the constructor opens a ROOT file with the name "Example.root" inside the histogram viewer. The BRoot class is a simple wrapper of the HistogramBrowser class from FreeHEP.

One can read ROOT files also without invoking a GUI browser. Reading 1D, 2D, and 3D histograms can be done using the HRoot class which opens a ROOT file and extracts histograms:

```
from jhplot  import *
from jhplot.io  import *

ro = FileRoot( "Example.root" )
print "Nr of histograms=",ro.getNKeys()
print "ROOT version=",ro.getVersion()
print "ROOT version=",ro.getTitle()
print "No of histograms=",ro.toString()

h1 = ro.getH1D("mainHistogram") # use keys
h2 = ro.getH1D("totalHistogram")
c1 = HPlot()
c1.setGTitle("ROOT histograms")
c1.visible()
c1.setAutoRange()
c1.draw([h1,h2])
```

Listing 9.24 Reading a ROOT file

Here we first read the key of the ROOT histogram and then convert it into the H1D histogram. Then, the object `h1` can be plotted with the usual `draw()` method of the `HPlot` class.

One can find more details about how to read ROOT data in Java in a more data analysis oriented book [12].

9.7.2   AIDA Files

Files created using AIDA can be read using the `FileAida()` class which is based on the FreeHep Java library [13]. One can read histograms and "Clouds", i.e., unbinned arrays of data, as shown in this example:

```
from jhplot  import *
from jhplot.io  import *

a= FileAida( "UsingJAIDAFromJython.aida" )
print a.getAllNames()
print a.getAllTypes()

c1d=a.get("Clouds/Cloud 1D")
c2d=a.get("Clouds/Cloud 2D")
h1=a.get("Histograms/Histogram 1D")

c1 = HPlot()
c1.setGTitle("Data from AIDA file")
c1.visible()
c1.setAutoRange()

c1.draw(c1d)
c1.draw(H1D(h1)) # convert to H1D
```

Listing 9.25   Reading AIDA files

The input AIDA file can be found in the directory "macros/examples/data/" of the DMelt installation.

9.8   Google's Protocol Buffer Format

Many situations require an analyzer to write structural data into a format which can be understood by variety of programming languages. Imagine an experimental apparatus producing a data stream. It is very likely that its code for data I/O is written in C/C++, since most hardware drivers are implemented in this system language.

But, at the end of the day, this is a user who should analyze the data and who may prefer a human-friendly language (like Python/Jython) for final analysis code. Therefore, we should find a way to read structural data produced by experiments using these higher level languages. Or, imagine an opposite situation: an analyzer produces data using Java or Jython, and an application implemented using a lower level language should read this data during communication with a hardware.

One way to deal with such kind of problems is to use a self-described file format, such as XML. But there are several problems with this format when dealing with large data volumes: (1) programs are slow for loading; (2) there is a significant penalty on program's performance; (3) Data files are large due to tags overhead.

Taking into account that the XML format is too cumbersome to use as an encoding method for large data files, one can use the ROOT format as an alternative. This may seem to be a heavy approach (installed ROOT takes several hundreds of megabytes!), and to read most recent ROOT files in Java is not too easy: in fact, the ROOT framework was not designed from the ground to be friendly for other programming languages.

The Google team has released the Protocol Buffers package [6] which deals with serializing structured data in a platform-neutral format. This data interchange format is used by Google for persistent storage of data in a variety of storage systems. It is also well tested on many platforms. The Protocol Buffers is a self-describing format, which is equally well supported by C++, Java, Python, and by other languages using third-party packages. In comparison with the XML format, the Protocol Buffers files are up to a factor 100 times faster to read, and file sizes are significantly smaller due to a built-in compression. This format is also very promising from the point of offered backward compatibility: new fields created by new protocol versions are simply ignored during data parsing.

Even more. The Protocol Buffers helps to abstract from a language-specific description of data structures. More specifically, this means that an analyzer only needs to produce a file describing his/her data records, and then the Protocol Buffers program generates a C++ or Java code for automatic encoding and parsing data structures.

Below we will discuss the Protocol Buffers in more detail. To develop applications in C++, the reader is assumed to install the Protocol Buffers (at least version 2.2) from the official web site [6].

9.8.1 Prototyping Data Records

For our next example we will assume the following experiment. We perform N measurements, each measurement is characterized by an identification number, a string (name), and arbitrary array with some other data. In each measurement, it is assumed that we observe n number of particles. This number is not fixed, and can vary from measurement to measurement. Finally, for each particle, we measure particle's energy and electric charge. We also assign a string with particle name.

Let us prototype an event record using the Protocol Buffers syntax, which is then can be used to generate a C++ or Java code. The code is shown below:

```
package proto;
option java_package = "proto";
option java_outer_classname = "Observations";

message Event {
  required int64   id = 1;
  required string  name=2;
  repeated double  data = 3 [packed=true];

  message Particle {
    required string   name=1;
    required sint32   charge=2;
    required double   energy=3;
  }
  repeated Particle particle = 4;
}
message Experiment {
  repeated Event event = 1;
}
```

Listing 9.26   Data prototype. File "experiment.proto"

This file contains the description of all objects in a platform-independent way, so later one can read or write data from a variety of languages. As one can see, the class Event envelops the entire data. Instead of using the class name, the Protocol Buffers uses the word "message". Each event has 4 records, id (integer), name (string), data (list of doubles), and the class Particle. Each such record (or "message") has numbered fields (in this example, running from 1 to 4). Such integer values are unique "tags" used in the binary encoding.

The value types can be numbers, Booleans, strings, or other message types, thus different messages can hierarchically be nested. The required field tells that such data field should always be present. One can also set the default value using the line [default=value], where value is a given default value. The message field can have other two types: optional—a message can have zero or one of this field (but not more than one), and repeated—a message can be repeated any number of times, preserving the order of the repeated values. The fields can include enumerators for tight association of a specific value to a variable name.

In the above example, the message data can be repeated (similar to a list), so we can append any number of double values in each record. The field [packed=true] is used for a more efficient encoding of multiple data entries. Finally (and this is very important), the message of the Particle type can also be repeated, since we can have multiple number of particles in events.

Table 9.1 Scalar value types used in .proto files and their Java and C++ equivalents

Scalar message field types	
int	int (Java and C++)
double	double (Java and C++)
int	int (Java and C++)
bool	Boolean (Java) and bool (C++)
string	String (Java) and string (C++)
fixed32	int (Java) and int32 (C++)
fixed64	long (Java) and int64 (C++)

As you may already have guessed, the message `Particle` keeps information about a single particle, such as its name, charge, and energy. This message is constructed in exactly the same way as the outer messages. The only difference is that now we should specify the fields which are the most appropriate for defining particle properties.

A scalar message field can have one of the following types shown in Table 9.1. More types are given in [6].

9.8.2   Dealing with Data Using Java

Now let us generate a Java code using the above prototype file. Assuming that the Prototype Buffer is properly installed, and assuming that we are working in the Unix/Mac environment, we generate the Java code as

```
protoc --java_out=.  experiment.proto
```

Listing 9.27   Generating a code based on prototype buffer

This creates a directory "proto" with the `Observations.java` file. The `--java_out` option tells to generate Java classes to be used for data encoding. Similar options are provided for other supported languages. The dot after this option tells to generate the output in the current directory. The output of this command is the Java file "Observations.java" located in the directory "proto", as specified in the original "experiment.proto" file (see the Java package statement).

Let us write a test code which generates 100 events with 10 particles in each event. We will create the file "WriteData.java" and copy it into the directory "proto" together with recently generated file "Observations.java":

```
package proto;
import proto.Observations.Experiment;
import proto.Observations.Event;
import java.io.FileOutputStream;

class WriteData {
```

```
public static void main(String[] args) throws Exception {
   Experiment.Builder exp = Experiment.newBuilder();
   FileOutputStream output =
                   new FileOutputStream("data.prod");
    for (int e=0; e<100; e++){
       Event.Builder ev = Event.newBuilder();
       ev.setId(e); ev.setName("collision");
       ev.addData(1); ev.addData(2);
          for (int i=0; i<10; i++){
             Event.Particle.Builder p =
                     Event.Particle.newBuilder();
             p.setName("proton");
             p.setCharge(1); p.setEnergy(1);
             ev.addParticle(p);
          };
     exp.addEvent(ev); } ;

   exp.build().writeTo(output);
   output.close();
  }
}
```

Listing 9.28   Writing data into a file. "WriteData.java" file

Next we need to compile all Java files and build a jar library. For this, make sure that Java CLASSPATH points to the DMelt library (or to "protobuf.jar" from the original Protocol Buffer library). Then compile and build the jar library:

```
javac −1.8 proto/*.java
jar −cf proto.jar proto/*
```

Listing 9.29   Compiling code with prototype buffer

This produces the file "proto.jar" with the compiled classes.

Let us test the above code. We will run the WriteData class in the usual Java fashion.

```
java −cp proto.jar proto.WriteData
```

Listing 9.30   Running WriteData program

The result of this command is the output file "data.prod" with our structural data written in a binary form. Let us verify what is written by adding the code "Read-Data.java" into the directory "proto".

```
package proto;
import proto.Observations.Experiment;
import proto.Observations.Event;
import java.io.FileInputStream;

class ReadData {
```

```
public static void main(String[] args) throws Exception {
  Experiment  exp=
    Experiment.parseFrom(new FileInputStream("data.prod"));
    for (Event ev: exp.getEventList()) {
      System.out.println("Event id:" +ev.getId());
      System.out.println("Event name:" +ev.getName());
          for (Event.Particle p: ev.getParticleList()) {
              System.out.println(p.getName());
              System.out.println(p.getEnergy());
      } }
} }
```

Listing 9.31 Reading data. "ReadData.java" file

It opens the file "data.prod" and fetches all objects in a loop, until all events are processed. For each event, the code extracts particles and their attributes. Let us compile this code by invoking the compilation command `javac` as was done before, assuming that it is located in the same directory "proto". This will produce a new jar file. Then, run the code using the command

```
java —cp proto.jar proto.ReadData
```

Listing 9.32 Running ReadData program

This code reads the data and prints all events with particle attributes.

9.8.3 Switching to Jython

In Sect. 16.2 we will discuss how to unwrap a Jython code into Java. In this example we will show the opposite case: We will rewrite our example code "WriteData.java" into a compact Jython script:

```
from proto.Observations import Experiment
from proto.Observations import Event
from java.io import FileOutputStream

exp=Experiment.newBuilder()
for e in range(100):
   ev = Event.newBuilder()
   ev.setId(e); ev.setName("collision")
   ev.addData(1); ev.addData(2)
   for i in range(10):
      p=Event.Particle.newBuilder()
      p.setName("proton")
      p.setCharge(1); p.setEnergy(1)
      ev.addParticle(p)
   exp.addEvent(ev)
exp.build().writeTo(FileOutputStream("data.prod"))
```

Listing 9.33 Writing data into a file. "WriteData.py" file

As you can see, it is much smaller than the equivalent Java program. To run this code, we must copy the "proto.jar" file to a place where Java can see it (for example, in the directory "lib/user") or to include its location into Java CLASSPATH. Then one can execute this script using DMelt IDE.

Similarly, one can rewrite the Java code used to read the data as

```
from proto.Observations import Experiment
from proto.Observations import Event
from java.io import FileInputStream

exp=Experiment.parseFrom(FileInputStream("data.prod"));
for e in exp.getEventList():
    print e.getId()
    print e.getName()
    for p in e.getParticleList():
        print p.getName(),p.getEnergy();
```

Listing 9.34 Reading data into a file. "ReadData.py" file

Again, run this code and you will see exactly the same output as in case of Java.

9.8.4 Adding New Data Records

Now let us illustrate how to add a new event record to the existing file. To make our codding shorter, we will use Jython and add an event (but without any particle) as

```
from proto.Observations import Experiment
from proto.Observations import Event
from java.io import FileOutputStream
from java.io import FileInputStream

exp=Experiment.newBuilder()
try:
    input=FileInputStream("data.prod")
    exp.mergeFrom(input); input.close()
except FileNotFound, e:
     print "The file was not found, going to backup file"

ev = Event.newBuilder() # add new event
ev.setId(99); ev.setName("new entry")
exp.addEvent(ev)

output=FileOutputStream("data.prod")
exp.build().writeTo(output)
output.close()
```

Listing 9.35 Adding new records. "AddData.py" file

In this example we simply read the file and then write a new file with the additional event.

9.8.5 Using C++ with the Protocol Buffers

Now we come to the main issue of how to generate the same file using a C++ program, so we can read it using Java, Jython, or any other language. Of course, we also need to test the opposite situation when reading the Java-generated file "data.prod" by a C++ program.

Below we will show the necessary steps to produce a C++ code which generates exactly the same data file as that shown in the previous section, and then how to read this file (or the file generated by the Java code above). We will use the same prototype file "experiment.proto" as before. Let us generate a C++ code to be used for data encoding:

```
protoc --cpp_out=. experiment.proto
```

Listing 9.36 Generating C++ code

Note the option cpp which tells to generated the C++ encoding code. After execution of this line, we should find two files, "experiment.pb.cc" and "experiment.pb.h" in the same directory. As in case of Java, they contain the necessary information to be used for data serialization.

Let us write a small test program which creates the output file.

```
#include <iostream>
#include <fstream>
#include <string>
#include "experiment.pb.h"
using namespace std;

int main(int argc, char **argv)
{
  GOOGLE_PROTOBUF_VERIFY_VERSION;
  proto::Experiment exp;

  for (int e=0; e<100; e++){
   proto::Event* ev = exp.add_event();
   ev->set_id(e); ev->set_name("collision");
   ev->add_data(1); ev->add_data(2);
     for (int i=0; i<10; i++) {
     proto::Event::Particle* p=ev->add_particle();
     p->set_name("proton");
     p->set_energy(1); p->set_charge(1);
     } };

   cout << "Write data.prod" << endl;
```

```
    fstream output("data.prod",
          ios::out | ios::trunc | ios::binary);
  if (!exp.SerializeToOstream(&output)) {
    cerr << "Failed to write address book." << endl;
    return −1; }

  output.close();
  google::protobuf::ShutdownProtobufLibrary();
  return 0;
}
```

Listing 9.37 Writing data into a file. "write_test.cc" file

For those who know C++ this code should look rather simple. Moreover, the code
logic is rather similar to that of the Java code. Since we declared the package "proto"
in the prototype file, one should use the corresponding C++ namespace for the class
declarations. The first statement in the main() method verifies that we have not
accidentally linked against the library which is incompatible with the version of the
headers.

Next, we will need to compile all these files. For this, we use GNU gcc compiler
installed by default on Unix/Linux systems

```
gcc write_test.cc experiment.pb.cc \
    −o write_test `pkg-config −−cflags −−libs protobuf`
```

Listing 9.38 Compiling the source codes

Finally, try to execute the program write_test. We will see the created file
"data.proto". Check its size. It should have the same size as in the case of Java
example. Surely, this is a good sign, since it likely indicates that the new file is
correctly produced and, in fact, is totally identical to that generated previously using
Java.

Let us move on and write a C++ code which will read the data from this file.

```
#include <iostream>
#include <fstream>
#include <string>
#include "experiment.pb.h"
using namespace std;

int main(int argc, char **argv)
{
 GOOGLE_PROTOBUF_VERIFY_VERSION;
 proto::Experiment exp;

 fstream input("data.prod", ios::in | ios::binary);
 if (!exp.ParseFromIstream(&input)) {
   cerr << "Failed to parse data file" << endl;
   return −1; }

 for (int i = 0; i < exp.event_size(); i++) {
```

```
    const proto::Event& ev = exp.event(i);
    cout << " ID: " << ev.id() << endl;
    cout << " Name: " << ev.name() << endl;

  for (int j = 0; j < ev.particle_size(); j++) {
    const proto::Event::Particle& p = ev.particle(j);
    cout << " - name: " << p.name() << endl;
    cout << " - energy: "   << p.energy() << endl;
  } }
  google::protobuf::ShutdownProtobufLibrary();
  return 0;
}
```

Listing 9.39 Reading data from a file. "read_test.cc" file

The file reads the data and loops over all event entries. We can compile all source files as usual:

```
gcc read_test.cc experiment.pb.cc \
    -o read_test `pkg-config --cflags --libs protobuf`
```

Listing 9.40 Generating C++ code

and run the executable program as `read_test`. We will see the print messages included for debugging.

As a final exercise, try to use the Java or Jython code developed in the previous subsection for reading file created by our C++ program.

9.8.6 Some Remarks

The Protocol Buffers has an easy-to-use format to organize sequential data in a platform-independent way. It is a good format for handling individual messages within a large data set which consists of large number of small pieces. Each piece can also be a structural data in the form of "messages".

We have already discussed that the class `PFile` (see Sect. 9.4) is completely based on the Protocol Buffers format. Each record is compressed and independently packed in the ZIP file format. The records are implemented using the Protocol Buffers file "HRecord.proto" located in the directory "macros/system". The file "HRecord.proto" can be used to construct applications in C++ which can read and write files generated by the `PFile` Java class.

9.9   Creating Excel Files

DMelt includes the Apache POI library [14] to work with Microsoft Office file formats. Our main goal is to show how to write and read Excel files. This gives a lot of benefits since one can build Excel files inside programs, i.e., creating easy-to-use spreadsheets which can be viewed with Microsoft Office, OpenOffice, or LibreOffice.

We should say that keeping large data in Excel files is not too efficient, and involves a lot of overhead related to the XML structure of this format. But, there one obvious advantage—one can view data in a convenient way using office suites for desktop and mobile applications. The POI library is consistent with Office OpenXML used in Microsoft Office 2007 and 2008, as well as with OpenOffice (LibreOffice).

Let us start by designing a simple code to create a file with one row and a few different cells:

```
from  org.apache.poi.xssf.usermodel import XSSFWorkbook
from  java.io import *

wb=XSSFWorkbook()
sheet=wb.createSheet("New spreadsheet");
row=sheet.createRow(0)                          # a new row
row.createCell(0).setCellValue(100)             # cell=0 has 100
row.createCell(1).setCellValue(200)             # cell=1 has 200
row.createCell(2).setCellValue("dmelt")         # cell=2 has "dmelt"
row.createCell(3).setCellValue(False)           # boolean
row.createCell(4).setCellFormula("SUM(A1:B1)")
out=FileOutputStream("table1.xlsx")
wb.write(out)
out.close()
```

Listing 9.41   Creating a simple excel spreadsheet

To check the output, you can choose whichever program you like, i.e., Microsoft Excel or OpenOffice.

How about making a large spreadsheet? We can do this too using our favorite approach based on random numbers: In addition to cells with float numbers, we will also add a row with the calculation of the sum of the cells, see the line with "SUM(A1:B1)":

```
from org.apache.poi.xssf.usermodel import XSSFWorkbook
from java.io import *
from java.util import Random

wb=XSSFWorkbook()
sheet=wb.createSheet("random numbers")
r=Random(1)
for i in range(50):
   row=sheet.createRow(i)     # create a new row
```

```
    for j in range(20):
        row.createCell(j).setCellValue(r.nextDouble())
row=sheet.createRow(50)
row.createCell(0).setCellFormula("SUM(A1:A50)")
out=FileOutputStream("table2.xlsx");
wb.write(out)
out.close()
```

Listing 9.42   Creating a large Excel file with random numbers

Now let us read this spreadsheet from the created file. We will fill our data to a Python list using Java iterators:

```
from   org.apache.poi.xssf.usermodel import XSSFWorkbook
from   org.apache.poi.ss.usermodel import Cell
from   java.io import *

f=FileInputStream(File("table2.xlsx"));
wb = XSSFWorkbook(f)
sheet=wb.getSheetAt(0) # first sheet from the workbook
elist=[]
row= sheet.iterator()
while (row.hasNext()):
   r = row.next();
   c = r.cellIterator();
   lrow=[]
   while(c.hasNext()):
        cell=c.next()
        if (cell.getCellType() == Cell.CELL_TYPE_NUMERIC):
            lrow.append(cell.getNumericCellValue())
   if (len(lrow)>0):elist.append(lrow)
print elist
```

Listing 9.43   Reading an Excel spreadsheet

Finally, we can also apply some styles to the table. Run this example and open the output Excel file using Microsoft Office or OpenOffice:

```
from org.apache.poi.xssf.usermodel import XSSFColor,
    XSSFWorkbook
from org.apache.poi.ss.usermodel  import CellStyle
from java.io import *
from java.util import *
from java.awt import Color

wb = XSSFWorkbook()
sheet=wb.createSheet("Custom fields")
helper=wb.getCreationHelper()
style=wb.createCellStyle()
style.setDataFormat(helper.createDataFormat().getFormat("m/d/yy
    h:mm"))
row=sheet.createRow(0)
cell=row.createCell(0)
```

```
cell.setCellValue(Date())
cell.setCellStyle(style)
# prepare next cell style
text= helper.createRichTextString("Test")
style=wb.createCellStyle()
style.setAlignment(CellStyle.ALIGN_CENTER)
font = wb.createFont()
font.setItalic(True)
font.setColor(XSSFColor(Color.red))
text.applyFont(font)
# create this cell
cell=row.createCell(1)
cell.setCellValue(text)
cell.setCellStyle(style)

out=FileOutputStream("table3.xlsx")
wb.write(out)
out.close()
```

Listing 9.44   Creating an Excel file after applying styles

You have now learned enough of the Java Apache POI library to start creating simple tables. Since this book is about numerical data, we will not go deeper to the subjects of how to design complex custom tables.

Of course, if you read the Excel files, you can use visually appealing office suites, instead of opening the data using scripts. The reader can check this by opening the created files in the dedicated office programs.

9.10   Non-SQL Object Databases

Data in DMelt can be organized in databases. The DMelt project has a broad support for in-memory and flat-file non-SQL databases, as well as for rational database.

This section discusses non-SQL databases that keep data as objects. Unlike relational databases that are table oriented, object databases are easier to use since all complex attributes and methods of data are preserved and can be restored using very little coding.

9.10.1   Nonsequential Input and Output

Data can be written in a nonsequential order using a simple file-based database. Such a database can be build using the HDataBase class.

A typical database created by this class associates a key of type `string` with
each written record. This has some similarity with Python dictionaries, but this time
we are dealing with truly persistent approach. A data record can consist of only
one "blob" of binary data. The file can grow and shrink as records are inserted and
removed. The database operations do not depend on the number of records in the
file, i.e., they are constant in time with respect to file accesses. It is a good idea
to design an index string such that is small enough for efficient loading into the
computer memory.

Let us give a brief example of how to create a small file-based database. This will
be a simple address book with two entries. First, we will create a binary database file,
inserting two entries with the names of people as keys. Then we read the database
again, check the number of entries, and print all entries inside the file using the keys:

```
from jhplot.io import *
from java.util import *

f=HDataBase("addressbook.db","w") # created a database
f.insert("steve", "Chicago")
f.insert("alexey","Minsk")
f.close()

f=HDataBase("addressbook.db","r") # read database
print "Number of records=",f.getRecords()
print "Is record exists?", f.isExists("alexey")
keys=f.getKeys()
while keys.hasMoreElements():
   next = keys.nextElement()
   print "key=",next," obj=",f.get(next)
```

Listing 9.45   A simple object database with addresses

The output of this script is shown below:

```
Number of records= 2
Is record exists? True
key= steve    obj= Chicago
key= alexey   obj= Minsk
```

The same approach can be used to store more complicated objects and associate
them with the keys in the form of strings. The example below shows how to build a
file ("data.db") with the object database which could be useful for writing arbitrary
event records:

```
from jhplot   import *
from jhplot.io import *
import os.path

def makeEvent(entry): # create 3 objects
    label="Event="+str(entry)
    p=P0D(label)
```

```
    p.randomUniform(10,0,1)
    h=H1D(label,10,-1,1)
    h.fill(i)
    return [label,p,h]

f=HDataBase("data.db","w")  # write database
Events=1000
for i in range(Events):
    event=makeEvent(i)
    if (i%100 == 0):
        print event[0]+" size=",os.path.getsize("data.db")
    f.insert(str(i),event)
f.close()

f=HDataBase("data.db","r")  # read the database
print "extract event 26"; event=f.get("26")
print "=== "+event[0]+"===="
print event[1].toString()
print event[2].toString()
f.close()
```

Listing 9.46 An object database: event records

Let us comment on this code. As before, for each `insert(key,obj)` method, we used the key defined by a string. Each string represents an event number but, of course, it can be any nonunique string. We access the event record number 26 using the random access feature of this database and then print its objects.

One can always remove an object from the file with the `remove(key)` method or update an entry with the `update(ob,key)` method. Look at the Java API documentation of this class.

9.10.2 Persistent Map

We will continue the discussion of simple file-based databases with random access using the class `FileHashMap`. The index file created by this class is saved in a separate file and can be kept in memory. However, the actual serialized objects (values) are stored in a separate file. Therefore, one can store a lot of objects before running out of the computer memory, assuming there is enough disk space. Typically, the index file has the extension ".ix", while on-disk data file with the serialized objects has the extension ".db".

Let us create a database based on the class `FileHashMap` that keeps a few DMelt objects:

```
from jhplot import *
from org.clapper.util.misc import FileHashMap

p1=P0D(); p1.randomNormal(1000,0,2)
```

```
p2=P0D(); p2.randomNormal(1000,1,2)
pp=P1D("test",p1,p2);   h1=H1D("OK",100,0,10)
ppp=PND(); ppp.add([1,2,3,4,5]); ppp.add([1,2,3,4,3])

fm=FileHashMap("mydata",FileHashMap.FORCE_OVERWRITE)
fm.put("d1",p1)
fm.put("d2",p2)
fm.put("d3",ppp)
fm.put("4",h1)
print "Database size=",fm.size()
fm.save()
fm.close()
```

Listing 9.47 Creating a file-based map with objects

As before, we use the string keys to identify the data in this file map. After execution
of this script, you will see two files in the current directory, "mydata.db" (data file)
and "mydata.ix" (index file). Of course, using this approach, one can store any Java
object that can be serialized. Let us restore data using the keys:

```
from jhplot import *
from org.clapper.util.misc import FileHashMap

map=FileHashMap("mydata",FileHashMap.RECLAIM_FILE_GAPS)
p1=map.get("d1")
p2=map.get("d2")
ppp=map.get("d3")
hh=map.get("4")
print hh
print ppp
print map.size()
map.close()
```

Listing 9.48 Reading the file map with objects

Now all objects have been restored as you can see from the printed messages.

9.10.3 MapDB Database

MapDB [15] is a database engine for collections backed by disk or memory storage.
As the name of this database says, it is a map. It keeps serialized data using a mechanism that mimics the standard Java serialization. Unlike the previously discussed
classes which store key-values entries in files, MapDB is a complete full-featured
database. It is designed for multi-core CPUs, has fine grained locks, concurrent
trees, consistency checks using checksums, encryption, and data compression.

   Let us give an example of writing one-dimensional array and a histogram into
the MapDB database:

```
from jhplot import *
from java.io import *
from  java.util.concurrent import *
from org.mapdb import DB,DBMaker

p1=P0D(); p1.randomNormal(1000,0,10)
h1=H1D("histogram",100,1,10)

dfile=File("mapdb.db") # open a new database
db=DBMaker.fileDB(dfile).closeOnJvmShutdown().make()
map = db.getTreeMap("data");
map.put("array",p1)
map.put("histogram",h1)
db.commit() # persist changes into disk
db.close()
```

Listing 9.49   Creating a MapDB database with numeric objects

Now we will retrieve the objects using their keys:

```
from java.io import *
from org.mapdb import DBMaker

dfile=File("mapdb.db");
db =DBMaker.newFileDB(File("mapdb.db")).make()
map = db.getTreeMap("data")
print map["array"]
print map["histogram"]
db.close()
```

Listing 9.50   Reading MapDB database

We will leave the reader here. Additional information can be found on the original web page [15] or by looking at the Java API.

9.10.4   NeoDatis Database

NeoDatis [16] is another full-featured database for storing data as objects, rather than as tables. Similar to previously discussed databases, it allows saving and retrieving native objects in a single line of code. Let us write a small code snippet which creates a database with some numeric data: a filled histogram and 2D array:

```
from jhplot  import *
from org.neodatis.odb import  *
import os

p1=P1D("X-Y points")
```

```
p1.add(10,20); p1.add(30,40)
h1=H1D("Histogram",20,0,20)
h1.fillGauss(100,10,2)

if os.path.exists("database.db"):
    os.remove("database.db")
odb = ODBFactory.open("database.db")
odb.store(p1)
odb.store(h1)
odb.close()
```

Listing 9.51   Creating a NeoDatis database with data

Now we can read the data back:

```
from jhplot   import *
from  org.neodatis.odb import   *
from  org.neodatis.odb.impl.core.query.criteria import *
from  org.neodatis.odb.core.query.criteria import *

odb=ODBFactory.open("database.db")
query=CriteriaQuery(P1D)
obj=odb.getObjects(query)
print "Number of P0D objects=",obj.size()
query=CriteriaQuery(H1D);
obj=odb.getObjects(query)
print  "Number of H1D  objects=",obj.size()

query=CriteriaQuery(H1D)
h1=odb.getObjects(query).getFirst()
print h1.toString()
query=CriteriaQuery(P1D,Where.equal("title", "X-Y points"))
p1=odb.getObjects(query).getFirst()
print p1.toString()
```

Listing 9.52   Reading objects from a NeoDatis database

One difference with the "map" type of databases is that we retrieve all objects of a given class, and then identify separate objects using their titles or other attributes.

This section concludes our discussion of object-based databases. Next in our plans is to discuss a table-like databases based on SQL which are more common for web applications.

9.11   Relational SQL Databases

We cannot avoid the discussion of relational databases based on the SQL standard, since such databases are are industry standard for web applications.

So far we have discussed object databases to keep complex data objects to be retrieved using keys. In contrast, a typical relational database stores simple data (strings, integers, etc.) in two-dimensional tables. An SQL-relational database has several advantages: one can implement a server–client mode, scalability (indexing of records), and concurrency when is required to synchronize multiple updates at the same time.

However, I should warn you: in many cases you do not need such SQL databases at all for data analysis where we access data sequentially, i.e., when we read data records from the top to the bottom all the way through. In many data analysis applications, we do not need to worry about random access, concurrency or client–server mode as it is implemented in many relational databases designed for web applications. A straight file access is usually faster than executing SQL queries, and also less memory consuming.

Thus, in terms of performance, a data analysis code may not benefit from switching to the SQL databases. Using a SQL database may or may not be worthwhile. Your decision should be based on complexity of the data access, where a database software needs to be installed, and many other factors.

9.11.1 Derby SQL Database

Below we will consider the Apache ("Derby") open-source relational database. More details about Java implementation of this database can be found elsewhere [17].

First of all, let us prepare a module that keeps common information about our SQL database. This module will be necessary for creation and retrieval information in our further examples. We will consider an embedded Derby database, i.e., a database for a simple single-user Java application. In this mode, only a single application can access the database at the time. No network access is required.

Let us prepare a file "openDB.py" with the information necessary for creation of a database, such as drivers, protocol, name of database table. The name of our database will be "derbyDB", which will keep a single table with the name "location". We also set a user name and a password for this database. Below is a script that has all the necessary information:

```
from java.sql import *
from java.lang import Class
from java.util import Properties

driver = "org.apache.derby.jdbc.EmbeddedDriver"
protocol = "jdbc:derby:"
dbName="derbyDB"
table="location"

Class.forName(driver).newInstance()
```

```
props = Properties();
props.put("user", "jhepwork")
props.put("password", "secret")
```

Listing 9.53 Common module: "openDB.py"

This file will be imported for all Jython modules to be discussed below.

Now let us create a database called "derbyDB". Here we will use the Python syntax to insert several SQL statements into the table "location", which will keep information about the street name (type string) and home number (integer).

The script below does the following: (1) loads the JDBC Java driver in order to start up the Derby database. (2) Creates a table inside the database, first checking whether this table is already created. If it does exist, we remove it (see the statement `"drop table"+table`). (3) Then we insert a few records with addresses.

```
from openDB import *

scon=protocol+dbName+";create=true"
conn = DriverManager.getConnection(scon, props)
s = conn.createStatement()

try:
  s.execute("drop table "+table)
except SQLException: print "no need to remove table"

s.execute("create table "+table+\
          "(num int, addr varchar(40))")
s.execute("insert into "+table+\
          " values (1956,'Webster St.')")
s.execute("insert into "+table+\
          " values (1910,'Union St.')")
s1="update "+table+" set num=?, addr=? where num=?"
ps = conn.prepareStatement(s1)
ps.setInt(1,180)
ps.setString(2, "Grand Ave.")
ps.setInt(3, 1956)
ps.executeUpdate()
print "Updated 1956 Webster to 180 Grand"
s.close()
conn.commit()

try:
 DriverManager.getConnection("jdbc:derby:;shutdown=true")
except SQLException:  print "all done"
```

Listing 9.54 Creating a SQL database

This example shows how to use two alternative methods to insert and update the database: (1) one is based on the standard SQL statements, and the second using a prepared statement based on the method `prepareStatement()` for fast query.

It is important to close the database properly using the line:

```
"jdbc:derby:;shutdown=true".
```

The `DriverManager` should raise only one exception: `SQLException`. The database should be shut down so it can perform a checkpoint and releases its resources.

After the execution of the script above, you will see a database directory "derbyDB" with the stored information.

Now let us read the database entries. As before, we will use exactly the same common module "openDB.py". This time, however, we will set `create=false` for the argument of `DriverManager.getConnection`. Using two SQL queries, we will print all database records and will search for a record with the string "Union St.". Finally, we close the database:

```
from openDB import *

scon=protocol+dbName+";create=false"
conn = DriverManager.getConnection(scon,props)
s = conn.createStatement()

s1="SELECT num, addr FROM "+table+" ORDER BY num"
rs = s.executeQuery(s1)
while rs.next():
  print "sorted="+rs.getString(1),rs.getString(2)

s2 = "SELECT * FROM "+table+" WHERE addr='Union St.'"
rs = s.executeQuery(s2)
while rs.next():
  print "Found=",rs.getString(1),rs.getString(2)
s.close()
conn.commit()

try:
  DriverManager.getConnection("jdbc:derby:;shutdown=true")
except SQLException: print "all done"
```

Listing 9.55   Reading a SQL database

Sometimes it is useful to create a read-only database and compress it into a jar file. This allows a database to be distributed as a single file instead of multiple files within a directory. In this case, we will not be able to modify it, since the database will be represented by a single self-contained file.

The operation discussed above can easily be done with the Jython module `os`. This module will be used to create the file "derbyDB.jar" with the database directory by calling the external `jar` command (it comes with the Java installation):

```
import os
cmd="jar cMf derbyDB.jar derbyDB"
print os.system(cmd)
```

Listing 9.56   Compacting a SQL database to a jar file

Now, how can we read the file "derbyDB.jar"? This can be done in same script, but adding the extra line after the `import` statement:

```
protocol ="jdbc:derby:jar:("+dbName+".jar)"
```

Try to execute the modified script. After this, you should be able to read the database stored in this jar file.

But now we want to do something else. Since this book is on numeric computations, we will show how to store binary data in such a database. And when we talk about binary data, we mean objects that are common for numerical computations, such as 2D arrays, or even histograms.

First, let us prepare a common module that helps write objects to bytes (the method `putObject(h)`), and reads bytes back (the method `getObject()`)

```
from java.io import *

def putObject(h): # convert to bytes
    bStr=ByteArrayOutputStream()
    objOut=ObjectOutputStream(bStr)
    objOut.writeObject(h)
    objOut.flush()
    objOut.close()
    bStr.close()
    return bStr.toByteArray()

def getObject(rs,key): # read an object using key
    data = rs.getBytes(key)
    bInStream=ByteArrayInputStream(data)
    objIn= ObjectInputStream(bInStream)
    return objIn.readObject();
```

Listing 9.57   A module to work with binary data

Now let us create a new database with a 2D array and a histogram after converting them to binary arrays using `putObject(h)` function. Of course, we should import the module "utilDB.py" before using this function. Our database has two fields: one is "name" (string type) and second is "data" (blob type with the maximum size of 16 MB).

```
from openDB import *
from utilDB import *
from jhplot import *
```

```
scon=protocol+dbName+";create=true"
conn = DriverManager.getConnection(scon, props)
s = conn.createStatement()

p1=P1D("test")
p1.add(1,2); p1.add(2,3)

h1=H1D("histo",10,-1,1)
h1.fillGauss(100,0,1)

try:
  s.execute("drop table "+table)
except SQLException: print "No need to remove table"

s.execute("create table "+table+\
          "(name varchar(40), data blob(16M))")
pstmt = conn.prepareStatement("Insert Into "+table+" Values
    (?, ?)")
pstmt.setString(1, "p1d")
pstmt.setBytes(2, putObject(p1))
pstmt.execute()
conn.commit()

pstmt = conn.prepareStatement("Insert Into "+table+" Values
    (?, ?)")
pstmt.setString(1, "histogram")
pstmt.setBytes(2, putObject(h1))
pstmt.execute()
conn.commit()
s.close()
conn.commit()

try:
 DriverManager.getConnection("jdbc:derby:;shutdown=true")
except SQLException: print "all done"
```

Listing 9.58   Writing a histogram and 2D array to a database

Now we can read the histogram and 2D array back using getObject() method:

```
from openDB import *
from utilDB import *
from jhplot import *

scon=protocol+dbName+";create=false"
conn = DriverManager.getConnection(scon, props)

query="SELECT * FROM "+table
s = conn.createStatement()
rs = s.executeQuery(query)
while (rs.next()):
     id = rs.getString("name")
     obj = getObject(rs,"data")
```

```
      print obj.toString()
s.close()

try:
  DriverManager.getConnection("jdbc:derby:;shutdown=true")
except SQLException: print "all done"
```

Listing 9.59   Reading a histogram and 2D array from the database

This script prints both objects.

So far we considered databases that exist in form of files, i.e., employ a disk storage. In order to create databases with fast response that is not limited by the disk I/O, one can create a database whose data is stored in the main computer memory. Working with data in memory is much faster than writing to and reading from a file system! In-memory databases are also useful for testing applications or processing temporary (transient) data.

We use the following connection URL to create an in-memory database:

```
protocol="jdbc:derby:memory:"
```

You can do this by changing the string "protocol" in the module "openDB.py". With this modification, the previous example does not shut down the database before reading the table.

9.11.2   HyperSQL Database

HyperSQL (or HSQLDB) [18] is another relational database engine written in Java. The library is small, fast, and multithreaded. It supports the standard SQL features and offers databases with in-memory and disk-based tables. This database is a part of many popular applications, such as Open Office (or Libre Office).

Let us make an example which is similar to the Derby engine.

```
from java.sql import *
from java.lang import Class
from java.util import Properties

# load HSQL Database Engine JDBC
Class.forName("org.hsqldb.jdbc.JDBCDriver")
db="db_file"
conn = DriverManager.getConnection("jdbc:hsqldb:"+db+";create=
    true","SA","")
s = conn.createStatement()

table="price"
try:
  s.execute("drop table "+table)
```

```
except SQLException:
  print "no need to remove table"
s = conn.createStatement()
s.execute("create table "+table+\
          "(num int, addr varchar(40))")
s.execute("insert into "+table+\
          " values (1956,'Webster St.')")
s.execute("insert into "+table+\
          " values (1910,'Union St.')")
conn.commit()
st=conn.createStatement();
# writes out to files and performs clean shuts down
st.execute("SHUTDOWN");
conn.close()
```

Listing 9.60   Creating and filling a HSQLDB database

We will stop here since you can read data exactly in the same manner as in the case of the Derby database.

9.11.3   SQLite Database

SQLite implements a self-contained transactional SQL database engine which can be embedded into client programs. It is not a clientserver database engine as Derby considered before. Rather, SQLite can be a part of an application, and it requires much less configuration than client–server databases.

When it comes to Java, the SQLJet library [19] can help you work with SQLite databases. This library does not support SQL queries, instead it has a lower level API to work with data.

Let us give an example to show this. First, we create an SQLite database with a table "Price" and two fields "name" and "value". Then we build an index "idx" based on these two fields. Then we add several entries (name, price):

```
from org.tmatesoft.sqljet.core.table import SqlJetDb
from org.tmatesoft.sqljet.core import SqlJetTransactionMode
from java.io import File

dbFile=File("data.sqlite")
dbFile.delete()
db=SqlJetDb.open(dbFile, True)
db.getOptions().setAutovacuum(1)
db.beginTransaction(SqlJetTransactionMode.WRITE)
db.getOptions().setUserVersion(1)
db.createTable("CREATE TABLE Price (name, value);")
db.createIndex("CREATE INDEX idx ON Price(name,value);")
table = db.getTable("Price")
table.insert("Kitaev", "100$")
table.insert("Steve", "200$")
```

```
table.insert("Jan", "1000$")
db.commit()
db.close()
```

Listing 9.61   Creating a simple SQLite database

Again we assume that the reader knows the basics of SQL language.

Now let us read this database. First we will print the number of rows (3) and fields (2), and then retrieve one entry using "low-level" query (i.e., without SQL language). For this, we use the index "idx" to retrieve the information about "Jan":

```
from org.tmatesoft.sqljet.core.table import SqlJetDb
from org.tmatesoft.sqljet.core import SqlJetTransactionMode
from java.io import File

dbFile=File("data.sqlite")
db = SqlJetDb.open(dbFile, False)
print db.getSchema()
db.beginTransaction(SqlJetTransactionMode.READ_ONLY)
table=db.getTable("Price")
cursor=table.open()                  # open table
print "Nr of fields=",cursor.getFieldsCount()
print "Nr of rows  =",cursor.getRowCount()
cursor=table.lookup("idx", "Jan") # info about Jan
print cursor.getString("name"),cursor.getString("value")
db.close()
```

Listing 9.62   Reading a SQLite database

Run this script and you will get a pretty good idea about how to work with such a database. Of course, refer the original documentation of this Java library for more details.

9.12   Miscellaneous Input–Output Topics

9.12.1   Building List of Files

Let us come back to the example given in Sect. 2.16 where we wrote a small Jython script to transverse all subdirectories in order to collect files with a given file extension. This time, the DMelt Java library will be used which significantly simplifies the code, reducing it to a single-line statement.

Instead of creating the function walk() shown in Sect. 2.16, one can use a static class from DMelt called FileList as in this example:

```
>>> from jhplot.utils import *
>>> files=FileList.get("dir",redex)
```

This creates a list with file names in the input directory, `"dir"`, and a Java regular expression string `redex` using the syntax from the standard `java.util.regex` package. For the example discussed in Sect. 2.16 we should use `redex=".dat$"`.

There are a few advantages in using this approach: (1) it is significantly faster than when using the Python loop implemented in the `walk()` function; (2) one can use a powerful Java regular expression engine; (3) Finally, it requires only one line of the code to scan over all directories to build the file list.

9.12.2   Reading Configuration Files

To facilitate efficient data processing, especially if one needs to run the same program multiple number of times with different initial conditions, it is often necessary to pass some initial values to this program at runtime. Similarly, this is necessary for numeric programs that read initial parameters for calculations from a file. Changing the conditions will only require editing text in the input file rather than editing and recompiling the source code.

It is very convenient to use the so-called Java configuration (or property) files. Unlike binary files, the configuration files should be readable by a human. This has a significant advantage over the Java serialization mechanism (unless you use the XML serialization, and know how to go around in editing XML files).

Below we will show several approaches to read the configuration files. One approach will be based on a pure-Python module, while the second one will use Java.

9.12.2.1   Configuration Files Using Jython

Let us create a small file with the name "jythonapp.conf" with several entries as in this example:

```
# this is a comment
# Nr of events to process
events = 1000
# release version
release = 1.1
# input file
input = data.txt
```

Listing 9.63 Example of a configuration file

Each line of this configuration file has the format: "name = value". The white spaces between these elements are ignored during reading such files. A configuration file can also contain comments: by default, the "#" character at the start of each line with a comment text.

How can you read such files in Python/Jython? We will use a small Python module based on the package `ConfReader` [20]. The function that reads conflagration files is in the module "Conf.py" which is located in the directory "system" inside the directory "macro". The module is rather flexible: you can change the default syntax of your configuration file rather easily.

The next example code reads the configuration file, parses all text lines and prints the input values and their types. Do not forget to import the module "Conf.py" before running this example. The easiest approach is to copy the file "Conf.py" to the directory where the example code is located.

```
from Conf import *
try:
  config = ConfReader("jythonapp.conf")
except IOError:
  print "Cannot read configuration file"

config.set("release", post=float)
config.set("events", post=int)
config.set("input", default="/tmp")

try:
    config.parse()
except ConfigMissingError, why:
    print "Missing config", why
except ConfigPostError, why:
    print "Postprocessing Error", why

print config.release, type(config.release)
print config.events, type(config.events)
print config.input,  type(config.input)
```

Listing 9.64   Reading configuration file

The parameters of the configuration file are retrieved with the help of the `set(str, option)` function. It takes two arguments. The first argument is a string which contains a parameter name. The second, `option`, should be in the form `post=value` (to be discussed below). The example contains several exception statements which have briefly been discussed in Sect. 2.14.

Running the script shown above prints:

```
1.1 <type"float">
1000 <type"int">
data.txt <type"str">
```

As you can see, the program correctly identifies the input values and their type.

Let us come back to the set (str, option) function. As mentioned, the second argument of this function should be in the form post=value, where the attribute value can have the following values:

default   the value returned if the config is not found;
post   post-processing function to use; can be a lambda form or any function.
required   set to 1 if config is required;
list   set to 1 if config should always be returned as a list. Multiple entries
           will be appended to the list, if this is not a list but a set; otherwise, the
           last entry will be used.

Optionally, the configuration file can be arranged in sections of data, which can be used to organize references to multiple resources. Please read the documentation given in the file "Conf.py".

9.12.2.2   Reading Java Configuration Files

One can also read the configuration files using the Properties class from the Java package java.util. The Properties can be used to save a property file in a stream. Each key and its corresponding value in the property list is a string.

Below we show an example illustrating how to read the configuration file created in the previous section. We will use the strings "release", "events" and "input" as the keys. Then we print the loaded strings in the ISO 8859-1 character encoding. Since we expect "events" to be an integer number, we convert it to the integer type and modify its value. Finally, we set a new property using "events" as the key and save it to a file with the optional comment "New settings" (one can use also "None" in case of no comments).

```
from java.util import Properties
from java.io  import *

p=Properties()
p.load(FileInputStream("jythonapp.conf"))
print  p.getProperty("release")
print  p.getProperty("input")
print  p.getProperty("events")

# save new file with x10 more events
ev=int(p.getProperty("events"))
ev=10*ev
p.setProperty("events",str(ev))
p.store(FileOutputStream("new.conf"),"New settings");
```

Listing 9.65   Reading a configuration file

One should note that, in the case of Java, the file extension of the property files is "properties" by convention, rather than that given in this example.

9.13 Summary

It may well be that you get confused by the significant number of I/O formats supported by DMelt. Below we will summarize the DMelt I/O and give some guiding tips about what exactly to use in certain situations.

9.13.1 Dealing with Single Objects

If we need to save one object at the time (like a histogram, function, or a P-type array), use their native methods. For example, to write a DMelt data object into a file, use the `toFile()` method, see Sects. 4.2.8 and 7.6. For reading data, one can use the corresponding constructors that accept the name of the file with input data for object initialization. This also includes a compression mechanism, see the method `readGZip()` in Sect. 4.2.8. The good thing about this approach is that: (1) it is fast; (2) data can be produced using other programming languages, as long as our codding strictly follows the native DMelt format.

A slower I/O approach would be to use the Java serialization, i.e., serializing separate objects with the methods of the class `jhplot.IO`. By default, files will be compressed. One can also write and read data stored in XML files using the method `writeXML()` and `readXML()`. This will be the slowest approach, but the advantage is that the XML files are human-readable and human-editable. Finally, it is easy to read such files using applications implemented in other programming languages.

9.13.2 Dealing with Object Collections

To read and save collections of DMelt objects, use the serialization methods of the class `jhplot.IO`. This can be done "in one go", if all Java objects are put either in a list or map. This approach also allows dealing with the XML-type formats. As we have already mentioned, this is a human-readable, self-describing, and platform-neutral protocol.

For large and complicated data structures, it makes sense to use the `HFile` class, see Sect. 9.3. In most cases, objects can be viewed using a GUI browser. Data in such files are compressed by default. As before, one can serialize data in XML using the class `HFileXML`.

An arbitrary collection of structural data can also be written using the Google's Protocol Buffer format discussed in Sect. 9.8. It is a compact way of encoding data in a binary format. The main advantage of this format is that data can be processed by programs implemented in other programming languages.

If we are interested in the DMelt objects only, use the class `PFile`. This approach has several advantages compared to the Java serialization discussed in Sect. 9.4. The class `PFile` writes data in the form of compressed records based on the Google's Protocol Buffer format. This approach encodes high-level DMelt objects (like histograms) in a platform-independent way. This means that data can be generated by C++ programs using the package CBook [7] and processed using the `PFile` class.

The Java class `PFile` deals with a rather broad range of data structured in "events" and "records". The `PFile` can be used to read events or records, without uncompressing the whole file. This feature is particularly useful for reading data in parallel using multi-core processors (this will be discussed in Sect. 12.4).

9.13.3 Text Files

The usage of plan text files makes sense if you need to deal with data that can be easily viewed. But they might not be the most efficient for large data volumes. It can be a problem with the disk storage too if no compression is used.

One can use a pure-Python I/O approach, or calling Java classes to write and read text files. See Sect. 2.15 describing the Python I/O modules. The pure-Python method is generally not recommended for large data files, but it is very useful when it comes to the simplicity.

The class `HBook` is mainly designed for an XML-like output, but it is more optimized to keep large data sets, removing unnecessary tags for array elements, compared to the usual XML serialization in the standard Java. The data can be structured. This class is useful for storing complex data objects from DMelt, such as histograms and arrays. No compression is used in this XML approach.

The format CSV is only useful for keeping data in tables. The only advantage of this approach is that it is simple, and data can easily be converted by spreadsheet applications.

9.13.4 Databases

As we discussed in this chapter, if one needs to find an association between a key and a data object, one can use databases. The usage of databases makes sense in the situations when we need to use indexing, multithreading, permissions, and network access. The databases can also update data easily, with data integrity checks and concurrent access.

In the simplest, nonpersistent case, use Python dictionaries or Java maps. For storing information in persistent state, one can use a database based on the class `HDataBase`, see Sect. 9.10.1, or the MapDB database. Note that one can use a similar database approach using the `PFile` class. As discussed earlier, it can also be used for a direct association between a stored object inside a file and its textual key (which is just object title). For complicated input queries, one can use the Apache Derby or HSQLDB SQL databases as discussed in Sect. 9.11.

References

1. Richardson C, Avondolio D, Vitale J, Schrager S, Mitchell M, Scanlon J (2005) Professional Java, JDK 5 edn, Wrox, Birmingham
2. Arnold K, Gosling J, Holmes D (2005) Java (TM) programming language, The (4th edn) (Java series), Addison-Wesley Professional, Boston
3. Flanagan D (2005) Java in a nutshell, 5th edn, O'Reilly Media Inc, Sebastopol
4. Eckel B (2006) Thinking in Java, 4th edn, Prentice Hall PTR, New Jersey
5. Bloch J (2008) Effective Java, 2nd edn, (The Java series), Prentice Hall PTR, New Jersey
6. Google protocol buffers project. http://code.google.com/p/protobuf/
7. Chekanov S, CBook histogram library. http://jwork.org/dmelt/cbook
8. EDN, extensible data notation format. https://github.com/edn-format/
9. Gonnerman C, Navy DIF file handler. http://newcenturycomputers.net/projects/dif.html
10. Brun R, Rademakers F, Canal P, Goto M (2003) Root status and future developments, ECONF C0303241, MOJT001
11. Brun R, Rademakers F (1997) Root: an object oriented data analysis framework. Nucl Instrum Meth A389:81. http://root.cern.ch/
12. Chekanov S (2010) Scientific data analysis using Jython scripting and Java (advanced information and knowledge processing). Springer, London
13. FreeHEP Java libraries. http://java.freehep.org/
14. Apache POI—the Java API for microsoft documents. https://poi.apache.org/
15. Kotek J, MapDB. an embedded database engine. http://www.mapdb.org/
16. Smadja O, NeoDatis. an object database. http://neodatis.wikidot.com/
17. Apache derby. http://db.apache.org/derby/
18. Toussi F et al, HyperSQL. a SQL relational database software written in Java. http://hsqldb.org/
19. TMate, SQLJet. a Java implementation of a popular sqlite database. http://sqljet.com/
20. Nottingham M (1998) Confreader—configuration file reading class

Chapter 10
Probability and Statistics

Statistical methods are used for solving a wide variety of problems in natural sciences and other fields. Statistics provides an important guidance in description and interpretation of collected data. Based on results of such an interpretation, one can discover patterns in data, determine what theoretical models can describe data and what predictions can be made to predict future events, or to find connections between seemingly independent events or trends.

This chapter discusses several computational aspects of the statistical methods frequently used for interpretation of data. As usual, the focus of this chapter is to show how to perform computations on the Java platform, rather than diving into detailed description of the mathematical concepts of statistics.

10.1 Descriptive Statistics

To describe the basic features of collected data, methods of descriptive statistics are used. They attempt a quantitative description of data in terms of basic statistical characteristics, such as mean, variance, standard deviations, and so on. Please refresh your knowledge of such terms before reading this chapter. As before, we assume that the reader knows about basic concepts of statistics, therefore, our role is to illustrate such concepts using programming codes. Refreshing your memory of the main statistical terms will help a lot in understanding many code examples that will follow after this chapter.

Let us turn from this abstract definition of descriptive statistics to concrete examples. Assume we have an array of numbers, say the Fibonacci integer sequence. We want to know about major statistical characteristics of this sequence. Let us make such calculations using the JythonShell:

© Springer International Publishing Switzerland 2016
S.V. Chekanov, *Numeric Computation and Statistical Data Analysis*
on the Java Platform, Advanced Information and Knowledge Processing,
DOI 10.1007/978-3-319-28531-3_10

```
>>> from jhplot import *
>>> p=P0D([0,1,1,2,3,5,8,13,21,34,55,89,144])
>>> print p.getStatString()
```

Run this script and you will get a detailed information about this distribution, such as the mean value, variance, standard deviation, standard error, geometric mean, skewness, kurtosis, moments up to sixth order, quantiles, etc. With some formatting, the output will look as

```
Size: 13
Sum: 376.0
SumOfSquares: 33552.0
Min: 0.0
Max: 144.0
Mean: 28.9230769
RMS:  50.8027861
Variance: 1889.7435897435898
Standard deviation: 43.47118
Standard error: 12.05673635
Geometric mean: 0.0
Product: 0.0
Harmonic mean: 0.0
Sum of inversions: Infinity
Skew: 1.52308282
Kurtosis: 1.14302725
Sum of powers(3): 3908764.0
Sum of powers(4): 5.0343876E8
Sum of powers(5): 6.8054635E10
Sum of powers(6): 9.4423980E12
Moment(0,0): 1.0
Moment(1,0): 28.92307
Moment(2,0): 2580.923
Moment(3,0): 300674.1
Moment(4,0): 3.872605E7
Moment(5,0): 5.234971E9
Moment(6,0): 7.263383E11
Moment(0,mean()): 1.0
Moment(1,mean()): -4.372570E-15
Moment(2,mean()): 1744.37869
Moment(3,mean()): 125120.318
Moment(4,mean()): 1.47952923E7
Moment(5,mean()): 1.60635741E9
Moment(6,mean()): 1.82476587E11
25%, 50%, 75% Quantiles: 2.0, 8.0, 34.0
```

```
quantileInverse(median): 0.5384615384615384
Distinct elements:
[0.0,1.0,2.0,3.0,5.0,8.0,13.0,21.0,34.0..]
Frequencies: [1,2,1,1,1,1,1,1,1,1,1,1]
```

The output of this simple command is impressive even for a professional statistician. This listing is generated by the package `cern.hep.aida` which is integrated in the `POD` Java class. Let us briefly remind a few definitions listed above: A moment of an order n for a list with N elements is given as

$$M_n = \frac{1}{N} \sum_i^N (\text{list}(i) - \text{mean})^n$$

A skewness is defined via the moments, $M_3/(M_2)^{1.5}$, while the kurtosis is $M_4/(M_2)^2$. A coefficient of variation can be computed the ratio of the biased standard deviation to the mean.

Although we have compressed all such information into a string, one should mention that one can access these statistical characteristics of data using the appropriate methods of the `POD` class designed to hold one-dimensional arrays (see Sect. 4.1).

Let us continue with this example and now we would like to return all statistical characteristics of the sample as a Python dictionary (see Sect. 2.6.3). Since this code is based on Java, and runs on the Java platform, the actual object that is returned as Python list is the Java map from the `java.util` package. We will remind that this object maps keys to values. We can do this exercise by appending the following lines that (1) create a dictionary "stat" with key/value pairs and (2) retrieve a variance of the sample using the key "Variance".

```
>>> stat=p.getStat()
>>> print "Variance=",stat["variance"]
```

The code prints "Variance = 1889.74". If not sure about the names of the keys, simply print the dictionary as "print stat". Here is a small example that prints all keys and values:

```
print stat.keys()
```

or, printing the keys together with the actual values:

```
for key in stat:
    print key , "=", stat[key]
```

This will print the following list of statistical descriptive properties:

```
sum = 376.0
mean_error = 12.0567363544
kurtosis = 1.14302725995
```

```
harmonicMean  =  0.0
moment_3_mean  =  125120.318616
sumOfInversions  =  inf
size  =  13.0
median  =  8.0
variance  =  1889.74358974
mean  =  28.9230769231
rms  =  50.8027861138
....
```

Now you can see what keys to use. The printed list is rather long, therefore we show only a few most common statistical characteristics (with a reduced numeric precision) of the one-dimensional array "p0".

Statistical characteristics of arrays can also be calculated with the Colt Java library previously discussed in Sect. 5.1.6. When it comes to the descriptive statistics, use the class DynamicDoubleBin1D for double values, or DynamicFloat Bin1D for float values. Here is a small example that analyzes an array of double values and calculates all major characteristics of the input data. We will write a few comments to help understand this code.

```
from hep.aida.tdouble.bin import *
from cern.colt.list.tdouble import *

N=1000000
M=DoubleArrayList()
for i in range(0,N):   # fill array
        M.add(i)
d=DynamicDoubleBin1D()
d.addAllOfFromTo(M,0,N-1)
print d.mean()
print d.rms()
print d.variance()
print d.standardDeviation()
print d.standardError()
print d.moment(2,10)   # moment of 2nd order with 10
print d.sum()
print d.geometricMean()
print d.harmonicMean()
print d.kurtosis()
print d.skew()
print d.sumOfLogarithms()
d.standardize(100,10)
d.sortedElements()     # sort elements
```

Listing 10.1 Descriptive statistics with Colt

As we discussed in Chap. 7, one can create the histograms that catch the most basic characteristics of data in multiple dimensions. The method getStat() discussed above also applies to the histogram classes. This is especially important if there is no particular reason to deal with large volume of data, and making "projections" of data into histograms makes more sense. Coming back to the example with the Fibonacci sequence, let us show how to extract statistical characteristics of a typical histogram:

```
>>> h=p.getH1D(10, 0, 200)    # h is histogram with 10 bins
>>> print h.getStat()         # look at statistics
>>> print h.toString()        # print on the screen
```

The code converts the array into a histogram object "h" with 10 equidistant bins in the range [0, 200], and then it prints the map with statistical characteristics. Please read Chap. 7 about how such histograms can be printed.

Now we will turn to a graphical example. We will create 100 random numbers distributed according to a normal (or Gaussian) distribution with the mean 0. The width (the standard deviation) will be set to 2. Then we will fill a histogram from this array to look at the shape of the distribution. Finally, we will create a box with statistical characteristics, mean, and RMS (the root mean square). The latter is used to calculate the spread of this distribution. Each histogram bin shows the statistical uncertainty with the vertical error bar, which is defined as the square root of the number of entries in that bin.

```
from jhplot import *

p0 = P0D()
# 100 numbers (Normal distribution)
p0.randomNormal(1000,0.0,2.0) # mean=0, sigma=2
h1=H1D("Normal",40,-10,10)
h1.fill(p0) # fill a histogram with 1D array

c1=HPlot()
c1.visible(); c1.setAutoRange()
c1.draw(h1)
c1.drawStatBox(h1) # show statistics
stat= h1.getStat()
print stat["mean"],"+-",stat["mean_error"]
```

Listing 10.2 Showing statistics of a histogram

The output of this script is shown in Fig. 10.1. In addition, the script prints the mean value and its standard error. As you will see, the mean value agrees with the expected value used to create the normal distribution.

10.1.1 Comparing Data

Several data sets can easily be compared using the `getStat()` method that performs calculations of descriptive statistics. Here is a small code snippet that shows how to find any small difference between two arrays by looking at statistical characteristics of these two arrays:

Fig. 10.1 Statistical
characteristics of a histogram

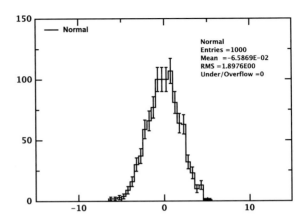

```
from jhplot import *
p1=P0D([0,1,1,2,3,5,8,13,21,34,55,89])
p2=P0D([1,1,1,2,3,5,8,13,21,34,55,89])
s1=p1.getStat()
s2=p2.getStat()
for key in s1:
  s=s1[key]-s2[key]
  if (s != 0):
      print "Diff for ",key,"=",s1[key]," vs  ",s2[key]
```

An observant reader will notice that the input arrays differ by one value, i.e., we
replaced "0" by "1" in the second array. Then we print descriptive characteristics
by calling the method getStat(). By running this example, you will see several
differences, i.e., the second array has larger mean, variance, and so on.

  If there are several histograms, it is common to compare them using the method
getStat(). This approach captures differences between data arrays using descriptive statistics. Section 10.5.5 shows how to run statistical tests on arbitrary data.

10.2   Statistical Analysis Using Python

In addition to the Java libraries provided by DMelt, one can also use third-party
libraries implemented in Python. The Python/Jython third-party packages are located
in the directory "python/packages." There are several advantages in using programs
implemented in Jython: (1) one can directly access the code with the implemented
numerical algorithms. (2) One can reuse the libraries in CPython programs.

  We remind that the directory "python/packages" is imported automatically when
using the DMelt IDE, so there is no need to worry about appending this directory
location to the "sys.path" variable. If one uses another editor, make sure that

Jython looks at the directory "python/packages" to import the third-party packages. For example, you can do it as

```
>>> import sys
>>> sys.path.append(SystemDir+"/python/packages")
```

where `SystemDir` is the directory with the installed DMelt. Below we assume that the reader uses the DMelt IDE, in which case there is no need for specifying this directory.

Here we will consider the module `"stats"` that provides basic statistical functions for Python collections. It allows calculations of simple characteristics from a Python list, such as the mean value and the standard deviation:

```
>>> from statlib.stats import *
>>> a = [1,2,3,4,5,6,7,8,10 ]
>>> print "Mean=",mean(a)
Mean= 5.1111
>>> print "Standard deviation=",stdev(a)
Standard deviation= 2.93446
```

Analogously, one can calculate other statistical characteristics, such as the median, variation, skewness, kurtosis, and moments. All such statistical characteristics can be assessed in one line of the code using the `describe(list)` function:

```
>>> a = [1,2,3,4,5,6,7,8,10]
>>> stat=describe(a)
>>> print "N=",stat[0]
N= 9
>>> print "tuple=",stat[1]
tuple= (1, 10)
>>> print "mean=",stat[2]
mean= 5.1111
>>> print "standard deviation=",stat[3]
standard deviation= 2.93
>>> print "skewness=",stat[4]
skewness= 0.199
>>> print "kurtosis=",stat[5]
kurtosis= 2.00
```

The output values have been truncated to fit the page width.

In the next example, we will extend this code snippet by including calculations of various characteristics of a list with 100 numbers. Moments about the mean for this list will be calculated in a loop (up to the order nine):

```
from statlib.stats import *

a=range(100)
print "geometric mean=",geometricmean(a)
print "median=",median(a)
```

```
print  "variation=",variation(a)
print  "skew=",skew(a)
print  "kurtosis=",kurtosis(a)
for  n  in  range(2,10):
     print  "moment  order  "+str(n)+":",moment(a,n)
```

Listing 10.3 Statistical analysis of lists

The output of this code snippet is given below:

```
geometric mean= 0.0
median= 49.005
variation= 58.315
skew= 0.0
kurtosis= 1.799
moment order 2: 833
moment order 3: 0.0
moment order 4: 1249
moment order 5: 0.0
moment order 6: 2230
moment order 7: 0.0
moment order 8: 4.335+12
moment order 9: 0.0075
```

Again, we have reduced the precision of the output numbers to fit the page width.

The module allows calculations of correlation coefficients between two lists. Let us generate two correlated lists using a Gaussian distribution and estimate their correlations using several tests:

```
from  statlib.stats  import  *

from  random  import  *
ran=Random()
mu,sigma=2.0,3.0

x,y=[],[]
for  i  in  range(100):
   t=ran.gauss(mu,sigma)
   x.append(t)
   y.append(t*2+ran.gauss(mu,sigma))

print  stdev(x),  "+/-",sterr(x)
print  mean(y),"+/-",sem(y)
print  "Covariance=",lcov(x,y)
print  "Pearson (correlation coeff. prob)=",lpearsonr(x,y)
print  "Spearman rank-order correlation=",lspearmanr(x,y)
```

Listing 10.4 Calculation of correlation coefficients

The output is shown below:

```
2.917 +/- 0.2917
6.610 +/- 0.6750
Covariance= 17.126
Pearson (correlation coefficient,prob)=(0.869,8.945e-32)
Spearman rank-order correlation=(0.848,8.361-29)
```

For clarity, and in order to fit the output to the page width, we truncated several numbers to three decimal digits.

The module "stats" has many statistical tests (including the Anova test) and functions. Please read the package description (and the Python source code) to learn more.

10.3 Random Numbers

A random number is a number chosen by chance from a specified distribution. It is an essential concept in many scientific areas, especially for numeric computations and simulations of complex systems using Monte Carlo methods.

For a set with random numbers, no individual number can be predicted from knowledge of any other number or group of numbers. However, sequences of random numbers in a computer simulation eventually contain repeated numbers after generation of many millions of random numbers. Thus, it is only a good approximation to say that the numbers are random, and the definition "pseudo-random" is more appropriate.

Another notion which is usually associated with a sequence of random numbers is the so-called "seed" value. This is a number that controls whether the random number generator produces a new set of random numbers after the code execution or repeats a certain sequence of random numbers.

For debugging of programs, it is often necessary to start generating exactly the same random number sequence every time you start the program. In this case, one should initialize a random number generator using the same seed number. In many code examples of this book, we use the Java class Random(L) to create a sequence of random numbers, where L is a long integer. This means we initialize the generator from the same seed, i.e., our sequence is reproducible every time we run the code.

The seed must be changed for each run if you want to produce completely different sets of random numbers every time the program is executed. Usually, this can be done by generating a new seed using the current date and time, converted into an integer value. This is also the case when calling the class Random() without arguments.

10.3.1   Using Random Numbers

This section is going to be short, since we have discussed this topic in Sect. 2.8 in the context of the Python language. The standard Python module implementing a random number generator is called "random". It must be imported using the usual statement import.

As before, it is advisable to use the standard Java libraries to create arrays with random numbers where possible, instead of filling lists with random values using Python loops. There are several reasons for this: (1) less chances that a mistake can be made; (2) programs based on the standard Java libraries can be faster; (3) code with calls to Java libraries can be reused by the standard Java programs, or by programs based on alternative scripting languages supported by Java.

First, we discuss the most common classes to generate random numbers in Java. Random numbers provided by the Java API have already been used in the previous sections. Let us remind that the class Random can be used to generate a single random number. Below we check the methods of this class:

```
>>> from java.util import *
>>> r=Random()       # seed from the system time.
>>> r=Random(100L)  # user defined seed=100L
>>> dir(r)
[.. "nextDouble", "nextFloat", "nextGaussian",
... "nextInt", "nextLong" ..]
```

In the first definition, the default seed comes from the computer system time. In the second example, we initiate the random sequence from an input value to obtain reproducible results for every program execution.

Below we describe the most common methods to generate random numbers. As usual, "i" denotes an integer value, "l" represents a long integer value, "d" means a double value, while "b" corresponds to a Boolean value.

i=r.nextInt(n)	random integer value ≥ 0 and $\leq n$
i=r.nextInt()	random integer value (full range)
l=r.nextLong()	random long value (full range)
d=r.nextDouble()	random double value ≥ 0.0 and ≤ 1.0
b=r.nextBoolean()	random Boolean value, true (1) or false (0)
d=r.nextGaussian()	random double value from the normal (Gaussian) distribution with the mean 0 and the standard deviation 1

To build a list containing random numbers, invoke a Python loop. For example, this code typed using the JythonShell creates a list with Gaussian random numbers:

```
>>> from java.util import *
>>> r=Random()
>>> g=[]
>>> for i in range(100):
>>> ... g.append(r.nextGaussian())
```

It was discussed earlier that Python loops are not particularly fast; therefore, it is recommended to use the predefined DMelt methods returning lists with calculations. In addition, using the predefined methods to build collections with random numbers guarantees that the code is sufficiently short and free of errors.

Now our goal is to illustrate that using high-level Java libraries is more efficient than using long loops in Jython codes. Look at the examples below: the first program is rather inefficient (and long), while the second code snippet is a factor five faster (and shorter). Both programs fill histograms with random numbers in the range between 0 and 100 and show them on a canvas. As usual, we write the program in Python, but the actual engine behind the code shown below is Jython/Java:

```
from java.util import Random
from jhplot   import *
from time import clock

start=clock()
h1 = H1D("Uniform distribution",100, 0.0, 100.0)
rand = Random()
for i in range(2000000):
    h1.fill(100*rand.nextDouble())
c1 = HPlot()
c1.visible()
c1.setAutoRange()
c1.draw(h1)
send=clock()
print "Time elapsed = ",send-start," seconds"
```

Listing 10.5 Creating random numbers in Python loops

The program below does the same, but it is faster by a large factor:

```
from jhplot   import *
from time import clock
from jhplot.math.StatisticSample import randUniform

start=clock()
h1= H1D("Uniform distribution",100, 0.0, 100.0)
h1.fill( randUniform(2000000,0.0,100.0) )
c1=HPlot()
c1.visible()
c1.setAutoRange()
c1.draw(h1)
send=clock()
print "Time elapsed = ", send-start," seconds"
```

Listing 10.6 Random numbers using high-level DMelt methods

The reason for this is that we use Java high-level methods instead of Python loops, so the engine behind the calculation is significantly more optimized for speed. Here we recall that the method `fill()` of the class `H1D` accepts not only separate values, but also arrays of different types—in this case, we build such array on the fly using the method `randUniform()`

10.3.2   Random Numbers in Colt

The Colt package provides a comprehensive list of methods to create random numbers. The Java classes necessary to build random numbers come from the package cern.jet. As example, let us consider a generation of reproducible random numbers using the MersenneTwister class from the sub-package random.engine. The macro below creates an array P0D with random numbers and then prints the statistical summary of a Gamma distribution:

```
from cern.jet.random.engine import *
from cern.jet.random import *
from jhplot import *

engine=MersenneTwister()
alpha=1;   lamb=0.5
gam=Gamma(alpha,lamb,engine)
p0=P0D()                       # fill 1D array
for i in range(100):
    p0.add(gam.nextDouble())
print p0.getStat()
```

Listing 10.7   Colt random numbers

Here we used the so-called "Mersenne-Twister" algorithm, which is one of the strongest uniform pseudo-random number generators. We did not specify any argument for the engine; therefore, the seed is set to a constant value and the output is reproducible next time you run the script. One can use the current system date for a seed to avoid reproducible results:

```
>>> import java
>>> engine=MersenneTwister(new java.util.Date())
```

Learn about all possible methods of this package as usual:

```
>>> import cern.jet.random
>>> dir(cern.jet.random)
```

The above command prints the implemented distributions:

```
Beta, Binomial, BreitWigner, BreitWignerMeanSquare,
ChiSquare, Empirical, EmpiricalWalker, Exponential,
ExponentialPower, Gamma, Hyperbolic, HyperGeometric,
Logarithmic, NegativeBinomial, normal, Poisson,
PoissonSlow, StudentT, Uniform, VonMises, Zeta
```

All these classes operate on a user supplied uniform random number generator.

Once you know which random number is necessary for your program, use the code assist or Java API documentation to learn more.

There is one special distribution you have to be aware of. One can generate an array of random numbers that follow a predefined probability distribution function (PDF). Such a distribution is called "Empirical". The PDF should be provided as an array of positive numbers. The function can be in the form of relative probabilities, but the absolute probabilities are also accepted. If LINEAR_INTERPOLATION constant is set, a linear interpolation within the bin is computed, resulting in a constant density within each bin. When NO_INTERPOLATION is passed, no interpolation is performed and the result is a discrete distribution. Let us see this:

```
from cern.jet.random.engine import *
from cern.jet.random import *
engine = MersenneTwister()
pdf=[1.,4.,4.,4.,3.,2.,1.,1.,1.]
enterpolation=Empirical.LINEAR_INTERPOLATION
em=Empirical(pdf,enterpolation,engine)
from jhplot import P0D
p0=P0D()
for i in range(100):
        p0.add(em.nextDouble())
print p0.getStat()
```

Listing 10.8 An empirical distribution of random numbers

Look also at the class EmpiricalWalker which implements the so-called Walker's algorithm.

Now let us put everything together. In the next example we create a few random distributions by filling histograms.

```
from jhplot import *
from cern.jet.random.engine import *
from cern.jet.random import *

c1=HPlot("Canvas",700,500,3,2)
c1.visible(); c1.setGTitle("Random Distributions")
eng=MersenneTwister()
N=100000 # number of random numbers

def plot(c1,h1,r,N): # density distribution:
    c1.setAutoRange()
    h1.setFill(True)
    w=1.0/(N*h1.getBinSize())
    for i in range(N):
        h1.fill(r.nextDouble(),w)
    c1.draw(h1)

c1.cd(1,1)
r=Gamma(1,0.5,eng); h1=H1D("Gamma",30,0,15)
plot(c1,h1,r,N)
```

```
c1.cd(2,1)
r=Binomial(10,0.5,eng);  h1=H1D("Binominal",15,0,15)
plot(c1,h1,r,N)

c1.cd(3,1)
r=Poisson(5,eng);  h1=H1D("Poisson",20,0,20)
plot(c1,h1,r,N)

c1.cd(1,2)
r=StudentT(5,eng);  h1=H1D("Student",20,0,10)
plot(c1,h1,r,N)

c1.cd(2,2)
h1=H1D("NBD",20,0,20);  r=NegativeBinomial(10,0.3,eng)
plot(c1,h1,r,N)

c1.cd(3,2)
r=Logarithmic(0.3,eng);  h1=H1D("Logarithmic",5,0,10)
plot(c1,h1,r,N)
```

Listing 10.9   Creating random density distributions in Colt

There is some repetition in this code, but making it shorter will lead to a more obscure code. The output of this code is shown in Fig. 10.2. The example is based on everything we know so far: We created the probability densities using random numbers and histograms. We recall Sect. 7.1.1 where it was said that the integral of a probability density is 1. The histograms are filled with the weights that reflect the bin sizes and the number of entries, so the plotted distribution is a density distribution, i.e., it is integrated to 1. We can check this by adding the statement:

```
>>> print h1.integral()
```

It will print "1" for the shown distributions.

10.3.3   Other Packages with Random Numbers

jMathTool [1] classes further extend the Java random number generators. This library is included into the package "jhplot.math".

The example below shows how to generate random numbers using the jMathTool package:

```
>>> from jhplot.math.Random import *
>>> r=rand()              # random number between 0 and 1
>>> i=randInt(min,max)   # integer in the range [min,max]
```

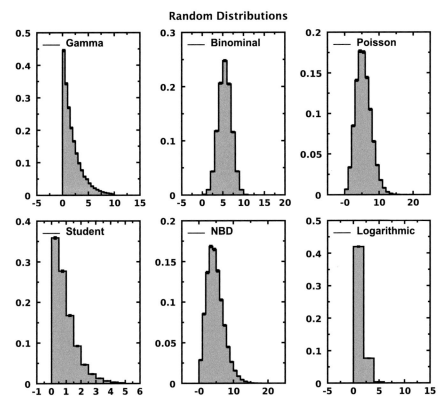

Fig. 10.2 Showing various density distribution functions using histograms filled from random numbers

Below we will show other possible options:

uniform(mi,ma)	a random number between mi and ma.
dirac(d[], p[])	a random number from a discrete random variable, where d[] array with discrete values, and p[] is the probability of each value.
normal(m,s)	a random number from a Gaussian (Normal) distribution with the mean ("m") and the standard deviation ("s").
chi2(i)	a random number from a ξ^2 random variable with "i" degrees of freedom.
logNormal(m,s)	a LogNormal random variable with the mean ("m") and the standard deviation ("s").
exponential(lam)	a random number from an exponential distribution (mean = 1/lam, variance = 1/lam**2).
triangular(mi,ma)	a random number from a symmetric triangular distribution

`triangular(mi,med,ma)`	a random number from a nonsymmetric triangular distribution ("med" means a value of the random variable with ma density)
`beta(a,b)`	a random number from a Beta distribution. "a" and "b" are the first and second parameters of the Beta random variable
`cauchy(med,s)`	a random number from a Cauchy distribution (Mean = Inf, and Variance = Inf). "med" is a median of the Weibull random variable, and "s" is the second parameter of the Cauchy random variable.
`weibull(lam,c)`	a random number from a Weibull distribution. "lam" is the first parameter of the Weibull random variable, and "c" is the second parameter of the Weibull random variable.

Finally, one can generate a random number from an analytic function using the known rejection method. This requires building an `F1D` function first and then passing its parsed object to the `rejection()` method. Below we show how this can be done:

```
>>> from  jhplot   import  *
>>> from  jhplot.math.Random  import  *
>>> f=F1D("x*exp(x)",1,2)
>>> p=f.getParse()
>>> print  rejection(p,15,1,2)
1.4
```

The method `rejection()` takes three arguments: a parsed function, a maximum value of the function (15 in this case), and a minimum and maximum value for the abscissa. The method returns a random number between 1 and 2, since these numbers have been specified as the arguments in the `rejection()` method.

Random numbers can also be generated using the Apache common math package. A random generator can be initialized using the Java class called `RandomData Generator()`. After the initialization, call its methods to generate a random number. The code below shows how to generate a single random number from different distributions:

```
>>> from org.apache.commons.math3.random import *
>>> r=RandomDataGenerator()
>>> dir(r)                     # check all methods
>>> d=r.nextBinomial(10,1)     # Binomial (trials=10, prob=1)
>>> d=r.nextUniform(0,1)       # Uniform number in [0,1]
>>> d=r.nextExponential(1)     # Exponential with the mean 1
>>> d=r.nextGaussian(0,1)      # Gaussian (mean=0 and sigma=1)
>>> d=r.nextGamma(1,2)         # Gamma (shape=1, scale=2)
>>> d=r.nextPoisson(1)         # Poisson with the mean 1
>>> s=r.nextHexString(10)      # hex string of the length 10
```

Check the corresponding API documentation for more options. One can reseed the random numbers using an integer argument `seed` of the constructor. This integer value sets the seed of the generator to the current computer time in milliseconds. One can also reseeds the random number generator with the supplied seed using the method `reSeed(i)`, where `"i"` is an arbitrary integer number.

The fact that one can define characteristics of random numbers in the same method that returns the random value is quite useful. For example, one can easily convolute several random distributions, and create a new random distribution. To be more specific, let us convolute a Poisson distribution with a Gamma distribution:

```
from org.apache.commons.math3.random import *
from jhplot import *

r=RandomDataGenerator()
c1=HPlot()
c1.visible()
c1.setRangeX(0,100)
h1=H1D("Gamma x Poisson",100,0,100)
tot=1000000             # total events
shape=10;   scale=3.0 # parameters of Gamma distribution
r=RandomDataGenerator()
for i in range(tot):
  mean=r.nextGamma(shape,scale)
  value=r.nextPoisson(mean)
  h1.fill(value)
c1.draw(h1)
```

Listing 10.10 Convoluting Gamma with Poisson random numbers

You can see that the mean of a Poisson distribution is distributed randomly using a Gamma distribution with fixed parameters. The resulting histogram is shown in Fig. 10.3.

There are also the so-called "secure" methods, much slower than those shown above. A secure random sequence has the additional property that knowledge of values generated up to any point in the sequence does not make it any easier to predict subsequent values. Such values are useful for a cryptographically secure random sequence. To see this, check the method `nextSecureHexString(10)` which generates a "secure" random string of size ten.

10.4   Random Sampling

When one needs to create a large array with random numbers obeying some probability distribution, it is inconvenient (and slow!) to create data containers using Python loops as shown at the beginning of this chapter. Instead, one should use the Java libraries implementing generations of arrays with random numbers.

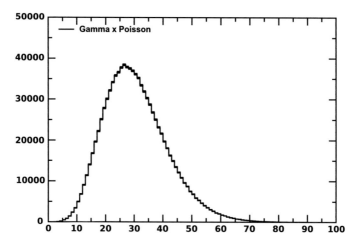

Fig. 10.3 A histogram filled with random numbers as the result of a convolution of a Poisson distribution with a Gamma distribution

We have discussed in Sect. 4.1 how to build a POD and populate it with random numbers:

```
>>> from jhplot import *
>>> p0= POD()
>>> p0.randomUniform(1000,0.0,1.0)  # Uniform distribution
>>> p0.randomNormal(1000,0.0,1.0)   # Gaussian distribution
```

In both cases we fill an array with 1000 numbers populated by either a uniform random generator (in the range [0–1]) or a Gaussian generator with the mean zero, and with the standard deviation equals unity.

Below we will consider several options to fill the POD container with custom random numbers.

10.4.1   Methods of 1D Arrays

Static methods of the package "jhplot.math" can be useful to generate arrays with random numbers. We remind that the package is based on the original class of jMathTools [1].

We will start from a simple example of how to create an array of size 10 with random integer numbers distributed between 0 and 100. After creation, we print out the array:

```
>>> from jhplot.math.StatisticSample import *
>>> a=randomInt(10,0,100)
>>> print a
array("i",[93, 19, 70, 36, 55, 43, 52, 50, 67, 38])
```

Analogously, `randUniform(N,min,max)` generates `"N"` random numbers uniformly distributed between `"min"` and `"max"`.

Below we show a list of methods designed to generate arrays with random numbers:

`randomDirac(N,d[],p[])`	a random array from a discrete random variable, where d[] is the array with discrete values and p[] is the probability of each value.
`randomNormal(N,m,s)`	a Gaussian (or normal) random number with the mean ("m") and standard deviation ("s")
`randomChi2(N,i)`	a random array with ξ^2 random numbers with "i" degrees of freedom.
`randomLogNormal(N,m,s)`	a random array with lognormal random numbers.
`randomExponential(N,lam)`	an array with an exponential random number. (mean = 1/lam, variance = 1 / lam^2).
`randomTriangular(N,m1,m2)`	an array with symmetric triangular random variable in the range [m1, m2]
`randomTriangular(N,m1,m,m2)`	an array from a nonsymmetric triangular random distribution in the range [m1, m2] ("m" indicates a value of the random variable with the maximal density)
`randomBeta(N,a,b)`	an array with Beta distribution, "a" and "b" are first and second parameters of the Beta random variable
`randomCauchy(N,m,s)`	an array from a Cauchy random distribution (Mean = Inf, and Variance = Inf). "m" is the median of the Weibull random variable, and "s" is second parameter of the Cauchy random variable.
`randomWeibull(N,lam,c)`	an array with Weibull random variables. "lam" is the first parameter of the Weibull random variable ("lam"), and "c" is the second parameter of the Weibull random variable.

In all these cases, the names of the methods are exactly the same as those shown in Sect. 10.3.3. The only difference is in the argument `"N"` which specifies how many random values should be generated in the output arrays.

Obviously, once a sequence of random numbers is generated, the next step would be to verify it. This can be done as follows:

- Convert the random array into a POD object;
- Display accumulated statistics with the method getStat() or convert it into a H1D histogram for visualization.

As an example, let us generate an array from the lognormal distribution, print statistics and plot it in the form of H1D histogram:

```
from  jhplot   import  *
from  jhplot.math.StatisticSample  import  *

a = randomLogNormal (1000,0,10)
p0 = POD (a)
print  p0.getStat ()
h = H1D ("LogNormal" ,40 ,-50 ,50)
h.fill (a)

c1  =  HPlot ()
c1.visible () ;  c1.setAutoRange ()
c1.draw (h)
```

Listing 10.11 Checking random numbers using a histogram

By running this script you will see a detailed statistical summary of the lognormal distribution, plus the shape of the distribution will be visualized by the histogram object.

10.4.2   Methods of 2D Arrays

The generation of 2D arrays (i.e., matrices) is rather straightforward. We will continue with the above example used to generate a random 1D array. This time, however, we will add an additional argument representing the number of rows in the matrix. This time our code snippet creates a matrix 3×2 with random integer numbers from 0 to 100:

```
>>> from  jhplot.math.StatisticSample  import  *
>>> a = randomInt (3,2,0,100)
>>> print  a
array([I,  [array("i",  [79,  92]) ,  array("i",  [78,  81]) ,
            array("i",  [92,  72])])
```

We should point out that all methods to generate random 2D arrays have the same names and the meaning, as for the 1D case. The only difference now is that all such methods have an additional argument representing the number of rows.

Now let us consider how to build arrays of random values in accordance with a functional form given by analytic functions. Below we give an example which shows how to (1) generate a vector with 1000 random numbers distributed between 1 and

2 using the analytic function $x * \exp(x)$; (2) fill a histogram with such numbers and plot them together with the function on the same canvas

```
from jhplot   import  *
from java.awt import  *
from jhplot.math.StatisticSample   import  *

c1 = HPlot()
c1.visible();  c1.setAutoRange()

f=F1D("x*exp(x) function","x*exp(x)",1,2)
f.setColor(Color.red)
c1.draw(f)

p=f.getParse()
a=randomRejection(80,p,20,1,2)
h=H1D("x*exp(x) histogram",10,1,2)
h.fill(a)
c1.draw(h)
```

Listing 10.12 Random arrays with arbitrary PDF

Figure 10.4 shows the plotted analytic function and the histogram filled from the array with random numbers. Note that the normalization of the histogram depends on the number of elements in the random array and the bin width.

Analogously, one can build a 2D array by adding an additional argument to the `randomRejection()` method representing the number of columns.

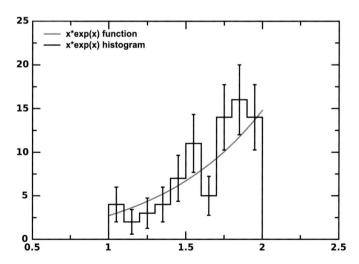

Fig. 10.4 Creating a random array in accordance with a given analytic function

10.4.3   Sampling Using the Colt Package

We have already considered how to generate separate random numbers using the Colt random engine in Sect. 10.3.2. Now we will learn how to use this package to create 1D and 2D random arrays.

First, let us discuss how to fill a POD container with random numbers using the Colt package. Look at the code below that creates the probability density function, or PDF, of a binomial distribution

```
from cern.jet.random.engine import *
from cern.jet.random import *
from jhplot import *

engine=MersenneTwister()
n,p=10,0.5
gam=Binomial(n,p,engine)
a=POD()
a.random(100,gam)
print a.toString()  # print random numbers
print a.getStat()   # print statistics
```

Listing 10.13 Statistical summary of the binomial PDF

The method to populate the POD with random numbers is useful since many predefined distributions are available. In addition, one can build an empirical distribution without making any assumptions about the functional form of the population distribution that the data come from. See the class Empirical() discussed in Sect. 10.3.2 for detail.

Analogously, one can fill the native Colt array called DoubleArrayList. This is exactly what we do in the next example: we change the random generator engine, fix the seed value to 99, and then create an array with the random numbers:

```
from cern.jet.random.engine import *
from cern.jet.random import *
from cern.colt.list import *
engine=DRand(99)
alpha,lamb = 1, 0.5
gam=Gamma(alpha,lamb,engine)
a=DoubleArrayList()
a.fillRandom(100,gam)
```

Listing 10.14 Creating random numbers using the Gamma function

Similarly, one can populate the array IntArrayList with integer random numbers.

10.5 Statistical Significance and Confidence Levels

10.5.1 Statistical Significance

Experimental measurements often have the disadvantage that they might have been caused by pure statistical accidents. In such cases we need to estimate the level of statistical significance (credibility) of observations, to ensure that a likelihood of such observations not happening by chance is small. One can approach this question from different directions by estimating the probability that the observed results have occurred due to statistical accident. Then, a large value of this probability would represent a small level of statistical significance (and vice versa). For a good recent review see [2].

The probability we are talking about is defined by the number called p-value. Mathematically speaking, it tells about the probability of a result being observed, given that a certain statement (the null hypothesis) is true. The p-value is close to 1 when results observed in measurements could have occurred by chance. Observations are statistically significant when the p-value is small, i.e., when one can demonstrate that the probability of obtaining observations by chance only is relatively low.

Let us give an example. Historical records on sunspot[1] activity suggest that solar activity is at the level of 80 sunspots per year. The population of sunspots is normally distributed with the standard deviation of $\sigma = 20$, and the records are large enough to assume the central limit theorem. During 1 year, however, 130 sunspots were observed. Is this finding significant? Does it indicate a beginning of unusual Sun activity?

To estimate the likelihood of the observation of 130 (or more) sunspots in a given year, we will calculate the area under the normal (Gaussian) probability density function from minus infinity to 130. The normal distribution should have the mean 80 and $\sigma = 20$. Then this value should be subtracted from 1 in order to calculate the probability for events with > 130 sunspots. This can easily be done as follows:

```
from org.apache.commons.math3.distribution import *
no=NormalDistribution(80, 20)
print "p-value=",1-no.cumulativeProbability(130)
```

Listing 10.15 Calculating the p-value for the normal distribution

The output probability is 0.0062. It states that the significance of the observation is very high: Only 6 years out of 1000 have occurrences with 130 sunspots due to statistical nature. This value was obtained using an analytic description of the normal distribution as implemented in the Apache Common Math library. In reality, the underlying distribution (and asked questions on significance) can be rather complex and, therefore, one needs numeric methods to estimate statistical significance.

---

[1] Sunspots are relatively dark areas on the surface (photosphere) of the Sun.

One can run an experiment where many events are generated using a normal distribution with the mean 80 and $\sigma = 20$, and counting a fraction of events with equal or more 130 values. We have considered a number of examples of how to generate random numbers in Sect. 10.3. Here is the code that numerically estimates the probability and plots a histogram with the null hypothesis.

```
from cern.jet.random.engine import *
from cern.jet.random import *
from jhplot import *

nr=Normal(80,20,MersenneTwister())
h1=H1D("Normal",100,0,200)
sum=0.;  total=100000
for i in range(total):
    va=nr.nextDouble()
    h1.fill(va)
    if (va>=130): sum +=1.0
c1=HPlot()
c1.visible();  c1.setAutoRange()
c1.draw(h1)
print "p value=",sum/total
```

Listing 10.16 Numeric calculation of the p-value

The code prints the expected p-value of $\simeq 0.0062$. The actual precision will depend on the number of random numbers generated. One should increase the number of random events to reduce statistical uncertainty on this estimate.

Now you know the main trick in the numeric estimation of the p-values. If you need to calculate the p-value numerically, first one should make multiple observations and determine the underlying distribution function, or your "null hypothesis." Then run numeric tests and estimate how many occurrences are expected for a total (large) number of observations, similar to those in the above example. The only difficult part is to find the correct null hypothesis. This can be done using analytic fits of data as explained in Chap. 11.

You may ask an opposing question. How many events should we see when we require a p-value less than 0.05, which is a common question in many fields of science (biology, psychology, physics, etc.). We should calculate the value, x, for which the area under the Gaussian probability density function (integrated from minus infinity to x) is equal to the argument:

```
from org.apache.commons.math3.distribution import *
no=NormalDistribution(80, 20)
print "p=0.05: ",no.inverseCumulativeProbability(0.95)
```

Listing 10.17 Calculating the number events for p-value of 0.05

This example returns 112.89.

Now let us consider one special case. If events are rare, i.e., when zero is a very likely number to observe, we should use a Poisson distribution instead of the normal distribution. In the example with the sunspots, one can ask question of how many sunspots occur with a certain size or type. For example, if we are talking about rare events with well-known mean 0.5, then an observation of three events can already be quite significant. Let us calculate the p-value in this situation:

```
from org.apache.commons.math3.distribution import *
no=PoissonDistribution(0.5)   # mean to 0.5
print "p-value=",1-no.cumulativeProbability(3) # 3 events
```

Listing 10.18 Calculating the p-value for the Poisson statistics

The code returns rather small probability of 0.0018.

In real life, expectations for experimental outcomes have some uncertainties. Let us calculate a statistical significance for an experiment with seven events, while the expectation is 3.1 ± 0.3, where 0.3 is uncertainty. In many situations, the uncertainty can be treated in the frequentist approach, i.e., we will assume that 0.3 is the standard deviation of a Gaussian distribution that has the mean value 3.1 (which is often true if the prediction is obtained using different probabilistic techniques). In order to account for uncertainty in the prediction, the mean value of the Poisson distribution should be obtained from a Gaussian distribution with the mean and width of the prediction [3]. And this is what we will do in the next code example using the random numbers discussed in Sect. 10.3.3:

```
from org.apache.commons.math3.random import *

r=RandomDataGenerator()
observed=7.          # observed number of events
bkg=3.1              # expected value
err=0.3              # expected error
tot=1000000; sum1=0;
for i in range(tot):
    mean=r.nextGaussian(bkg, err)
    value=r.nextPoisson(mean)
    if (value>=observed): sum1 +=1.0
print "p=", sum1/tot
```

Listing 10.19 p-value for the Poisson statistics with uncertainty

This code returns 0.041. Note that the above example is not very abstract—it was taken from the real-life situation of the top-quark discovery when the prediction was called "expected background rate," while the number 7 was the total number of observed top-quark candidates [4].

If you are interested in a more conservative estimate of the p-value, and have doubts about the statistical nature of the uncertainty, remove the Gaussian distribution and shift the mean value of the Poisson distribution by the uncertainty, i.e., evaluate the Poisson distribution with the mean 3.4.

10.5.2   Discovery Sensitivity

It should be said that, in many areas of science, events in a counting experiments are considered to be statistically significant when the p-value is less than 0.05. This means that an observation has more than 95 % chance of being true, i.e., not caused by pure chance. In some scientific fields, such as particle physics, a discovery is announced only when the p-value is less than 3×10^{-7} (assuming one-tailed p-value). This value corresponds to the five-sigma (5σ) observation, i.e., this probability can be obtained by integrating the tail of the normal distribution above five standard deviations. In other words, the observation would occur by chance alone as rarely as a value sampled from a Gaussian distribution would be five standard deviations from the mean. The Gaussian significance, which will be denoted as Z, is related to the p-value as $Z = \Phi^{-1}(1 - p)$, where Φ^{-1} is the inverse of the cumulative distribution.

Let us see how we can convert the p-value into the Z value in the expression $Z \cdot \sigma$:

```
from org.apache.commons.math3.distribution import *
p=NormalDistribution()
print "Z=",-p.inverseCumulativeProbability(3E-07)
```

Listing 10.20 Converting the p-value 3×10^{-7} to Z

You will see a value Z close to 5, i.e., we can say "5σ" observation.

If the p-value is smaller than 0.003 (which corresponds to three sigma, 3σ), one can report an evidence for a new phenomenon (or a particle) being observed. The reason for such a strong requirement by particle physicists is that new laws of physics should be discovered with much stronger evidence than what is usually required for the p-value of 0.05, which still leaves good room for statistical fluctuations and various instrumental anomalies. In addition, the five-sigma rule is a good protection against problems in estimating the p-value in real life when the background model (or the null hypothesis) is not known well, and measurements can be effected by uncertainties that are hard to foresee. Such uncertainties are not of statistical nature, and are usually called systematic uncertainties.

Now let us come back to the previous example and calculate the p-value that corresponds to three sigma (3σ) and five sigma (5σ). Then we estimate the expected numbers of events we should observe to claim 3 and 5 σ discoveries:

```
from org.apache.commons.math3.distribution import *
mean=80; sd=20
no=NormalDistribution(mean, sd)
p1=no.cumulativeProbability(mean+3*sd)
p2=no.cumulativeProbability(mean+5*sd)
print "p-value for 3 sigma:",1-p1
print "Events:",int(no.inverseCumulativeProbability(p1))
print "p-value for 5 sigma:",1-p2
print "Events:",int(no.inverseCumulativeProbability(p2))
```

Listing 10.21 Statistical significances for 3 and 5 σ

This prints the answer:

```
p-value for 3 sigma: 0.00135
Events: 139
p-value for five sigma: 2.86651E-07
Events: 180
```

As you can see, the claim of a new phenomena that causes high Sun activity requires the observation of 180 sunspots in a single year assuming the 5σ approach. Observing 140 sunspots in a single year is good enough for evidence of unusual Sun activity assuming the 3σ criteria. As explained before, simple situations in which the null hypothesis is well established and systematical variations (i.e., those that do not occur by chance) are well understood do not require the 5σ extreme, i.e., values of $p = 0.05$ and $p = 0.01$ should be usually sufficient.

10.5.3 Confidence Interval

A confidence interval is an estimated range of values which is likely to include an unknown population parameter. Common choices for the confidence level are 90, 95, and 99 %. They correspond to percentages of the area of the normal probability distribution. The confidence interval is usually symmetric around the mean. For a normal distribution, the confidence interval is expressed using its standard deviation denoted by σ. For example, $1.64485 \times \sigma$ for 90 %, $1.95996 \times \sigma$ for 95 % and $2.57583 \times \sigma$. For the example with sunspots in the previous section, a 95 % confidence interval is $20 \times 1.95996 = 39.3$ around the mean and, therefore, the probability of observing a value outside this area is less than 0.05.

In many practical applications, the mean and standard deviation are unknown; therefore, the standard deviation is replaced by the estimated standard deviation from a sample. In the case of unknown mean and standard deviation, a confidence interval for the mean is $\bar{x} \pm ts/\sqrt{n}$, where t is the upper $(1 - C)/2$ critical value for the t-distribution with $n - 1$ degrees of freedom (n is the size of the sample). The t-distribution is similar to the standard normal distribution, except that it has larger tails due to the increased variability associated with the sample standard deviation.

Let us show how to calculate a 95 % confidence interval for an array of values with unknown mean and standard deviation.

```
from org.apache.commons.math3.distribution import *
from org.apache.commons.math3.stat.descriptive import *
import math

level = 0.95
data = [100, 203, 272, 102, 302, 120, 221]
stats = SummaryStatistics()
for i  in range(len(data)):
```

```
      stats.addValue(data[i]);
t=TDistribution(stats.getN()-1)
val = t.inverseCumulativeProbability(1-(1-level)/2)
ci=stats.getStandardDeviation()/math.sqrt(stats.getN())
lo=stats.getMean()  -  ci*val
up=stats.getMean()  +  ci*val
print "95% Confidence Interval (low,upper):",lo,up
```

Listing 10.22 95 % confidence interval using the *t*-distribution

The code returns

```
95% Confidence Interval (low,upper): 112.0   265.1
```

As an exercise, replace 0.95 by 0.99 to see how the values change for 99 % confidence
level.

10.5.4   Confidence Levels for Small Statistics

Often, in physics and other disciplines of knowledge one needs to calculate upper
limits, or the so-called exclusion limits, on some value. In this section we will consider small statistics counting experiments with outcome observations distributed
according to the Poisson statistics.

Consider the case of finding $N = N_b + N_s$ events where N_b is the number of
events from known processes (we will call this "background"), and N_s represents
the number of events from a new theory ("signal"). In the case of small statistics,
the background has known mean b, while the signal has the mean s. Therefore, the
total number of events N_b is distributed according to the Poisson distribution with
the mean $s + b$.

For a counting experiment, one can introduce a confidence level CL_b that specifies the probability that the expected number of events from the background-only
hypothesis is smaller than or equal to the observed number, $CL_b = P_b(X \leq N_{obs})$.
In the presence of signal, poor compatibility of the observation with the background-
only hypothesis is indicated by CL_b being close to one. This definition might be
confusing, but in reality, $1 - CL_b$ is a *p*-value discussed in the previous sections.
This value quantifies our confidence in the background hypothesis. If the *p*-value is
small, this would indicate either a large statistical fluctuation, or a discovery of a
new theory that leads to incompatibility of the background hypothesis with observed
data. As discussed before, $1 - CL_b$ is required to be no more than 2.87×10^{-7} (or
twice that in the case of two-sided Gaussian tails) for a 5σ discovery.

Analogously, one can define our confidence in the background plus signal hypothesis. The confidence level for excluding the possibility of simultaneous presence
of the new theory and known, "background" (the $s + b$ hypothesis), is $CL_{s+b} =$

$P_{s+b}(X \le X_{obs})$. This probability assumes the presence of both signal and background at their hypothesized levels, and shows that the test statistic would be less than or equal to that observed in the data.

Then one can define [5]

$$CL_s = \frac{CL_{s+b}}{CL_b}$$

which specifies our confidence in the signal (or in a new theory). At this stage, we will not discuss problems that you might discover with this quantity, such as lack of the usual frequentist explanation—we will direct the reader to more specialized books. Below we will show how to perform calculations of CL_s using numeric experiments.

When excluding a new theoretical model, 95 % exclusion is usually used. This means CL_s should be computed with large enough precision to tell that an observed outcome is less probable than about 5 % of the time assuming the presence of a signal.

Our code for calculation of the confidence levels will be based on a Monte Carlo technique. This method is well suited for computing confidence levels when combining a large number of counting experiments. To compute CL_{s+b} and CL_b probabilities, many pseudo-experiments need to be created, and then each possible outcome should be checked.

10.5.4.1 Exclusion Limit

Let us show how to set an exclusion limit. To do this, we will need to calculate CL_s. For 95 % confidence level, a signal hypothesis is excluded if $CL_s = 0.05$ assuming that signal is present.

We have already discussed the statistical significance for the Poisson statistics in Sect. 10.5.1, where we assumed frequentist uncertainties on predictions. We will use again the example from the previous section: Assume an experiment detected 10 unusual sunspots in a certain year. Scientists have a well-established model that expects exactly five events of such type, with an uncertainty of two events. The uncertainty on theory is considered using the Bayesian viewpoint. For example, the uncertainty is estimated from an analytic calculation where statistical methods based on large statistical samples cannot be applied. The prediction 5 ± 2 will be called background since, from the point of view of a possible discovery, there is nothing unusual in such a prediction. Given that events are rare, and assuming no uncertainty on the prediction, one can numerically calculate the statistical significance of the observation, $1 - CL_b$, or the p-value, using the Poisson statistics:

```
from jhplot.math   import  *

expected=5   # number  of  expected  events
observed=10  # number  of  observed  events
sum=0.0;  total=1000000
nr=Poisson(expected)
```

```
for  i  in  range(total):
     if  (nr.next()>=observed):  sum  +=1.0
print  "p-value=",sum/total
```

Listing 10.23 Calculation of *p*-value for the Poisson statistics

As it can be seen, the *p*-value is 0.032. It is not too small. If one takes into account
the upper value of the theory uncertainty, the *p*-value will be 0.16—simply replace 5
by 7 to see this. This would be the proper answer for a conservative estimate of the
p-value.

But now we want to ask a different question. Imagine we have developed an
alternative theory that predicts a larger number of unusual events, and it has some
free parameter, say λ, that determines the number of events in such a theory. What
we want to do is to set an upper limit on this parameter λ at a 95 % confidence level.
This means that a theory with a given value of λ that predicts X number of events
is excluded with a 95 % confidence level. We need to find this unknown X, and this
would lead to the limit on λ.

We will build a simple calculator to do this. We will calculate the probability for
a background event with the mean 5 to give more than nine events. Numerically,
we calculate arrays of probabilities for $5 + X$ events to give more than nine events,
scanning X from 1 to 20:

```
from  jhplot.math   import  *

observed=10.  #  observed  number  of  events
background=5  #  expected  background
tot=1000000;  sum1=0;
nr=Poisson(background)

for  i  in  range(tot):
  if  (nr.next()>=observed):  sum1  +=1.0
for  X  in  range(1,20,1):
  sum2=0
  for  i  in  range(tot):
      if  (nr.next(background+X)>=observed):  sum2  +=1.0
  CLb  =1-sum1/tot
  CLbs=1-sum2/tot
  CLs  =CLbs/CLb
  print  "X=",X,"CLb=",CLb,  "  CLbs=",CLbs,  "  CLs=",CLs
```

Listing 10.24 Calculation of CL_b using a Monte Carlo method

The code uses the method $next(background+X)$ that returns a value from a Poisson
distribution with the mean background+X where $X = 1 - 20$. As you will see,
CL_s hits the value of 0.05 if X value is between 11 and 12. This value of X gives us
an upper limit on a parameter λ of this theory. Make finer steps in this calculation to
get more precise value.

In reality, there is no need to code homemade calculators to set exclusion limits.
There are well-established tools to do such calculations. They take into account

uncertainty on backgrounds and even can add multiple counting experiments. One can use a calculator described in Ref. [6].

Let us use this tool in our next code. We will set an arbitrary number of guessed events (denoted by "X"), and set an experiment with 10 observed events and 5 ± 2 "background" events. We will remind that the uncertainty ± 2 does not have the usual frequentist reasoning since this is expected theoretical uncertainty. The output of this calculator is the p-value estimated using a Monte Carlo method. In this example, use the approach called Cousins–Highland [7] (other choices are also possible in this tool).

```
from jhpro.stat.limit import *

guess=12.7 # initial guess of events
s=StatConfidence(guess,500000)
efficiency=1; error=0
# 10 observed, 5+-2 expected
exp=ExpData(10,5,2,efficiency,error)
s.addData(exp)
s.runUpper()
print "P(Cousins+Highland)=",s.getProbabilityCH()
```

Listing 10.25 95 % confidence interval in the Cousins–Highland method

The number of guessed events was set to 12.7. This roughly leads to the correct p-value of 0.05 when using 500,000 Monte Carlo iterations defining the precision on the final answer. The calculator also accepts the observation efficiency and its uncertainty on the "background" which, for simplicity, was set to 1 and 0, respectively. The reason why we used 12.7 in the above code is simple: we ran the code several times with different initial values until we found the value 12.7 that corresponds to $p = 0.05$.

Thus, the above code gives the answer: if a new theory predicts the number of events that exceeds 13 events, it will be excluded with 95 % confidence level, since the 12.7 observations are excluded at this level. One can run this script several times, iterating toward whatever limit type you choose, e.g., for a 99 % upper limit one should run limit guess until you see the probability of 0.01. For the above example, this corresponds to a value of 16. If you need to run this code for a lower limit, replace runUpper() with runLower().

It is worth mentioning that a typical experiment has uncertainties on the measured efficiency. And the efficiency itself may not be 100 %. Let us assume that the efficiency is 0.8, with its uncertainty of 0.1. Replacing 1 and 0 by 0.8 and 0.1, respectively, we will see that 95 % upper limit excludes 18.9 events, thus we can exclude a new theory if it predicts (together with the established theory) more than 19 events.

Finally, assume that a second group of scientists did a similar measurement using different techniques. They have recorded 13 events, with a theory prediction of 7 ± 2 (where 2 is the theory uncertainty). We can extend the above code by inserting the lines:

```
exp=ExpData(13,7,2)
s.addData(exp)
```

before the "runUpper()" statement. This changes the 95% confidence limit to about 14.5 events.

10.5.4.2 Discovery

Let us come to the discovery case again. As we discussed, $1 - CL_b$ indicates the probability that the background could have fluctuated to produce a distribution of candidates at least as signal-like as those observed in the data. We again remind that for a 5σ discovery, $1 - CL_b$ is required to be no more than 2.87×10^{-7} (or twice of that). This means that Monte Carlo calculations for discoveries are more CPU intensive compared to the exclusion limits. As discussed previously, evidence for a new phenomena (or for a new particle) requires 3σ, which corresponds to $1 - CL_b = 1.3 \times 10^{-3}$ (or twice that).

The discovery situation can be simulated using the calculator shown above. You should set the number of expected events to a larger value, and to increase the number of Monte Carlo iterations (say, to 100 million). The 5σ discovery level can be obtained if one observes about 33 events assuming the predicted 5 ± 2 and an experimental efficiency of 1. An observation of about 20 events corresponds to 3σ "evidence".

It should be said that a measure of expected discovery significance in the Poisson counting experiments is given by the expression $s/\sqrt{b}$, where s is the number of signal events and b is the number of "background" (or expected) events. If background events have some systematic uncertainty characterized by a standard deviation σ_b; the expected discovery significance becomes $s/\sqrt{b + \sigma_b^2}$. Both expressions are widely used in particle and nuclear physics, especially for designs of new experiments when one needs to estimate the needed numbers of events to be able to say about the possible discovery.

10.5.5 Statistical Tests

Suppose we have a histogram or X–Y data points with known statistical uncertainties (or standard deviations, σ) on the bin height (or values of Y in the case of X–Y arrays). We would like to measure how well the data agree with some theoretical distribution, given by another set of values. Such a comparison is often done by calculating the χ^2 value:

$$\chi^2 = \sum_{i=1}^{n} \frac{(y_i - y_i^t)^2}{\sigma_i^2} \tag{10.1}$$

where σ are the standard deviations associated with each measurement y_i, and y_i^t are expected (theoretical) values. The χ^2 value is used by the Pearson's χ-squared test (see (11.1) for a review), a statistical test that evaluates how likely it is that any

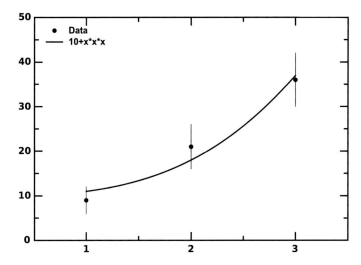

Fig. 10.5 Comparing $X-Y$ data with an analytic function

observed difference between two data sets is due to chance. The χ^2 value can be considered as a distance between two histograms. A large value of χ^2 indicates that the histograms are drawn from distinct underlying distributions.

Let us assume that the underlying (theoretical) values are given by the analytic function $10+x^3$. Our $X-Y$ data points are $(1, 9)$, $(2, 21)$, and $(3, 36)$, with uncertainties on the Y values of 3, 5, and 6, respectively. Now we will run a small test by comparing these three experimental data points with uncertainties on each point with the analytic function:

```
from jhplot   import  *

p=P1D("Data")
p.add(1,  9,  3);  p.add(2,  21,  5);  p.add(3,  36,  6)
f=F1D("10+x*x*x",1,3)
print p.compareChi2(f)
c=HPlot()
c.visible();  c.setAutoRange()
c.draw(p)
c.draw(f)
```

Listing 10.26 Calculation of a χ^2 test for X–Y data

The output plot is shown in Fig. 10.5. This example also prints a map with the χ^2, the number of degree of freedom (ndf) and the p-value:

```
chi2:  0.83,      ndf:  3,     p-value:  0.84
```

According to the rule of thumb, a theory (or expected function) agrees with data well when χ^2/ndf is less or close to 1, and the p-value is close to unity. This is true for the above example. The data and predictions are identical if χ^2 is zero and the p-value is 1 exactly. If the p-value is small, this indicates that our test fails and data and predictions are different.

But how is the χ^2-statistic used to calculate the p-value? This value comes from the χ^2 distribution. One can use the Apache Common Math package to calculate the p-value as in this example:

```
from org.apache.commons.math3.distribution import  *
chi2D=ChiSquaredDistribution(3)  # ndf=3
print "p-value=",1-chi2D.cumulativeProbability(2)
```

Listing 10.27 Calculation of the χ^2 test

Here we use 3 for the number of degrees of freedom, and calculated the p-value for $\chi^2 = 2$. The output of the code is 0.572.

If experimental data and predictions in Eq. 10.1 are given by histograms, we can use a similar approach. Data in the first histogram represent our experimental values, while the second histogram can be derived from a theoretical prediction using a Monte Carlo technique, or it can represent an alternative set of data that needs to be compared to. If the histograms are denoted as "h1" and "h2", run the χ^2 test in as in this code:

```
d=h1.compareChi2(h2)
```

The method returns a map with the χ^2 value, the number of degrees of freedoms ("ndf"), and the probability of similarity ("p-value"). The χ^2 between two histograms reflects the difference between heights of the bins taking into account statistical uncertainties. The number of degrees of freedoms corresponds to the number of bins. The values χ^2 and ndf are used as inputs for the cumulative distribution to calculate the p-value. Histograms are identical if χ^2 is zero and the p-value is one. If the p-value is small, this indicates that our comparison test fails and the histograms are different.

Let us show an example where we compare two histograms filled with Gaussian random numbers. We will make the second histogram slightly narrower. The value 0.9 in this example reduces the Gaussian σ by 10% for the second histogram.

```
from jhplot import  *
from java.util import  *

h1  =  H1D("H1",20,  -1,  1)
h2  =  H1D("H2",20,  -1,  1)
r=Random(1L)  # make reproducible output
for i in range(100):
    d=r.nextGaussian()
    h1.fill(d);  h2.fill(d*0.9)
d=h1.compareChi2(h2)
```

```
print "chi2 / ndf =",d["chi2"]/d["ndf"]
print "p-value=",d["p-value"]
```

Listing 10.28 Comparing two histograms using the χ^2 test

The final answer is $\chi^2/ndf = 0.67$ and the p-value is 0.86, with ndf being 20 (i.e., equal to the number of bins). This test shows that the histograms are nearly identical.

One can also try several other approaches to assess the similarity of two histograms, such as Kolmogorov–Smirnov, Anderson–Darling, and Goodman tests. For these tests, use the `hep.aida.util` package from FreeHep. Let us show how to run the Kolmogorov–Smirnov statistical test (see, for example, Ref. [8]):

```
from hep.aida.util.comparison import *
r=StatisticalComparison.compare(h1.get(),h2.get(),"KS")
print "ndf=",r.nDof()  #  = number of bins
print "Kolmogorov-Smirnov method=",r.quality()
```

where "h1" and "h2" are histograms defined in the previous example. The histograms are identical if the method `r.quality()` returns 1. The actual value obtained in this example is close to 1, which indicates again that the histograms are nearly identical. This test returns zero when histograms are very different.

10.5.6 Confidence Levels for Distributions

The previous two sections, Sects. 10.5.3 and 10.5.4, described limit settings for situations with certain number of events from counting experiments. In real life, we are typically dealing with distributions of events that can be represented, for example, by histograms. Section 10.5.5 considered statistical methods to test compatibility of several histograms. One histogram can describe experimental data, while the other can represent a theoretical expectation.

This time we will study the computation of confidence levels for distributions of events. As in the previous section, we will consider the Poisson statistics, i.e., when the rate of occurrence of events is small, and events are independent of each other. Here we will follow the discussion in Ref. [9] and illustrate the confidence level computation using several code examples.

As for Sect. 10.5.4, our code for the calculation of the confidence levels will be based on a Monte Carlo technique. To compute CL_{s+b} and CL_b probabilities, we will generate many pseudo-experiments and then each possible outcome will be checked. For distributions represented by histograms, this requires generating Poisson random numbers in each histogram bin according to our expectation.

10.5.6.1 Exclusion Limit

Let us show how to set an exclusion limit. To do this, we will need to calculate CL_s. We remind that a signal hypothesis is excluded at the 95 % CL if $CL_s = 0.05$ and at more than the 95 % CL if $CL_s < 0.05$, assuming the presence of a signal.

Let us generate three histograms using one-tailed Gaussian distributions. The first histogram will represent "experimental" data with 1000 events distributed according to one-tailed Gaussian distribution with the standard deviation 1. Then we will create two other histograms. One histogram will represent our expectation for a known theory ("background"), while the second histogram will represent a new (possible) theory. The latter histogram will be called "signal". What we want to know is how likely the new theory, or signal, is true for the given experimental data. Assuming that theoretical uncertainties are small for the latter histogram, we will use a trick: We will fill these two histograms with large number of events, but apply weights (0.01 and 0.0005), respectively. In this case, the normalization of the background histogram is 1000 (10^5 times 0.01), while the signal has an area that corresponds to 50 events (10^5 times 0.0005). Then we will combine the background and signal histograms, creating a histogram that represents our expectation for data assuming a contribution from a new theory. The code below initialize and fills all our histograms and writes data to a file:

```
from java.awt import *
from java.util import *
from jhplot import *

data=H1D("Fake data",30,0.0,4.0)
data.setColor(Color.black)
data.setStyle("p")

backg=H1D("Backg",30,0.0,4.0)
backg.setColor(Color.green)
backg.setFill(1); backg.setErrAll(0)
backg.setFillColor(Color.green)

signal= H1D("Signal",30,0.0,4.0)
signal.setFill(1)
signal.setFillColor(Color.red)
signal.setColor(Color.red)

r=Random(2L) # make random numbers reproducible
for i in range(100000):
   backg.fill(abs(r.nextGaussian()),0.01)
   signal.fill(1+abs(r.nextGaussian()),0.0005)
   if (i<1000): data.fill(abs(r.nextGaussian()))
sigback=backg.oper(signal,"Signal+Backg","+")
sigback.setErrAll(0)
IO.write([backg,signal,sigback,data],"data1.jser")
```

Listing 10.29 Creating histograms with "experimental" data for limit settings

Next, we will use the data created in the above example and set an exclusion limit by calculating CL_s. The code shown below calculates and prints CL_s:

```
from jhplot import *
from jhpro.stat import *

data=IO.read("data1.jser")
print "Wait. Calculating .."
datasource = DataSource(data[1], data[0], data[3])
climit = CLimits(datasource,1000000)
conf = climit.getLimit()
print "CLs     :",conf.getCLs()
print "CLb     :",conf.getCLb()
print "CLsb    :",conf.getCLsb()
print "expected <CLs>  :",conf.getExpectedCLs_b()
print "expected <CLb>  :",conf.getExpectedCLb_b()
print "expected <CLsb> :",conf.getExpectedCLb_b()
excl=(1-conf.getCLs())*100
print "Signal hypothesis is excluded at level (%)",excl
c1 = HPlot()   # show the results
c1.visible();  c1.setRange(0,3,0.0,200)
c1.draw(data)
```

Listing 10.30 Calculating exclusion limit using histograms

The code uses the `CLimits` Java class that reads the input data and the number of iterations (1,000,000) for the Monte Carlo technique. The above code leads to CL_s close 0.02. This means that our new theory is excluded with more than 95 % confidence level or, to be more precise, it is excluded at 98 %. If you want to set the upper limit on the signal exactly at 95 % confidence level, you should change the number of signal events until $CL_s = 0.05$. The output of this example code is shown in Fig. 10.6.

10.5.6.2   Discovery Case

Let us remind that $1 - CL_b$ indicates the probability that the background could have fluctuated to produce a distribution as observed in the data. For a 5σ discovery, $1 - CL_b$ is required to be no more than 2.87×10^{-7}.

Now we will illustrate a 5σ discovery case using a real-life situation. Let us create "experimental" data for a "discovery" of the Higgs boson, an elementary particle in the Standard Model that was first observed at the CERN laboratory in Geneva (Switzerland). Our code will estimate the discovery significance, or the p-value, from which we can conclude that the observation is real and is inconsistent with random fluctuation because of the statistical nature of the experimental distribution. This discovery case is best illustrated in Fig. 10.7 released by the CERN laboratory. It features a peak in the invariant mass of two photons created in proton–proton collisions at the LHC experiment at CERN. This peak corresponds to about 7σ

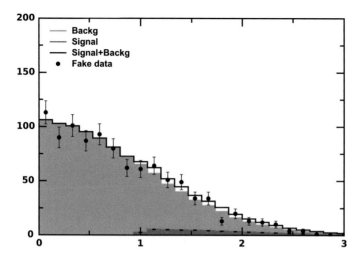

Fig. 10.6 An illustration of the exclusion limit calculation for distributions. The signal distribution is shown as the *red* histogram, while the *green* histogram shows our expectation for "background". Symbols illustrate our hypothetical "experimental" data. According to the test described in the text, the data are fully consistent with the expectation given by the background histogram (color figure online)

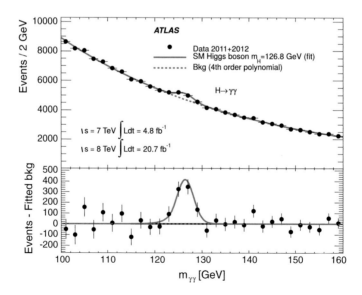

Fig. 10.7 The invariant-mass distribution [10] of two photons recorded by one of the two collaborations at the CERN laboratory. A bump near 125 gigaelectronvolts indicates the presence of the Higgs particle. The image is released under the CC-BY-SA-4 license. Credit for this image goes to CERN

discovery. We will try numerically "reproduce" it by creating fake data that resemble the real data shown in this figure, and then we will run a statistical significance code similar to that shown in the previous example.

Let us prepare a data sample assuming 250 events near the 125 region of a histogram (assuming arbitrary units). Then we will initialize and fill histograms the signal peak and background, making sure that statistics is large enough to populate these histograms.

```
from java.awt import *
from java.util import *
from jhplot import *

backg=H1D("Expected background",30,110,140)
backg.setColor(Color.blue)
backg.setFill(1); backg.setErrAll(0)
backg.setFillColor(Color.cyan)

signal= H1D("Higgs signal",30,110,140)
signal.setFill(1)
signal.setFillColor(Color.red)
signal.setColor(Color.red)

data=H1D("Observed data",30,110,140)
data.setColor(Color.black)
data.setStyle("p")

r=Random(1L) # make reproducible distribution
for i in range(5000000):
   backg.fill(abs(100*r.nextGaussian()),0.02)
   signal.fill(125+1.2*r.nextGaussian(),0.00005)
   if (i<100000): data.fill(abs(100*r.nextGaussian()))
   if (i<250):    data.fill(125+1.2*r.nextGaussian())
sigback=backg.oper(signal,"Signal+Backg","+")
sigback.setErrAll(0)
IO.write([backg,signal,sigback,data],"data2.jser")
```

Listing 10.31 Creating fake "Higgs" data using random numbers

Now we will run a code that calculates the *p*-value using the `CLimits` method discussed before:

```
from jhplot import *
from jhpro.stat import *

c1 = HPlot(); c1.visible()
c1.setRange(110,140,0.0,1000)

data=IO.read("data2.jser")
c1.draw(data)
print "Wait. Calculating.."
datasource = DataSource(data[1], data[0], data[3])
climit = CLimits(datasource, 10000000)
```

```
conf  =  climit.getLimit()
print  "CLs       :"  ,conf.getCLs()
print  "CLb       :"  ,conf.getCLb()
print  "CLsb      :"  ,conf.getCLsb()
print  "expected  <CLs>  :",conf.getExpectedCLs_b()
print  "expected  <CLb>  :",conf.getExpectedCLb_b()
print  "expected  <CLsb>:"  ,conf.getExpectedCLb_b()
print  "p-value:",  1-conf.getCLb()
```

Listing 10.32 Calculating 5σ significance for "Higgs discovery"

Depending on the CPU, the calculation can take up to 30 minutes on a modern Intel's Core i7 processor. You should increase the precision of the CL_b calculation using more iteration steps. The likely conclusion you will arrive is that the signal significance is larger than 5σ, i.e., $1 - CL_b$ is smaller than 2.87×10^{-7}. The output of this code is shown in Fig. 10.8.

As a bonus, you may try to use the created histograms to run the peak-finder algorithms discussed in Sect. 14.4.

10.5.6.3   Statistical Significance from Fits

Looking at Fig. 10.8, the immediate thought is to ask for the number of events above the expected background, pretending that we do not know how many signal events were included when we created the signal histogram. There are several ways to do this. One possibility is to subtract the known background histogram from the "data" histogram (shown with the symbols) using the usual mathematical operations of the

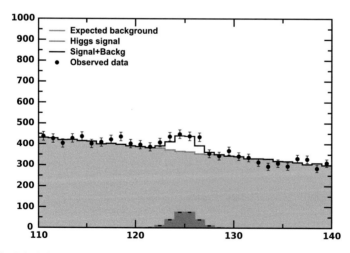

Fig. 10.8 Calculation of the discovery significance using expected background and signal histograms. The experimental distribution is shown using symbols. The signal plus background hypothesis is shown by the *black line* (color figure online)

histogram class `H1D`. But if you are not sure about the background histogram, one can perform a fit of the data with a function which has a background-like component and a signal-shaped peak. You can read about this in Chap. 11, so do not be afraid to skip this section.

We will fit the data using two analytic functions: one function is a first-order polynomial, $a + b \cdot x$ (a and b are constants), that represents the background shape, while the second is a Gaussian function for the peak description. The goal is to fit the data and determine the peak parameters. We will determine the number of events under the background from the extracted amplitude of the Gaussian peak, and then we will estimate the width and the peak position.

And there is a good bonus for doing all of this. One can also estimate the uncertainty on the number of events contributing to the signal. Why do we need to do this? This again comes to the significance calculation discussed previously. One can define a significance of the peak (this time we will call it S) as $S = N/\delta N$, where δN is a statistical uncertainty on N. Since the standard deviation, σ, provides the best estimate for the statistical uncertainty, one can quote the statistical significance in terms of $S \cdot \sigma$. One can calculate the numbers N and δN using the fit function, so the significance can be estimated in a simple and intuitive way.

How this statistical significance is related to the p-value and $Z\sigma$ value has been discussed before. In fact, they are directly related, but not in a simple way. But what is important for our numerical case is this: the calculation of the statistical significance using fit functions takes almost no computational time, while the calculation of the p-value using many millions of Monte Carlo events takes a substantial amount of the CPU time. The only nontrivial part of the fit procedure is to find a correct function that is believed to describe the observed data.

The last comment is important. When extracting the significance from an analytic-function fit, one must be sure that the fit hypothesis is correct, i.e., calculated χ^2 per degree of freedom is smaller or close to one. The fit is a powerful way to estimate statistical significance if we know the shape of signal and background, and if we are sure that among other possible signal and background shapes, the chosen function is the most plausible. Signal and background functions, which (a) are simplest, (b) have the smallest number of free parameters and, at the same time, (c) lead to smallest χ^2 per degree of freedom, are preferential for the statistical significance calculation and for extracting the number of signal events. Systematic uncertainties can be checked using alternative background and signal shapes, and/or redoing the fit. As an additional check, one should vary histogram fit ranges and bin sizes, or perform fits using an alternative procedure for the function minimization.

Let us put together a code to extract the number of signal events under the background distribution:

```
from jhplot    import *
from java.awt import Color

data=IO.read("data2.jser")
h1=data[3] # histogram with fake data
```

```
p1="p0+p1*x[0]"
gauss="(0.3989/s)*exp(-0.5*(m-x[0])*(m-x[0])/(2*s*s))"
f=HFitter() # fit data with gaussian+P1
f.setFunc("p1+gauss",1,p1+"+N*"+gauss,"p0,p1,N,s,m")
f.setPar("p0",1000);    f.setPar("p1",-5.0)
f.setPar("m",125);   f.setPar("s",1)
f.setPar("N",50)
f.setRange(100,150)
f.fit(h1)

c1=HPlot()
c1.visible();  c1.setRange(110,140,0.0,1000)
result=f.getResult()
par=  result.fittedParameters()
err=  result.errors()
n=  result.fittedParameterNames()
print "Fit results:"
for i in range(len(n)):
    print n[i]+": %.2f   +-   %.2f"%(par[i],err[i])
print "Chi2/Ndf=",result.quality(),  " S=",par[2]/err[2]
f2  =  F1D("P1+Gauss fit",f.getFittedFunc(),110,140)
f2.setColor(Color.blue)
c1.draw(f2);   c1.draw(h1)
```

Listing 10.33 Fitting data with a signal plus background function

The code defines a background and Gaussian function with the amplitude "N". The Gaussian, or the normal distribution, has two parameters: mean ("m") and the width ("s" $= \sigma$). Note that we added the factor $1/(\sigma\sqrt{2\pi})$ for the Gaussian function to make sure the correct normalization of this density function. Please look at Chap. 11 describing how to select fit functions for data fits.

The output of this code is shown in Fig. 10.9. It also prints the fit quality (χ^2/ndf), which is close to 0.86 (good fit!), and also the number of the signal events with statistical uncertainty entering the Gaussian amplitude $N = 233 \pm 46$. The last number leads to 5.1σ from the fit (i.e., $S = 234/46$). So the signal strength is above 5σ, assuming that the background is well understood and other checks discussed above have been done before claiming the "discovery". The fit also returns the correct peak width 1.07 ± 0.21, which is consistent with that used to create such a peak.

To extract the number of signal entries under the peak, it is important to make the correct normalization of the fit function that describes the signal shape. This means that the function should be normalized to 1. For this, one needs to multiply N by the bin width. In the above example, the bin width was conveniently set to 1 when we created the input histograms.

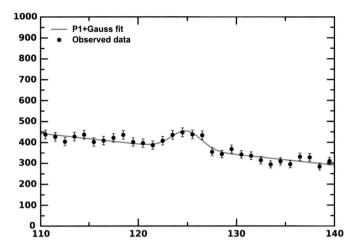

Fig. 10.9 An illustration of the signal extraction using a χ^2 fit with the first-order polynomial function for the background description. The Gaussian function is used for the description of the peak ("signal")

10.6 Error Propagation

In statistics, propagation of errors (or, better to say, statistical uncertainties) is a topic that deals with determination of statistical uncertainties on a complex observable constructed from several quantities with uncertainties. If two quantities (A and B) with their individual uncertainties (ΔA and ΔB) are measured, the main question is how to determine the uncertainty on an arbitrary function $F(A, B)$ constructed from these two quantities and their uncertainties. For counting statistics, the statistical uncertainty on a quantity is typically defined by the standard deviation σ (or by the square root of variance). For example, if A is the number of counts recorded during some time interval, then the uncertainty on this number is $\sqrt{A}$ assuming the Poisson statistics.

You already probably know the answer to the question of how to combine statistical uncertainties for simple expressions. If $F(A, B)$ is the sum (or difference) of A and B, then the total uncertainty is $\sqrt{\Delta A^2 + \Delta B^2}$, assuming A and B are uncorrelated variables.

For the general case, an analytic error propagation for a complex function $F(A, B, \ldots)$ can be expressed as the Taylor linearization based on the first-order partial derivatives:

$$\Delta F = \sqrt{\left(\frac{\partial F}{\partial A}\delta A\right)^2 + \left(\frac{\partial F}{\partial B}\delta B\right)^2 + \cdots},$$

assuming that the variables in this expression are not correlated. Otherwise, the above expression requires a modification that includes covariance, i.e., a degree of correlation between the variables.

DMelt has a handy library to deal with statistical uncertainties. Let us consider the value 10 with uncertainty 2 that has to be transformed to $\log_{10}$. Here is a small example of how you can make such a transformation:

```
>>> from jhplot.math import *
>>> x=ValueErr(10,2)  # 10 +- 2 (error)
>>> ylog=x.log10(x)   # transform to log10
>>> print ylog
1.0 +- 0.086
```

The Java package `jhplot.math.ValueErr` can perform various mathematical transformations with values and associated errors, as well as the usual mathematical operations. The library also supports error propagation with correlations. Let us consider an example with several measurements:

```
from jhplot.math import *

A=ValueErr(30,3)
B=ValueErr(100,20)
C=ValueErr(50,10)
AB=A.times(B)        # A multiplied by B
ABC=AB.plus(C,0.5)  # A*B+C with 50% correlation
print AB," ",ABC
```

Listing 10.34 Propogation of statistical uncertainties

To determine correlations between measured values, you will likely need to look at multiple measurements, and calculate the correlation coefficient (see, for example, Sect. 10.2).

When you have multiple measurements, it is convenient to use the P1D array (discussed in Sect. 4.2) which has useful uncertainty-propagation methods:

```
from jhplot import *

A=P1D("Array A")
A.add(1, 8, 3)   # error on 8 is 3
A.add(2, 9, 2)   # error on 9 is 2

B=P1D("Array B")
B.add(1, 5, 2)   # error on 5 is 2
B.add(2, 3, 1)   # error on 3 is 1

A.oper(B,"*")  # A*B
print A
```

Listing 10.35 Propogation of statistical uncertainties for arrays

The method `oper()` supports addition ("+"), multiplication ("*"), subtraction ("−"), and division ("/").

10.6.1 Propagation Using Monte Carlo Technique

The propagation of uncertainties discussed above used the analytical linearization method. This has its limitation: complex functions are difficult to evaluate since this approach requires the computation of derivatives. Their evaluation can be quite complicated in certain cases. However, there is an alternative to the analytic approach. One can randomly generate values using normal distributions, and look at the standard deviation of the outcomes. Such a Monte Carlo approach can be used for rather complex situations. It can reduce uncertainties associated with linearization, and it can even take into account correlations between variables.

To illustrate this Monte Carlo technique, let us numerically estimate the uncertainty on a complex function containing two parameters with uncertainties, 100 ± 10 and 30 ± 4. As before, we assume that the uncertainties are distributed according to a normal distribution (Gaussian). Our complex F function is

$$F = x^2 * \sqrt{x} * A + B \quad \text{if } x \geq 5,$$
$$= 0 \text{ if } x < 5,$$

where A and B are the input values with their uncertainties. We define this function using the FNon class designed to create mathematical functions (see Sect. 3.7). Our goal is to calculate the value of the function and its uncertainty at a given fixed point, say $x = 10$. We will run 10,000 Monte Carlo events and calculate the standard deviation with the method `getStd()` of this function at the required value:

```
from java.lang import *
from jhplot import *
from jhpro.upropog import *

class MyFunc(FNon):
  def value(self, x):
      y=x[0]*x[0]*Math.sqrt(x[0])*self.p[0]+self.p[1]
      if (x[0]<5): y=0
      return y

p1 = MyFunc("Function",1,2)
val=[100,10]  # define values
err=[30,4]    # define errors
x=10          # calculations at x=10 (with zero error)
ff=UPropagate(p1,val,err,[x],[0],10000)
p1.setParameter("par0",val[0])
p1.setParameter("par1",val[1])
val=p1.value([x])
print "val=",val," +- ", ff.getStd()
```

```
c1 = HPlot (); c1.visible () # plot histogram
c1.setAutoRange ()
c1.draw (ff.getHisto ())
```

Listing 10.36 Uncertainty propagation using a Monte Carlo method

The code prints the uncertainty for the complex function defined above. In addition, you will see a histogram with values from the Monte Carlo simulation. The histogram can be rather useful since it shows possible biases: In fact, the output distribution may be asymmetric, or even not normal-shaped, and this should be taken into account when quoting the final uncertainty.

The code shown above uses the class `UPropogate`, but in fact, one can write a more custom code that does not use this class at all. Our next example will make the explanation of the Monte Carlo technique to the uncertainty propagation more intuitive. Let us rewrite the previous example using the histogram class `H1D` and a Gaussian random number generator:

```
from java.lang import *
from jhplot import *

class MyFunc (FNon):
  def value (self, x):
      y = x[0] * x[0] * Math.sqrt (x[0]) * self.p[0] + self.p[1]
      if (x[0] < 5): y = 0
      return y
p1 = MyFunc ("Function",1,2)
val,err = [100,10],[30,4]   # define values and errors
x = 10                       # calculations at x=10
p1.setParameter ("par0",val[0])
p1.setParameter ("par1",val[1])
y = p1.value ([x])

from java.util import Random
rd = Random ()
hh = H1D ("error",50,y-y,y+y)
for i in range (10000):
  p1.setParameter ("par0",val[0] + err[0] * rd.
    nextGaussian ())
  p1.setParameter ("par1",val[1] + err[1] * rd.
    nextGaussian ())
  hh.fill ( p1.value ([x])   )
stat = hh.getStat ()
print "val=",y," +- ",stat ["standardDeviation"]
c1 = HPlot (); c1.visible (); c1.setAutoRange ()
c1.draw (hh)
```

Listing 10.37 Uncertainty propagation using random numbers

Run this example, and you will see the correct answer for the final uncertainty. This code snippet uses random numbers to fill the histogram with the outcomes from the function, and then the standard deviation that represents the output errors is calculated. Note that if you know that the input uncertainties, for some strange reasons, are not distributed using the normal distribution, you should replace the Gaussian random generator with the appropriate distribution. The only difficult part in this code is to define the ranges for the histogram. Here we assumed that the histogram range is symmetric, and the number of the bins is 50. The histogram range should be adjusted if you will notice that the plotted distribution is outside the plotted range.

References

1. jMathTool Java Libraries. http://jmathtools.berlios.de
2. Ziliak S, McCloskey D (2009) The cult of statistical significance. University of Michigan Press, USA
3. Cousins RD, Highland LH (1992) Nucl Instrum Methods A320:331
4. Sinervo PK (1992) Proceedings of the conference "Advanced Statistical Techniques in Particle Physics", Durham, England. arXiv:hep-ex/0208005. Accessed 18–22 March 2002
5. Read AL (2002) Presentation of search results: the CLs technique. J Phys G: Nucl Part Phys 28(10):2693. http://stacks.iop.org/0954-3899/28/i=10/a=313
6. Barlow R (2002) A calculator for confidence intervals. Comput Phys Commun 149(2):97–102
7. Cousins R, Highland V (1992) Nucl Instrum Methods A320:331
8. Stuart A, Ord K, Arnold S (1999) Classical inference and the linear model. Kendall's advanced theory of statistics.Arnold, London
9. Junk T (1999) Confidence level computation for combining searches with small statistics. Nucl Instrum Methods A434:435–443
10. ATLAS (2013) Invariant-mass distribution of diphoton candidates after all selections of the inclusive analysis for the combined 7 TeV and 8 TeV data, ATLAS-PHO-EVENTS-2013-001. Released under the CC-BY-SA-4.0 license. http://cds.cern.ch/record/1605822/

Chapter 11
Linear Regression and Curve Fitting

Interpretation of empirical data using mathematical equations or functions is widely used technique to describe data, explain relationships between variables, or compare data with theoretical expectations. Such an approach is also used for forecasting of future trends.

We start this chapter with the simplest linear case, and then consider curve fitting using arbitrary functions. At the beginning of this chapter we will discuss the elements of the linear regression that has the goal of determining a relationship between two variables by fitting a linear equation to observed data. Then, in the following sections, we will discuss a nonlinear regression, which is often called curve fitting. Finally, we will consider complex fitting procedures with many variables and fitting techniques based on functions which cannot be easily defined by simple analytical expressions.

11.1 Linear Regression

Linear regression is a method of finding a linear correspondence between two data sets. The method is based on a fitting data set by the simple linear function, $y = A + Bx$. The variable x is considered to be an explanatory variable, while the other variable, y, is considered to be a dependent variable. Sometimes, the variable A is called the intercept, and B is the slope of the line.

Before going into the heart of the matter, it is useful to remind that it is always a good practice to visualize the data under consideration before attempting to apply the fit procedure with the linear equation. For example, in the case of two-dimensional data, one can make a contour plot (see Sect. 4.2.4) in order to determine the strength of the relationship between the variables x and y.

We start the discussion of this topic using our usual approach: first, we simulate the input data for linear regression, and then we illustrate how to apply this technique to the simulated data.

© Springer International Publishing Switzerland 2016   399
S.V. Chekanov, *Numeric Computation and Statistical Data Analysis
on the Java Platform*, Advanced Information and Knowledge Processing,
DOI 10.1007/978-3-319-28531-3_11

11.1.1 Creating Input Data

First, let us create a data set in order to illustrate a linear regression analysis. We assume that an explanatory variable x is distributed in accordance with a Gaussian distribution. We also assume that the "dependent" variable y is a function of the explanatory variable plus an additional "noise" simulated using random Gaussian numbers. The noise effect will be added to make our example as realistic as possible. Look at this code:

```
from jhplot   import *
from java.util import *

p1 = P1D ("data")
r = Random (1L) # create reproducible dataset
for i in range (100):
  x = 10 * r.nextGaussian ()
  y = 20 * x + 200 * r.nextGaussian () + 50.0
  p1.add (x, y)
IO.write (p1, "data.jser")
```

Listing 11.1 Creating data for the linear regression analysis

In this example, we fill a P1D using random values x and y (the latter depends on x) and store it in a serialized file "data.jser" for use in the next section. Note that the random numbers were initiated with the seed "1L," so the results can be reproduced every time you run the example.

11.1.2 Performing a Linear Regression

For linear regression analysis of the created data, we use the Java class LinReg from the package jhplot.stat which provides all necessary tools for the linear regression fits. To initialize the linear regression calculation, we should pass our data in the form of P1D to the LinReg() constructor:

```
>>> from jhplot.stat import *
>>> r = LinReg (p1d)
```

We should note that one can also use the usual Java arrays or Python lists for x and y inputs; but here, we concentrate only on the P1D case.

After calling the constructor LinReg with the input P1D object, the calculation is already done and the next step is to extract the linear regression results. First of all, we are interested in the intersect and the slope values of the fit. Both values, including their statistical uncertainties, can be extracted using the following methods:

```
>> print r.getIntercept (), "+/-", r.getInterceptError ()
>> print r.getSlope (), "+/-", r.getSlopeError ()
```

The next step is to display the fitted values. This can be done using the method
`getResult()` that returns an `F1D` function with the results of the linear regression.

```
>>> f=r.getResult()
```

Finally, the prediction and the confidence levels of the linear regression can be
extracted as

```
>>> d=r.getCorrelation()
>>> p1d=r.getResiduals()
```

The first method returns the usual correlation coefficient

$$\rho = \frac{cov(x, y)}{\sigma_x \sigma_y}$$

where $cov(x, y)$ is the covariance. σ_x and σ_y are the standard deviations of x and
y random variables. We remind that the correlation coefficient varies from -1 to 1,
with -1 being for a perfect negative correlation between x and y values, and 1 for a
perfect positive correlation.

The second method returns a `P1D` array with the residuals, i.e., the vertical distances of each point from the regression line.

In addition, one can also calculate the 95 % confidence interval for the regression
result given by the method `getResult()`. This band encloses the true best-fit linear
regression line, leaving a 5 % chance that the true line is outside the confidence
interval boundaries. The confidence interval is the area that has a 95 % chance of
containing the true regression line. This typically means that many points could be
far away from the confidence-interval band.

Finally, we can calculate the 95 % prediction interval. The prediction interval is
the area where 95 % of data points are located. The prediction interval is further away
from the best-fit line than the confidence bands.

Both the prediction and the confidence intervals can be obtained using the following two methods:

```
>>> a1=r.getPrediction()
>>> a2=r.getConfidence()
```

These two methods return two `P1D` objects representing the lowest and the highest
levels for the prediction and confidence level. We remind than one can easily plot
them with the method `c1.draw(obj)`, where `obj` is either `a1` or `a2` shown above,
since this method can be used for displaying lists of `P1D` objects. There are several
other convenient methods that can be found by reading the API documentation of
this package.

Now, let us illustrate how to analyze the data set generated in the previous section. First, we read the data array from the serialized file. Then, we plot the data, the linear regression fit result, the confidence interval and, finally, the predictions:

```
from jhplot   import *
from jhplot.stat import *
from java.awt import Color

p1=IO.read("data.jser")
c1 = HPlot("Canvas",700,500,2,1)
c1.visible(); c1.setAutoRange()
c1.setGTitle("Linear regression")

r = LinReg(p1) # perform linear regression
print r.getIntercept(),  "+/-",r.getInterceptError()
print r.getSlope(),"+/-",r.getSlopeError()
print "Correlation=",r.getCorrelation()

c1.cd(1,1)
c1.draw(p1)
c1.draw( r.getResult()  )
c1.draw( r.getConfidence() )

c1.cd(2,1)
c1.setAutoRange()
c1.draw(p1)
c1.draw(  r.getResult()  )
c1.draw( r.getPrediction() )
```

Listing 11.2 Linear regression analysis

The output of the script is shown below:

```
30.89 +/- 22.120
19.47 +/-  2.149
Correlation= 0.675
```

The output numbers are printed here with a reduced precision. The script also generates the plot shown in Fig. 11.1.

11.2   Curve Fitting

Now, let us come to a general, nonlinear case, when the relationship between x and y values cannot be explained by the linear behavior. In such situations we need to fit data points using an arbitrary mathematical function.

Moreover, each data point can represent a group of events or measurements, thus points can have statistical uncertainties. In addition, the data points can have systematic uncertainties related to the nature of the measurement apparatus. There-fore, points do not have same statistical weights during the fit procedure: points with larger uncertainties should be taken into account with smaller weights.

Let us consider the case when data are represented by a histogram as shown in Fig. 7.1. A histogram height in a certain bin contains a statistical uncertainty. In the case of a simple event-counting measurement, such an uncertainty is related to a variance σ_i^2 in a bin i, where σ_i is the standard deviation, a measure of how spread out numbers are. The numerical value of σ_i is proportional to the square root of the number of entries in each histogram bin for a Poisson process. Thus, one can introduce a weight factor w_i which reflects the degree of influence of each data point on the final parameter estimate. Typically, the weight w_i for the bin i is expressed in the terms of the variance as $w_i = 1/\sigma_i^2$.

Data points in x and y can also be represented by a P1D array with arbitrary statistical or systematic uncertainties. If they are known, one can perform a fit by attributing a weight factor to each point. Of course, errors can be arbitrary, i.e., they do not need to be exactly the same as in the case of counting experiments.

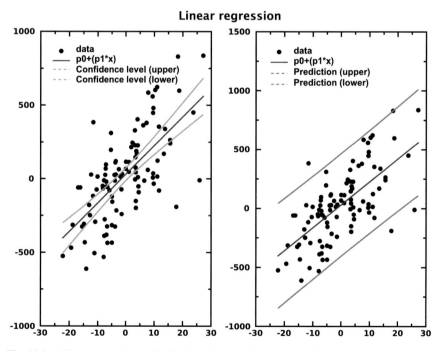

Fig. 11.1 A linear regression analysis using the Java class LinReg. The data are shown as *dots*, while the *lines* show the 95 % confidence level (*left*) and the prediction interval (*right*). The line in the middle of these intervals indicates the result of the linear regression (color figure online)

To start the curve fitting, we should define:

- A mathematical function that is expected to describe the fitted data. The function can correspond to a theoretical model, or to our expectations for trends seen in the data.
- A fit method defining a distance measure between data points and the analytical function.

To find a proper function that is expected to describe data after adjusting its free parameters is a nontrivial task. First of all, we should start with an initial guess for the analytic expression and initial values of free parameters. The number of free parameters should be as small as possible, and the resulting fit quality should be sufficiently satisfactory, so one can claim that the analytical model describes the data. The degree of the fit quality is usually characterized by the χ^2 value:

$$\chi^2 = \sum_{i=0}^{N} \left[\frac{y_i - f(x_i)}{\sigma_i} \right]^2, \tag{11.1}$$

where $f_i(x)$ is a function with free parameters found after the fitting procedure, y_i is a data value in Y, and σ_i is the standard deviation of data entries in the bin i. The fitting procedure tries to find the values of free parameters of the input function, such that the χ^2 has the minimum possible value.

As mentioned before, it is usually a good idea to plot data first to see how the fit function may look like and what initial parameter values may have sense to produce the desired result.

11.2.1   Preparing a Fit

First of all, let us initialize the curve fitting package by creating an instance of the HFitter Java class.

```
>>> from jhplot   import *
>>> f=HFitter()
>>> print f.getFitMethod()
["chi2","cleverchi2","uml","bml","leastsquares"]
```

The example shows how to print the available fitting methods. The most popular method is the so-called "chi2". This method, which is based on the minimization of Eq. (11.1), is the default method used by the HFitter() class if it is instantiated without passing arguments. The χ^2 method works the best if data are represented by discrete points with Gaussian uncertainties. The main assumption is that the points y_i are normally distributed around a function $f(x_i)$, uncorrelated, and have variances σ_i^2. If the number of events in each bin is larger than a certain minimum number (say, several hundred entries), then the distribution of the expected events per bin is approximately Gaussian.

Unlike the χ^2 method, the method `cleverchi2` does not include the variance of experimental data points. This approach is based on the minimization of the following function:

$$\chi^2 = \sum_{i=0}^{N} \left[\frac{(y_i - f(x_i))^2}{|f(x_i)|} \right] \tag{11.2}$$

As before, y_i is the value of a data point, or the height of the bin i if the data are represented by a histogram, x_i is the center of the ith bin, and $f(x_i)$ is the value of the function calculated in the ith bin.

The other two options, `uml` and `bml`, are unbinned and binned maximum likelihood fits, respectively. If the dataset is binned (i.e., when we deal with a histogram), a binned maximum likelihood should be used.[1]

The maximum likelihood method should be used if data are represented by integer numbers of events which follow the Poisson statistics, i.e., for situations with event counts. Experimentally, if you see a distribution with a small number of events in each bin, the best bet would be to try to use the maximum likelihood method. If the amount of data is very small, a larger bin size should be tried. In this case, it is preferable to avoid data binning at all: the fit can be done using the unbinned maximum likelihood method.

The binned maximum likelihood method minimizes the quantity:

$$bml = \sum_{i=0}^{N} [f(x_i) - y_i \ln(f(x_i))]$$

while the unbinned method is based on:

$$bml = -\sum_{i=0}^{N} [\ln(f(x_i))]$$

Finally, the `leastsquares` method assumes that the best-fit curve is the curve that has the minimal sum of the deviations squared (least square error) from a given set of data. Unlike the methods discussed above, this method does not use errors for data points, i.e.,

$$ls = \sum_{i=0}^{N} [y_i - f(x_i)]^2 . \tag{11.3}$$

This method was considered for the linear regression and it will not be discussed anymore.

---

[1]Note that, for a small number of events, the binning can result in a loss of information and large statistical uncertainties for the parameter estimates [1]. On the other hand, a benefit of the binning is that it allows for the goodness-of-fit test.

11.2.2   Creating a Fit Function

Let us assume that we know what method should be used for the curve fitting.
The next question is how to find a function that can describe our data. The package contains several pre-built functions in the HFitter catalog, which can be
printed as

```
>>> from jhplot   import *
>>> f=HFitter()
>>> print f.getFuncCatalog()
["e", "g", "g2", "lorentzian", "moyal",
 "p0", "p1", "p2", "p3", "p4", "p5",
 "p6", "p7", "p8", "p9", "landau", "pow"]
```

This command prints the names of the implemented fit functions as described in
Table 11.1.

There are several ways to create a function to be used for the fit. The simplest
approach would be to pick up a predefined function from the catalog. Alternatively,
one can use a simple script to define any custom function.

If you need to include a custom function to the catalog, use the method addFunc
(name, func), where func is a function initialized by the Java class IFunction,
which was briefly discussed in Sect. 3.7. To use this method is rather straightforward
when working with Java: one needs to implement a Java class using the examples
located in jhplot.fit package. We will discuss this topic later.

Table 11.1 The list of
implemented fit functions of
the HFitter class

Name	Definitions
e	Exponential
g	Gaussian
g2	Double Gaussian
lorenzian	Lorenzian
moyal	Moyal
landau	Landau
pow	Power-law, $a * (b - x)^c$
p0	$y = a$
p1	$y = a + b * x$
p2	$y = a + b * x + c * x^2$
pn	Polynomial of nth order

11.2.2.1 Built-in Fit Functions

Let us create a Gaussian fit function using the predefined key "g" and print all attributes of this function:

```
>>> f.setFunc("g")
>>> func=f.getFunc()
>>> print func
BaseModelFunction:   Title=g, name=g
Dimension: 1, number of parameters: 3
Codelet String: codelet:g:catalog
Variable Names: x0,
Parameters: amplitude=1.0, mean=0.0, sigma=1.0,
Provides Gradient:true,Provides Parameter Gradient:true
Provides Normalization: false
```

One may find this information a bit cryptic, but the important information can easily be found: this output tells that the function is implemented in one dimension and it has three free parameters (named "amplitude", "mean", and "sigma"). One can also see the initial values assigned to these parameters during the initialization.

It should be noted that the object func, which is returned by the method getFunc(), belongs to the class IFunction. Let us illustrate several methods of this class:

```
print "Metatype and implementation:",func.codeletString()
print "Parameter names:",func.parameterNames()
print "List parameters", func.parameters()
print "No of free parameters",func.numberOfParameters()
print "Function title:", func.title()
print "Function value:", func.value([1])
```

The last method is important: it accepts a list of the values at which the function is evaluated. In the case of one-dimensional functions, the list contains only one value. Usually, the printed information is sufficient to understand the function implementation. If you are not sure how the function is implemented, evaluate it at a fixed value.

One can also set free parameters to certain values. For example, one can set the mean of a Gaussian to 10 using this method:

```
>>> func.setParameter("mean",10)
```

One can also set all three parameters by using a list:

```
>>> func.setParameters([100,10,1])
```

which sets the values for "amplitude," "mean," and "sigma" to 100, 10, 1, respectively.

The real power of the pre-built fit functions comes when one needs to go beyond of the simple functional expressions that already exist in the catalog. We can easily construct new functions using simple operations. For example, one can build a Gaussian function plus a second-order polynomial using the string "g+p2", i.e.,

```
>>> f.setFunc("g+p2")
>>> func=f.getFunc()
>>> print func.title()+" has:  ",func.parameterNames()
"amplitude", "mean", "sigma", "p0", "p1", "p2"
```

Now, you can add this function to the catalog as shown in this code:

```
>>> from jhplot import *
>>> f=HFitter()
>>> f.setFunc("g+p2")
>>> func=f.getFunc()
>>> f.addFunc("g+p2",func)
>>> print f.getFuncCatalog()
array(java.lang.String, [ ... "g+p2" ..])
```

11.2.2.2 Building Functions from a String

One can also create a custom IFunction object using an analytic expression, function dimension and the list of parameters. Let us illustrate how to build a parabolic function from a string:

```
>>> f.setFunc("parabola",1,"a*x[0]*x[0]+b*x[0]+c","a,
    b,c")
>>> func=f.getFunc()
>>> print func.title()+" has:  ", func.parameterNames
    ()
parabola has:   ["a", "b", "c"]
>>> print func.codeletString()
parabola:verbatim:jel:1:a*x[0]*x[0]+b*x[0]+c:a,b,c:
```

The setFunc() method takes several arguments: the function title, the number of dimensions, a string representing the function, and the names of the free parameters. The only independent variable of this function is "x[0]". In the case of 2D functions, one can add a second variable "x[1]". The method codeletString() prints the implementation details of this function.

11.2.2.3 Functions from a Script

One can also build a `IFunction` using Jython scripts as discussed in Sect. 3.7. In this case, one can incorporate any complex logic for your functional expression into its definition (like `if-else` statements), or even call external Java libraries to access special functions. We will remind that one should create a custom class which is based on `ifunc` as shown in this example::

```
>>> from shplot import *
>>> class [name](ifunc):
>>>    def value(self, v):
>>>       [equation]
>>>       return [value]
```

where `[name]` is the function name, `[equation]` its functional form. The purpose of the method `value()` is to return the calculated function value. Then, one can instantiate an object of this class which now has all the properties of the `IFunction` class. This object can be used for the fits as usual.

11.2.2.4 Preparing a Fit Function

During the fit procedure, the fit optimization program tries to find the best possible parameters of the function which is expected to describe input data. The minimization starts with some initial values, and proceeds until the best distance measure is found. It is always a good practice to set initial values to some sensible numbers. This can be done using two methods discussed before, either the method `setParameters(list)` or the method `setParameter(name,v)`, where `"v"` is an input value. One can also use the method `setPar(name,v)` of the class `HFitter` (there should not be any confusion here, since only one function is allowed for the fit).

```
>>> f.setPar("mean",10)
```

This sets the parameter `"mean"` to the value 10, assuming that `setFunc("g")` has been called first as shown previously (i.e., `"f"` represents an instance of a Gaussian function).

During the minimization, one can restrict parameter variations to a certain range:

```
>>> f.setParRange("mean", -10, 10)
```

This sets the range $[-10, 10]$ for the Gaussian mean.

Finally, in some situations, one may constrain certain fit parameters. Let us give an example by creating a double-Gaussian function. During the fit, we want to keep the mean of the first Gaussian exactly to be the same as for the second one:

```
>>> f.setFunc("gauss2",1,"N*( a*exp(-0.5*(mean0-x[0])\
>>> *(mean0-x[0])/(s0*s0))+(1-a)*exp(-0.5*(mean1-x[0])\
>>> *(mean1-x[0])/(s1*s1) ))",\
>>> "N,a,mean0,s0,mean1,s1")
>>> f.setParConstraint("mean0=mean1")
```

Here, we had to break up the long string defining the function. The method
setParConstraint() does the trick. Make sure that you have correctly assigned
the names of the parameters during function creation.

11.2.3   Displaying Fit Functions

Once the function is defined, it should be able to display in the usual way. This can
be done using either F1D or F2D functions, after passing an IFunction object to
the function constructor:

```
>>> ff=F1D(f.getFunc(),min,max)
>>> c1.draw(ff)
```

where min and max are the minimum and maximum values used to draw the function,
and c1 represents the HPlot canvas. How to display an F1D function was discussed
in Sect. 3.3. Analogously, one can build and display functions in 2D.

11.2.4   Making a Fit

Next, we will fit our input data with the prepared function. This part is easy: if you
have P1D, H1D, H2D, or PND containers filled with data, simply execute the method
fit(obj), where the obj is one of the objects mentioned above.

   In the case of 1D data, you may also restrict the fitting range with the method
setRange(min,max) applied to the object f. After the fit, one can get the resulting function after the minimization as

```
>>> func=f.getFittedFunc()
```

The method returns a IFunction function with the parameters determined by the
fit minimization.

   Another way to retrieve the results is to use this method:

```
>>> result=f.getResult()
```

The object `result` is sophisticated as it keeps essentially everything you need for retrieving the final fit results. Below, we list the most important methods of this class and comment on some of them:

```
>>> result.fittedParameters()
>>> result.errors()
>>> result.constraints()
>>> result.covMatrixElement(0,1)
>>> result.engineName()
>>> result.errors()           # parabolic errors
>>> result.errorsMinus()      # error -
>>> result.errorsPlus()       # error +
>>> result.fitMethodName()
>>> result.fitStatus()
>>> result.fittedFunction()
>>> result.fittedParameter("mean")
>>> result.fittedParameterNames()
>>> result.fittedParameters()
>>> result.ndf()              # number of degrees of
    freedom
>>> result.quality()          # chi2/ndf for the Chi2
    method
```

We should point out that the quality of the fit can be obtained via the method `quality()`, which represents the χ^2 per degree of freedom. The smaller its value is, the better the fit minimization.

It is time now to give a small example before going any further. Below we generate random data that follow a Gaussian distribution and then we fit the data with a Gaussian function.

```
from jhplot  import *
from java.util import Random

f=HFitter()
f.setFunc("g")
f.setPar("amplitude",50)

h1 = H1D("Data",50, -4, 4)
h1.setPenWidthErr(2)
h1.setStyle("p")
h1.setSymbol(4)
h1.setDrawLine(0)

r = Random()
for i in range(1000):
   h1.fill(r.nextGaussian())

c1 = HPlot("Canvas")
c1.visible()
c1.setAutoRange()
c1.draw(h1)
```

```
f.setRange(-4,4)
f.fit(h1)
ff=f.getFittedFunc()

r=f.getResult()  # get fitted results
Pars    = r.fittedParameters()
Errors = r.errors()
Names   = r.fittedParameterNames()
print "Fit results:"
for i in range(ff.numberOfParameters()):
   print Names[i]+" : "+str(Pars[i])+" +- "+str(Errors
   [i])

mess="&chi;^{2}/ndf="+str(round(r.quality()*r.ndf()))
mess=mess+" / "+str(r.ndf())
lab=HLabel(mess, 0.12, 0.69, "NDC")
c1.add(lab)

f2  = F1D("Gaussian",ff,-4,4)
f2.setPenWidth(3)
c1.draw(f2)
print "Quality=",r.quality(), " NDF=",r.ndf()
```

Listing 11.3 Performing a fit using a Gaussian function

The output of this program is shown below:

```
Fit results:
amplitude : 62.578 +- 2.465
mean : -9.484E-4 +- 0.032
sigma : 0.986 +- 0.023
Quality= 0.990   NDF= 34
```

In addition, the script generates Fig. 11.2 and shows the χ^2/ndf value inside the interactive label.

Fig. 11.2 Curve fitting example using a Gaussian function

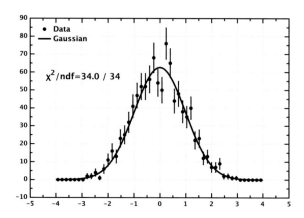

11.3 Real-Life Example. Signal Plus Background

With the curve fitting class in our hand, we now show a realistic example of how to perform a fit of data using a Gaussian plus a background function. As usual, we will do it step-by-step, dividing the example into two parts: in one part, we will generate a histogram with data, and in the second part, we will read this histogram and perform a fit.

11.3.1 Preparing a Data Sample

Let us prepare the input data for the curve fitting. Our data will be in the form of two histograms, H1D and H2D. The 1D histogram is filled with two Gaussian random numbers, one for a signal and the second for the background. The signal distribution is modeled also by a Gaussian distribution after setting its width to a large value. Also, we fill a 2D histogram using the Java class H2D. We use random Gaussian numbers, but this time, we shift their mean values from zero and increase the widths. This histogram will also be used in the example that will be considered a bit later. We will store these histograms in a dictionary, in which a string will be used as a key. Then, we write the dictionary into a binary file "data.jser":

```
from java.util import Random
from jhplot   import *

h1 = H1D("Data",50, -7, 7)
h2 = H2D("3D Data",40,-10,10,40,-10,10)
r= Random(1L)
for i in range(10000):
   if (i<5000): h1.fill(r.nextGaussian())
   h1.fill(5*r.nextGaussian()+5)
   h2.fill(2*r.nextGaussian(),2*r.nextGaussian()+3)
d={"h1":h1,"h2":h2,"description":"Gaussian+background"}
IO.write(d,"data1.jser")
```

Listing 11.4 Preparing histograms for curve fitting

Later we will read the "data.jser" file, and will use the keys "h1" and "h2" to retrieve the histograms from the dictionary.

11.3.2 Performing Curve Fitting

Let us now fit the data prepared in the previous example. After reading the serialized file, we will retrieve the histograms and fit them using a Gaussian function with a first-order polynomial describing the background under the signal peak:

```
from jhplot   import  *
from java.util import Random
from java.awt import Color

d=IO.read("data1.jser")
h1=d["h1"];  name=d["description"]

f=HFitter()
print f.getFuncCatalog()
f.setFunc("g+p1")
f.setPar("p0",10);  f.setPar("amplitude",100)

c1 = HPlot()
c1.visible();  c1.setAutoRange()
h1.setPenWidthErr(2)
h1.setStyle("p")
h1.setSymbol(4)
h1.setDrawLine(0)

c1.draw(h1)
f.setRange(-7,7)  # set  ranges
f.fit(h1)            # perform  the  chi2  fit

ff=f.getFittedFunc()
r=f.getResult()
fPars   = r.fittedParameters()
fErrs   = r.errors()
fNames = r.fittedParameterNames()
print "Fit  results:"
for i in range(ff.numberOfParameters()):
   print(fNames[i]+":"+str(fPars[i])+" +- "+str(fErrs[
     i]))
print "Chi2/Ndf=",r.quality()*r.ndf(),"/",r.ndf()

f2 = F1D("Gaussian",ff,-7,7)
f2.setColor(Color.blue)
f2.setPenWidth(3)
c1.draw(f2)
```

Listing 11.5 Signal + background fit of data

This code snippet prints the fit results:

```
Fit results:
   amplitude : 568.3 +- 11.3
   mean : 0.04 +- 0.02
   sigma : 1.01 +- 0.02
   p0 : 123.6 +- 2.0
   p1 : 17.4 +- 0.4
   Chi2/Ndf= 114/45
```

Fig. 11.3 Fitting a complex histogram with a signal peak together with a broad background using the χ^2 minimization (color figure online)

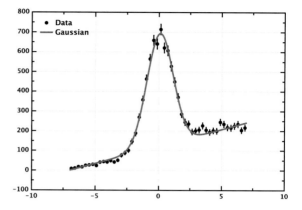

The output plot is shown in Fig. 11.3. The quality of the fit is not particular good for the right side of the peak, but this is just a simple example to illustrate the method.

11.3.3 Fitting Multiple Peaks

This time we will learn how to fit multiple peaks. For this, we will prepare a histogram with three Gaussian peaks plus a background. Unlike the example discussed before, we will create the background distribution using random numbers generated according to the function $10 + 10 * x$ (see Sect. 10.4 for details). Below we show how to do this:

```
from java.util import Random
from jhplot    import *
from jhplot.math import StatisticSample

xmin,xmax=0,20
h1 = H1D("Data",100,xmin,xmax)
f=F1D("10+10*x",xmin,xmax)
p=f.getParse()
max=f.eval(xmax)
r= Random(1L)
for i in range(10000):
  a=StatisticSample.randomRejection(10,p,max,xmin,xmax)
  h1.fill(a)
  h1.fill(0.3*r.nextGaussian()+4)
  h1.fill(0.6*r.nextGaussian()+10)
  h1.fill(0.8*r.nextGaussian()+15)
IO.write({"h1":h1},"data2.jser")
```

Listing 11.6 Creating a histogram with multiple peaks

Note that, for each iteration, the object a is an array with 10 numbers between xmin and xmax. The Gaussian peaks are located at 4, 10, 15 units (note the additive factors) and have different widths, given by the scaling factors, 0.3, 0.6, 0.8. The histogram is saved in a serialized file using a dictionary with the key h1.

Now let us fit this histogram. First, we read the histogram using its key and plot it. Then we fit a region around the first Gaussian, making sure that we fit only the specified range before the second peak starts. Then we add a Gaussian to a new fit function and use the results of the previous fit for initialization of a new function. The code that fits the Gaussian peaks and the background is given below:

```
from jhplot   import  *
from java.awt import Color

d=IO.read("data2.jser")
h1=d["h1"]

c1  = HPlot();
c1.setRange(0,20,0,5000)
c1.visible();  c1.draw(h1)

f=HFitter()
f.setFunc("p1+g")  # first fit
func=f.getFunc()
f.setPar("mean",4);  f.setPar("amplitude",100)
f.setRange(0,7)
f.fit(h1)
ff=f.getFittedFunc();   r=f.getResult()
fPars   = r.fittedParameters()

f.setFunc("p1+g+g")  # Next Gaussian
func=f.getFunc()
func.setParameters(fPars.tolist()+[500,10,0.5])
f.setRange(0,13)
f.fit(h1)
ff=f.getFittedFunc();   r=f.getResult()
fPars   = r.fittedParameters()

f.setFunc("p1+g+g+g")  # Next Gaussian
func=f.getFunc()
func.setParameters(fPars.tolist()+[500,15,0.5])
f.setRange(0,20)
f.fit(h1)

f2  = F1D("Fit function",f.getFittedFunc(),0,20)
f2.setPenWidth(2)
f2.setColor(Color.blue)
c1.draw(f2)
```

Listing 11.7 Fitting multiple peaks

Fig. 11.4 Fitting multiple peaks and a background using the χ^2 minimization

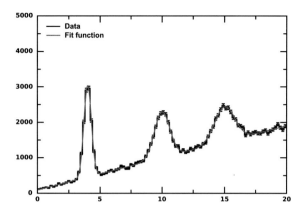

In this example, the list [500,10,0.5] simply specifies the initial parameters for the new Gaussian function, and it is added to the list created from the array fPars holding the parameter values from the previous fit. Similarly, we fit the third peak. Finally, we extract the resulting function and plot it. If you are not sure which parameter names are necessary to use at each step, print the outputs of the methods func.parameterNames() and func.parameters().

The result of the fit is shown in Fig. 11.4.

One can access the fit results using the same approach as in the previous examples: insert the line r=f.getResult() at the end of your code and use the methods of the object r to print the fit quality (χ^2/ndf), fitted parameter values and their statistical uncertainties.

11.3.4 Fitting Histograms in 3D

Let us turn to the H2D histogram filled with data in Listing 11.4, which was not yet analyzed. This time, we have to prepare a fit function with two independent variables. As before, we fit the 2D function using the HFitter class and extract the resulting function using the same methods as for the 1D case. The output should be shown on the HPlot3D canvas. Look at the example below:

```
from jhplot   import *

d=IO.read("data1.jser")
h1=d["h2"]

c1 = HPlot3D("Canvas",800,400,2,1)
c1.visible()

f=HFitter()
f.setFunc("g2D",2,  "N*(exp( -0.5*((mu0-x[0])*\
```

```
            (mu0-x[0])+0.5*(mu1-x[1])*(mu1-x[1]))\
            /(s0*s0) ))","N,s0,mu0,mu1")
f.setPar("N",100);    f.setPar("s0",1.0)
f.setPar("mu0",0.0);  f.setPar("mu1",1.0)
f.fit(h1)

ff=f.getFittedFunc()
r=f.getResult()
Pars    = r.fittedParameters()
Errors  = r.errors()
Names   = r.fittedParameterNames()
print "Fit results:"
for i in range(ff.numberOfParameters()):
   print Names[i]+": "+str(Pars[i])+" +- "+str(Errors[i])

print "chi2/ndf="+str(round(r.quality()*r.ndf()))\
               +"/",r.ndf()

c1.setGTitle("Fitting 2D data")
c1.cd(1,1)
c1.setRange(-10,10,-10,10,0,200)
c1.draw(h1)

c1.cd(2,1)
c1.setRange(-10,10,-10,10,0,200)
c1.draw(F2D(ff))
```

Listing 11.8 Fitting a 2D histogram using an analytic function

The output figure is shown in Fig. 11.5. As you can see, the resulting fit function resembles the original data. The output of the fit procedure also tells about the found parameters and the fit quality:

```
Fit results:
N: 103.6530 +- 1.6990
s0:  1.5142 +- 0.0095
mu0: 0.0575 +- 0.0240
mu1: 3.0047 +- 0.0191
chi2/ndf=1505.0/ 522
```

As you can see, $\chi^2/ndf = 2.9$. This is not an ideal fit, but the value is sufficiently small to declare a reasonable agreement of data with the fit function.

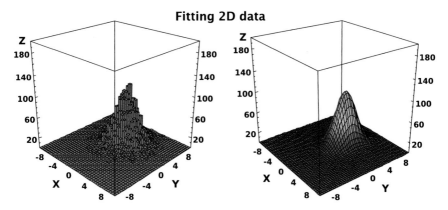

Fig. 11.5 The example shows a 2D histogram used as input for an χ^2 minimization. The resulting fit function after the minimization procedure is shown on the *right*

11.4   Interactive Fit

11.4.1   HFit Method

Data containing two variables, *x* and *y*, which are typically represented by H1D histograms or P1D arrays, can be used for interactive curve fitting. The fit can be done using the HFit Java class from the package jhplot. Assuming that c1 represents the HPlot canvas and h1 is an H1D histogram, one can start this fitter program as

```
>>>a=HFit(c1,h1)
```

The execution of this line brings up a fit dialog that can be used to select the necessary fit function (it must exist in the catalog). One should mention that one can also pass a P1D object instead of the histogram.

The fit panel allows a user to add custom functions, fit methods and perform the fit. One can also set up the initial parameters for the selected fit function using the [Settings] button. Once the fit is acceptable, one can generate the output fit parameters, which will be inserted directly to the editor in a form of Jython code. One can also automatically generate the JAIDA source code which corresponds to the fit. Once inserted, it can further be corrected using the editor. Then, HFit() statement can be removed and the generated JAIDA code can be run manually.

One can also specify a user-defined function and add it to the HFit dialog. This is an example of how to insert a first-order polynomial function to be used for the curve fitting:

```
>>>a=HFit(c1,h1)
>>>a.addFunc("User1","Tool tip","a*x[0]+b","a,b")
```

After script execution, a new function with the name `User1` will be added to the list of known functions. Select it with the mouse and click the button [add]. One can add several functions for the fit. Then, by pressing the button [fit], you will see the fit result.

Let us give a complete example of how to fit data points with statistical errors using a custom function interactively:

```
from java.awt import Color
from jhplot  import *

c1 = HPlot()
c1.visible(); c1.setAutoRange()
c1.setGTitle("&chi;^{2} curve fitting",Color.blue)
c1.setGrid(0,0)
c1.setGrid(1,0)

p1 = P1D("Linear fit")
p1.add(1, 2, 1) # x, y and dy
p1.add(2, 2, 1)
p1.add(3, 5, 1)
p1.add(4, 5, 1)
p1.add(5, 7, 2)
p1.setErr(1)
p1.setErrFillColor(Color.yellow,0.3)
c1.draw(p1)

a=HFit(c1,p1) # run interactive curve fitter
a.addFunc("User1","My fit function", "a*x[0]+b","a,b"
   )
```

Listing 11.9 Fitting $x-y$ data with a second-order polynomial

The select a function, say P2 (a second-order polynomial, $a + bx + cx^2$) from the left dialog. The result of this script, after pressing the button [add] followed by [fit], is shown in Fig. 11.6.

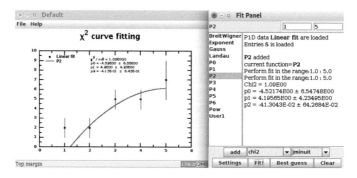

Fig. 11.6 Fitting data with statistical uncertainties using an interactive fitter

11.4.2 JAS Method

There is an alternative method for fitting data interactively. This method is based on JAS [2, 3], a program that was developed for high-energy applications. The original program was redesigned in order to merge with the API of DMelt.

Let us consider an example with three data sources for the `HPlotJas` plot function implementing the the JAS fitter. Run this code and you will see a pop-up window:

```
from   java.util import  *
from jhplot   import  *

h1  = H1D("1D  histogram",20,  -2,  2.0)
h1.fillGauss(500,  0,  1)

p0=P0D("Normal  distribution")
p0.randomNormal(1000,0.0,10.0)

p1=P1D("data  with  errors")
p1.add(1,10,3)
p1.add(2,5,1)
p1.add(3,12,2)
a=ArrayList([h1,p1,p0])
c=HPlotJas("JAS",ArrayList([h1,p1,p0]))
```

Listing 11.10 An interactive fit of $x-y$ data

Then select the data source (say, the histograms) using the left menu and plot it with the mouse click. Next, again using the mouse menu, select a function, and press the "fit" button. You will see the result of the fit. The attractive feature of this program is that one can use the mouse to adjust the initial parameters of the fit function, so the fit will converge quicker, and the result will be closer to what you would expect from the visual examination of data. Figure 11.7 shows the result of the fit using a Gaussian distribution.

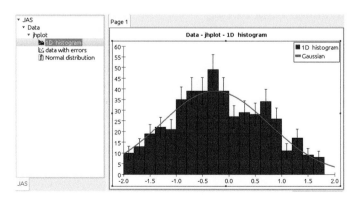

Fig. 11.7 Fitting a histogram with a Gaussian distribution using the `HPlotJas` class (color figure online)

11.5 Polynomial Regression

A polynomial regression is a method to fit data using a polynomial function of nth degree. This type of regression is a special case of the nonlinear regression discussed in Sect. 11.2. The reason why we consider the polynomial regression separately is because there is a simple and intuitive method to fit $x-y$ data or histograms using the Java class called `PolySolve` [4].

How easy is it? Let us see this in the example shown below. First, we create an $x-y$ array, and then we use it as the input for the `PolySolve` class.

```
from jhplot import *
from jhplot.math.polysolve import *
p=P1D()
p.add(10,20);    p.add(20,30)
p.add(30,80);    p.add(40,130)
p.add(50,70);    p.add(60,430)
pl=PolySolve(p)
```

Listing 11.11 Performing a polynomial regression

This code brings up a window with the plotted data. Now you can perform a polynomial fit by selecting a polynomial degree and repeating the fit. Each time you do this, you will see how the fit converges to the designed function that gives the best description of the input data. The output of the code is the found polynomial function. For the above example, the answer is $f(x) = -680 + 164x - 13.7x^2 + 52.5x^3 - 9.16 \times 10^{-3}x^4 + 5.91 \times 10^5 x^{-5}$. Of course, polynomial regressions with smaller number of polynomial degrees are more preferred compared to large-degree polynomials. The complex fit shown above is just for your exploration of this class.

Note that the input for this function can be the usual arrays, or histograms of the class `H1D`. At the time this book was written, the statistical uncertainties on data points were not supported.

11.6 Advanced Data Fitting

Now we will study more sophisticated cases of data fitting. This time we create a "custom" function (see Sect. 3.7) which cannot easily be defined by a single string using the usual approach. Then, we define a "distance measure" between data points and the function values, and then we will apply a minimization procedure using the Minuit package (see Sect. 3.9). We start from the χ^2 method, and then we consider the maximum likelihood method (Sect. 11.2). One of our goals is to show the result of the χ^2 minimization for each iteration step.

Let us prepare the data as shown below:

```
from org.freehep.math.minuit import *
from java.lang.Math import *
from jhplot import *
from java.awt import Color
data=P1D("data")
data.add(5,4,2);          data.add(10,8,3)
data.add(20,10,3);        data.add(30,12,3)
data.add(40,18,4);        data.add(70,20,4)
data.add(90,26,5);        data.add(150,37,6)
data.add(200,50,6);       data.add(300,60,7)
data.add(500,90,9);       data.add(700,120,10)
```

Listing 11.12 Data preparation for fitting

This code creates an array with $x-y$ values and uncertainties on the y values. We also import the needed packages that will be used in the following scripts.

Now, we will define two classes: one class `PowerLawFunc` will contain a custom power-law function used for the fit. The second function, `PowerLawChi2FCN`, defines the distance measure between data and the function. It should return the χ^2 that should be minimized. This code should be appended to the above code with the data definition and package imports:

```
class PowerLawFunc(FNon):
    def value(self, x):
        return  self.p[0]*pow(x[0],self.p[1])
class PowerLawChi2FCN(FCNBase):
  def __init__(self, meas, pos, mvar):
    self.meas=meas; self.pos=pos; self.mvar=mvar;
  def valueOf(self, par):
    pl = PowerLawFunc("PowerLaw",1,2)
    pl.setParameters(par)
    chi2 = 0
    for n in range(len(self.meas)):
      delta = pl.value([self.pos[n]]) - self.meas[n]
      sigma2=self.mvar[n]*self.mvar[n]
      if (sigma2>0): chi2=chi2+(delta*delta)/sigma2
    print "Chi2=",chi2
    return chi2;
```

Listing 11.13 Creating a fit function

Now, let us make another step of defining the minimization algorithm. We will use the MIGRAD algorithm from the Minuit package. The MIGRAD algorithm is considered the most efficient algorithm for function minimizations with the possibility for inclusion of arbitrary free parameters. Let us show how to call it:

```
x=data.getArrayX()
y=data.getArrayY()
err=data.getArrayErr()
```

```
theFCN=PowerLawChi2FCN(y, x, err)
upar=MnUserParameters()
upar.add("p0",1.0,0.1)
upar.add("p1",0.2,0.1);
migrad =MnMigrad(theFCN, upar)
vmin = migrad.minimize()
if vmin.isValid()==False:
    print "Try Alternative strategy"
    migrad = MnMigrad(theFCN, upar, 2)
    vmin = migrad.minimize()
state=vmin.userState()
output=state.params()
print "Fit details:",vmin
```

Listing 11.14 Fitting data using the MIGRAD algorithm

This code should also be appended to the two code snippets given above. The program minimizes the χ^2 values and prints all the needed information about the χ^2 values for each iteration step and the quality of the final minimization procedure.

We also need to visualize the data and the fit result. To do this, append the next code snippet to the above script and run it:

```
c1  = HPlot()
c1.visible()
c1.setRangeX(0,800)
c1.setRangeY(0,150)
c1.draw(data)

pl  = PowerLawFunc("PowerLaw",1,2)
pl.setParameters(output)
ff=F1D(pl,0, 800) # define the range for the function
ff.setPenWidth(2)
ff.setColor(Color.blue)
c1.draw(ff)
```

Listing 11.15 Visualization of the fit results

Fig. 11.8 Fitting data with a custom power-law function using the MIGRAD minimization algorithm (color figure online)

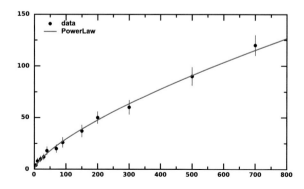

This will produce a plot showing the data and the result of the nonlinear fit with the power-law function. The plot is shown in Fig. 11.8.

Alternatively, one can fit the data using the maximum likelihood method. What we need to do is to redefine the distance measure as

```
class PowerLawLogLikeFCN(FCNBase):
   def __init__(self, meas, pos, mvar):
      self.meas=meas;
      self.pos=pos;
      self.mvar=mvar;
   def valueOf(self, par):
      pl = PowerLawFunc("PowerLaw",1,2)
      pl.setParameters(par)
      logsum = 0
      for n in range(len(self.meas)):
         mu = pl.value([self.pos[n]])
         k=self.meas[n]
         logsum = logsum+(k*log(mu) - mu);
      print "Log likehood=",-1*logsum
      return -1*logsum;
```

The above function calculates the sum of $y_i \log(e_i) - e_i$ and minimizes it. Note the minus sign in the front of this expression. It means that the likelihood sum should be maximal.

Create the object theFCN of this function exactly as in the previous example and run the code. As before, the script prints the value of the maximum likelihood and returns the most optimal description of the data.

We will leave the reader in this place. If you need to learn more, please look at the code of the FreeHEP Java library [5] which includes the MIGRAD algorithm for data fitting.

11.7 Fitting Using Parametric Equations

DMelt offers Java classes for fitting 2D data using a parametric representation of curves. For example, such curves can be eclipses and circles. Typically, such shapes can be specified using parametric equations and functions. In this section, we show how to fit data using ellipses.

Let us randomly generate points around an ellipse given by a parametric equation. The code shown below illustrates how we do this assuming that the ellipse is centered at $(0, -10)$, with a and b being the semi-major and semi-minor axis. We can also define a rotation angle *theta*:

```
from java.awt import *
from jhplot import *
from jhplot.shapes import *
from java.util import Random
```

```
from math import *
from jhpro.fit import *

p1=P1D("Random ellipse")
r=Random()
a=7; b=2; yshift=-1; theta=3.14/3
xshift=0; yshift=-1; step=0.05
ra=[v*step for v in range(0,int(6.28/step))]
for t in ra:     # fill random ellipse.
  x=a*cos(t)*cos(theta)-b*sin(t)*sin(theta)+xshift
  y=b*cos(theta)*sin(t)+a*sin(t)*cos(t)+yshift
  p1.add(x+0.2*r.nextGaussian(),y+0.3*r.nextGaussian())
efit= FitEllipse2D(p1) # initialize fitter
print "Mean distance=", efit.getDistance()
```

Listing 11.16 Step 1: Data preparation

This code imports a number of packages that will be needed for the next step. The above code tries to find the best equation of the ellipse which fits the data. The approach implemented in this code uses the least-squared method discussed in Ref. [6].

Now, we can visualize our results. Simply append the code shown below to the previous example and run the whole program:

```
c1=HPlot("Canvas",500,500)
c1.setLegend(0)
c1.setRange(-10.0, 10.0, -10, 10.0)
c1.visible(1)

ele1=efit.getEllipse()
ele1.setFill(0)
ele1.setColor(Color.red)
ele1.setTransparency(0.6)
c1.add(ele1)

c1.draw(p1)
```

Listing 11.17 Step 2: Fit results

The program will try to minimize the distance measure between data and the predefined ellipse parametric function. Figure 11.9 shows the result of the minimization procedure, and well as the original data. In addition, the code computes the quality of the fit, such as the average distance of points to the ellipse function.

Fig. 11.9 Fitting data using
an ellipse given by a
parametric equation

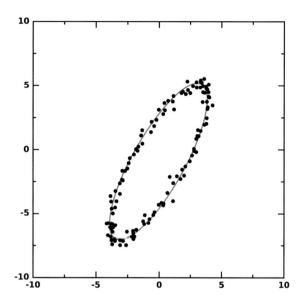

DMelt also contains Java classes to fit data with circles (which should not represent a challenge after showing how fit data with ellipses).

11.8 Symbolic Regression

While reading the previous sections, probably you have been asking one question—if we have some data, how do we know what function should be chosen for curve fitting? Perhaps, you can make some guesses about the function that can describe the data just by examining the data visually. But, often, it is not easy to find the best fit function.

In fact, computers can choose the best fit function using the so-called "symbolic" regression, i.e., an algorithm that searches mathematical expressions to find the analytic model that best fits a given dataset. For such algorithms, mathematical expressions are formed by randomly combining building blocks consisting of functions and constants. The symbolic regression is typically based on the genetic programming [7] that utilizes the same methodology of natural selection found in biological evolution.

We will illustrate a Java algorithm [8] to perform a symbolic regression. The code shown below will attempt to find a target function $f(x) = x^2 + 10 * x$ from input data. Let us create a test $x-y$ data set and build a symbolic regression using several mathematical operations, such as "+," "−," "*". We also assume that the function should include constants. The code shown below runs the symbolic regression algorithm against the input data. Then, it prints the found function, and writes a list with the input data and the functional expression (in the form of Unicode string) to a file. We use 200 iterations for this algorithm (this number should be increased to improve the precision).

```
from com.lagodiuk.gp.symbolic import *
from com.lagodiuk.gp.symbolic.interpreter import *
from jhplot import IO

x=[0,1,2,3,4,5,6]; y=[0,11,24,39,56,75,96]
data=[]
for i in range(len(x)):
    data.append(Target().when("x", x[i]).targetIs(y[i
    ]))
ADD=Functions.ADD  # +
SUB=Functions.SUB  # -
MUL=Functions.MUL  # *
VARIABLE=Functions.VARIABLE
CONSTANT=Functions.CONSTANT

fit=TabulatedFunctionFitness(data)
engine =SymbolicRegressionEngine(fit,["x"],
             [ADD,SUB,MUL,VARIABLE,CONSTANT])
engine.evolve(200)

bestSyntaxTree = engine.getBestSyntaxTree()
currFitValue = engine.fitness(bestSyntaxTree)
print "Iterations=",engine.getIteration()
print "Fit value=",currFitValue
fun=str(bestSyntaxTree.print())
print "Function="+fun
IO.write([fun,x,y],"function.jser")
```

Listing 11.18 Performing a symbolic regression using input data

The output of the code is

```
Iterations= 199
Fit value= 1.67079e-09
Function=(x + ((9.0 + x) * x))
```

As you can see, the analytic function was found correctly, but it is not nicely formatted as the original function used to create test data. Now, let us read the list with the function and data from the saved file, and plot the answer:

Fig. 11.10 Original data and the analytic function found by the symbolic regression algorithm (color figure online)

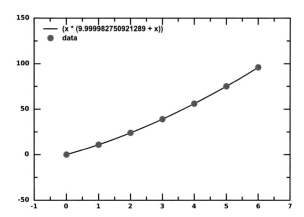

```
from jhplot import *
from java.awt import *

c1=HPlot()
c1.visible(); c1.setAutoRange()

d=IO.read("function.jser")
ff=F1D(d[0],0,6)
p1=P1D("data",d[1],d[2])
p1.setColor(Color.blue)
p1.setSymbolSize(10)
c1.draw(ff)
c1.draw(p1)
```

Listing 11.19 Drawing results of symbolic regression

Figure 11.10 shows the output of the symbolic regression, illustrating a perfect agreement between the original data and the found function.

The algorithm supports the following mathematical blocks for function searches: CONSTANT, VARIABLE, ADD, SUB, DIV, SQRT, POW, LN, SIN, COS (the names are self-explanatory). Generally, the found function can be rather long and confusing. Therefore, in certain cases, one may try to run the symbolic simplification algorithm discussed in Chap. 6.

DMelt also includes an alternative algorithm for the symbolic regression called JGap [9]. We will not discuss it here. How to use this algorithm can be found on the original JGap web page.

References

1. Eidelman S et al (2004) Review of particle physics. Phys Lett B 592:1
2. FreeHep, Jas, Java analysis studio. http://jas.freehep.org/jas3/
3. Donszelmann M, Johnson T, Serbo V, Turri M (2008) JAIDA, JAS3, WIRED4 and the AIDA tag library: experience and new developments. J Phys Conf Ser 119:032016

4. Lutus P, Polynomial Regression Data Fit. Java version. http://www.arachnoid.com/polysolve/
5. FreeHEP Java Libraries. http://java.freehep.org/
6. Fitzgibbon AW, Pilu M, Fisher RB (1999) Direct least-squares fitting of ellipses. IEEE Trans Pattern Anal Mach Intell 21:476
7. Koza J (1992) Genetic programming: on the programming of computers by means of natural selection (complex adaptive systems). Massachusetts IT, USA
8. Lagodiuk Y, Genetic-programming Java package. https://github.com/lagodiuk/genetic-algorithm
9. JGap, Java genetic algorithms package. http://jgap.sourceforge.net/

Chapter 12
Data Analysis and Data Mining

Now, we are equipped with all necessary software tools for data analysis, a systematic process of understanding surrounding our world by means of collecting data, inspecting data, and discovering useful information. Data mining is a part of the process of data analysis. Unlike the latter, data mining focuses on data classification, knowledge discovery, and predictions.

Typically, data analysis includes several phases of scientific research:

- Data gathering, digitization, and transformation to a necessary format. Usually, data come from experimental apparatus.
- Reduction of data volume, structuring, and cleaning erroneous entries where possible.
- Data description, which can usually be done via statistical analysis of data. At this stage, data summary is an important computational task to proceed further.
- Data mining focuses on knowledge discovery and predictions. It aims to identify and classify patterns in data. Data mining is usually an exploration of data that involves artificial intelligence techniques, machine learning, and statistics methods.
- Comparison of data with other data sets, finding interdependence or similarities, and drawing conclusions based on such studies.
- Confronting data with theoretical expectations that can be numerical or analytic. Numerical data modeling and simulation of experimental apparatus can be used when an analytic description is impossible. At this stage, data can also be compared to other data.
- Data visualization, extraction of relevant results, and data interpretation.

As you can see, the topic of data analysis is very broad and cannot easily be covered in a single book. We do not plan to do this. As the reader has already noted, the approach of this book is different. There are plenty of books that go into the depth of certain data-analysis subjects. In this book, I give numerical recipes and complete code snippets which illustrate essentially all the phases in data analysis that are discussed above.

© Springer International Publishing Switzerland 2016
S.V. Chekanov, *Numeric Computation and Statistical Data Analysis
on the Java Platform*, Advanced Information and Knowledge Processing,
DOI 10.1007/978-3-319-28531-3_12

This chapter focuses on rather general programming techniques used for data analysis, with a few (quite realistic) examples at the end of this chapter. A more detailed discussion of data-mining topics based on machine learning and other numerical algorithms will be discussed in the subsequent chapters.

12.1 First Steps in Data Analysis

This section discusses the common analysis techniques used to process and analyze data before final visualization. We will gain insight into how to transform data, reject unnecessary records, add additional information needed for further processing, remove duplicate records, sort data, and many other common operations. We will also learn how to process data in parallel using multiple processors.

From the programming point of view, our discussion is heavily based on the knowledge gained in the previous chapters.

A first step toward understanding of general analysis strategy is to abstract from a particular type of data and focus on generic steps for data processing. They consist of:

Measurement	a determination of numerical values of some observables and collection of experimental results into data files. A group of measured values forms an "event record," or simply, "events." Each event can include a set of records, each consisting of a set of objects (strings, values, multi-dimensional arrays, histograms, etc.). The event record may represent various characteristics of a single observation.
Transformation	a conversion of the event records into the most appropriate format for further data manipulation. This can be done at any analysis step, depending on a concrete situation. If the event records are too big for disk storage, they can be converted into a more compact format. But keep in mind that we may pay for this later with a lower processing speed during data analysis. If data requires a lot of processing and the file storage is not an issue, one can keep data in a less compact format.
Checking integrity	a process of assuring the accuracy of data. If data files have to be moved between different file storages using the network, one should check data integrity. The main question is whether the data records have been corrupted during the process of copying. Data files can also be altered at various stages of data manipulation (intentionally or unintentionally), thus one should always verify if they are valid and contain the original information.
Data skimming	a reduction of data volumes by rejecting uninteresting events. First of all, one should reject events which do not pass selection requirements or do not convey useful information, events

with duplicate records, and so on. The reason for this is simple: runtime for most data analysis algorithms is roughly proportional to the data volume. In addition, one can save disk space when saving data persistently. The choice of skimming criteria is often a balance between many factors.

Data slimming   removing uninteresting information from event records. This means that, instead of dealing with complete event records, one may keep only the most interesting objects to be used for the final analysis. The reason is pretty much the same as for the skimming step.

Data thinning   a reduction of information characterizing separate objects inside remaining event records after the skimming step. Again, the reason for this step is the same as before: it guarantees that one deals with a meaningful part of data records in future.

Data inclusion   adding necessary information to the event records which can simplify data interpretation. This additional information can be derived from the remaining data stored in the event records (or from the rejected information before the previous "slimming" or "thinning" steps). For example, one can add new collections that represent statistical summaries of the event records by imposing some selection criteria.

Building metadata   a process of constructing short event records that characterize either the entire event or distinct objects inside event records. The metadata are usually necessary in order to quickly find necessary events without reading the entire events, if data should be processed more than once. The metadata information can be inserted either inside the event records (using the data inclusion process discussed above) or can be included into a separate file for easy manipulation.

Final preparation   a step that is necessary for the final knowledge-discovery step. It includes a construction of statistical summaries for final data interpretation and classification. Typically, this stage includes a data projection to lower dimensions for data visualization. It can also include data sorting, removal of duplicate entries, or transformation to the most useful format for analysis.

The prepared event records can be used for meaningful predictions on the basis of discovered pattern. For instance, data can be visualized to draw certain conclusions, or used as inputs for machine learning techniques. The prepared data can be processed by the algorithms that discover various interesting patterns. The data can also be used to compare with other data, to make conclusions about similarity of the underlying processes that create such data. Finally, the selected data can be compared with a theory, or used to confirm or refute a hypothesis.

Data skimming, slimming, and thinning are not new concepts in data analysis. However, in recent years, there has been growing interest in such analysis techniques at the Large Hadron Collider constructed by the international community in Geneva (CERN, Switzerland), where such concepts have been reshaped and refined [1].

In this book, our approach will remain the same: instead of being abstract, we will illustrate the above analysis techniques using numerical examples.

12.2   Real Life Example. Analyzing a Gene Catalog

In this section, we analyze a published gene catalog of human chromosome 11 and illustrate most of the analysis steps discussed above. Other aspects of data analysis are discussed in the following sections.

Of course, we cannot discuss the actual measurement of the gene catalog; instead, we will copy the available data from a public domain and then perform the necessary operations using Python scripting. The analyzed data file can be copied from a web page given in Ref. [2]. For easy manipulation, we have transformed the original file into a CSV file. This can be done with any spreadsheet program.

Let us download a CSV file with the gene catalog and display it in a spreadsheet using the Python language:

```
from jhplot.io.csv import *
from jhplot import *
from java.net import *

http="http://jwork.org/dmelt/examples/data/"
xfile="nature04632-s16-2.csv"
r=CSVReader(URL(http+xfile),",")  # read data from URL
SPsheet(r)                         # open in a spreadsheet
```

Listing 12.1 Reading a CSV file from the web

A visual study of this file reveals that each gene symbol is characterized by several records. In fact, each row representing a gene can be considered an "event record" using the terminology adopted above. Each gene has its symbol, name, category, location, length, and so on. We will use this file as a starting point for our examples.

Here is a final comment. Instead of opening the CSV file directly, one can copy it to the local disk using the static class called `jhplot.Web`. This code shows this:

```
from jhplot.io.csv import *
from jhplot import *

http="http://jwork.org/dmelt/examples/data/"
xfile="nature04632-s16-2.csv"
print Web.get(http+xfile)  # print download status
r=CSVReader(xfile,",")     # read local file
SPsheet(r)                 # open in a spreadsheet
```

Listing 12.2 Downloading a CSV file and showing it in a spreadsheet

Now, the CSV file should be downloaded to the directory where you ran this script.

12.2.1 Data Transformation

Now, we will show how to transform the gene event records into a machine-readable form. For many practical applications, this is a necessary step to save the disk space while storing data. This also can lead to the best processing performance while reading data.

The next example illustrates how to read the CVS file (see Sect. 9.6.2.1) and convert the event records, line by line, into a compressed serialized file (see Sect. 9.3):

```
from jhplot.io.csv import *
from jhplot.io import *

r=CSVReader("nature04632-s16-2.csv",",")
f=HFile("nature.jser","w")
i=0
while 1:
  line= r.next()
  if line == None: break
  if (i%100 == 0): print i
  f.write(line)
  i +=1
f.close()
r.close()
```

Listing 12.3 Transforming data records to a binary file

The code looks quite simple: we loop over all rows in the CSV file and serialize them into the file "nature.jser" which keeps the data in a compressed form. The size of the output file is typically three times smaller than the original ASCII file. This Jython code invokes two Java classes, CSVReader and HFile.

12.2.2 Data Skimming

Let us turn to the skimming procedure during which the data volume is reduced. At this stage, we are going to reject uninteresting event records, i.e., we will assume that some rows in the CSV file convey no useful information for further analysis. In particular, we would like to skip all genes with undefined gene names ("symbol"). By examining the CSV file using a spreadsheet program, one can find that a gene symbol is not defined if the first column has the string "undef". So, let us rewrite the file by rejecting such entries:

```
from jhplot.io import *

f1=HFile("nature.jser","r")        # read file
f2=HFile("nature_skim.jser","w")   # write a new file
while 1:
  row=f1.read()
  if row == None:   break
  if row[0] == "undef" or len(row)<18:  continue
  f2.write(row)
f1.close()
f2.close()
```

Listing 12.4 Skimming data records

The output will be saved into the file "nature_skim.jser". Note that the above code includes a simple check that verifies the correct length (18) for the processed rows.

12.2.3   Data Slimming

This time, we are not interested in storing all information within each event record. For example, we would like to remove two last columns in each row, which do not convey any useful information for us, but keeping the same number of rows as before. It takes only a few lines of the code to perform this manipulation

```
from jhplot.io import *

f1=HFile("nature_skim.jser","r") # read input
f2=HFile("nature_slim.jser","w") # write a new file
while 1:
    row=f1.read()
    if row == None:   break
    s=len(row)
    del row[s-1]
    del row[s-2]
    f2.write(row)
f1.close()
f2.close()
```

Listing 12.5 Slimming data records

12.2.4   Data Sorting

Next, we would like to sort the event records using some criteria. Our event records are not simple because they are constructed from several fields. Therefore, the standard use of the sort() method from Python lists will not work. If each row contains several records, sorting rows should be done using values stored in a certain column, thus we must use the standard Python method `sort()` in a somewhat different way.

Sorting data will be performed in a computer memory in order to make the entire process fast. For this example, we will sort the records based on the first column. We will read data into a list, and use the `sort()` method together with the `cmp()` function which provides an effective sorting algorithm. After the sorting procedure, we will write the event records into a new file:

```
from jhplot.io import *

f1=HFile("nature_slim.jser","r")     # read the file
f2=HFile("nature_sorted.jser","w")   # write a new file
data=[]
while 1:
    row=f1.read()
    if row == None: break
    data.append(row)
f1.close()

# compare first ellements
def cmp1(a,b): return cmp(a[0],b[0])

# sort based on first column
data.sort(cmp1)
for k in data:
    f2.write(k)
f2.close()
```

Listing 12.6 Sorting records

In our example, the `sort()` method uses a custom `cmt1()` function which calls the standard Python `cmt()` function. As you may guess, if you want to sort the table using the second column, use this line

```
def cmp1(a,b): return cmp(a[1],b[1]) # 1 means sorting using
    2nd column
```

One can design a totally custom function `cmp1()`. This depends on what you expect from this algorithm. For example, one can construct the `cmp1()` function used in the above example as:

```
def cmp1(a,b):
    if a[0]<b[0]:  return -1
    if a[0]==b[0]: return  0
    if a[0]>b[0]:  return  1
```

Listing 12.7 A custom sorting function

One can also define the function `cmt1()` before the `sort()` command, and include any logic for sorting.

We do not need to use the Python-type data collections as shown in the above example for all data-analysis tasks. One can also use pure Java collections to store and sort elements. Section 2.7 shows how to use the `List` class to store data, and how to sort records using the `Collection` class.

12.2.5 Removing Duplicate Records

Removal of duplicate event records requires a small preparation step. First, let us write a simple function that accepts a list containing other lists and sort the list in accordance with an input index. Then, we scan from the end of the list, deleting duplicate entries based on the input index as we go

```
def unique(s,inx):
    def cmp1(a,b): return cmp(a[inx],b[inx])
    s.sort(cmp1)
    last = s[-1]
    for i in range(len(s)-2, -1, -1):
        if last[inx] == s[i][inx]:
            del s[i]
        else:
            last = s[i]
    return s
```

Listing 12.8 Module "unique.py"

The function returns a list of elements without duplicate entries. We pass a variable `inx` which defines the column number used for the duplicate removal.

Now, we should test this function. Assume we have a list in which each element is another list. We will remove duplicate entries based on the first index (0), and then remove duplicates based on the second index

```
from unique import *

L=[[1,2],["a",10],["a",100],["b",1],["c",2]]
print unique(L,0)
print unique(L,1)
```

Listing 12.9 Testing duplicate removal module

The output of this code is given below:

```
[[1, 2], ['a', 100], ['b', 1], ['c', 2]]
[['b', 1], ['c', 2], ['a', 100]]
```

As you can see, after the first call, we have removed `['a', 10]` (based on the first index), while the element `[1, 2]` was rejected after the second call based on the second index.

Now, we are ready to remove duplicate records in our example of the gene catalog. We will remove duplicate rows based on the second column (index = 1):

```
from jhplot.io import *
from unique import *

f1=HFile("nature_sorted.jser","r")
f2=HFile("nature_unique.jser","w")
data=[]
while 1:
    row=f1.read()
    if row == None: break
    data.append(row)
f1.close()
unique(data,1)
for k in data:
    f2.write(k)
f2.close()
```

Listing 12.10 Removing duplicate entries

12.2.6 Sorting and Duplicate Removal in Java

We should remind that there is no need for the Python-type approach for removing duplicate objects while working with Jython. On can always convert data into the collection Set from the Java package java.util discussed in Sect. 2.7. We have shown that one can use the class HashSet implementation of the interface Set which, by construction, cannot contain duplicates. Moreover, the Java class SortedSet always maintains its elements in an ascending order, so there is no need to worry about sorting and removing duplicates. The reason for switching to Java classes in Jython code depends on many factors, but one obvious reason is that your analysis code can easily be converted into other scripting languages, or to the original Java code.

So, plan the analysis beforehand: if you think that the final data output should not contain the same elements in the output container, just find an appropriate container to keep data records.

While doing data analysis, you may find yourself in a typical situation: one needs to read a huge data sample and remove duplicate records based on some value. One cannot keep all records in the computer memory, but one can still store values used for duplicate removal. In this case, there is no need for loading a full event into the computer memory: just read the file and write the event back as you go. Use the class HashSet to skip the event records with exactly the same description (in our case, this is the first element). When you add a description into the HashSet object, it returns "1" (success) if no duplicate is found and "0" if a duplicate is found. Using this approach, one should be able to process big data files without too much load on the computer memory.

In the next example, we show how to sort and remove duplicate records using Java classes. In particular, the class `TreeMap` will be used for sorting and removal of duplicate objects at the same time. This example is similar to that shown in Sect. 2.7.6. We will extract the first column with the description from the rest of the event record. We use the description of genes as the keys for the `TreeMap` object which removes duplicates and sorts the keys as we fill it in. Finally, we loop over all sorted keys and restore the event records by combining the keys.

```
from jhplot.io import *
from java.util import *

f1=HFile("nature_slim.jser","r")
map=TreeMap() #  map with 1st element  as  a  key
while 1:
    row=f1.read()
    if row == None: break
    map.put(row[0],  row[1:len(row)])
f1.close()

f2=HFile("sorted_unique.jser","w") # write sorted record
for i in map:
    row=map[i]
    row.insert(0,i) # prepends the key
f2.close()
```

Listing 12.11 Sorting and removing duplicates using `java.util`

At this stage, we cannot use the gene catalog created before for our further examples. Our remaining gene records are not sophisticated enough to perform thinning or building a metadata file. The latter techniques will be illustrated later using an appropriate data sample.

12.3 Metadata

We will continue with our discussion of data analysis techniques by diving into the notion of metadata. Metadata is especially useful for analyzing large data files with numerous complex records. To illustrate this concept, we should create a new data set. We cannot use the data sample from the previous section since it lacks the necessary volume and complexity with which we usually have to deal with when the metadata records become necessary for efficient data access.

As mentioned before, metadata is a short data entry that captures the basic characteristics of the entire event or objects inside records. Such characteristics included in small additional record appear useful when one needs to find a necessary information as fast as possible, without reading all data. The metadata concept is especially important when one needs to find a few rare events without reading data again and again. In this case, one can gain a lot in terms of program performance when multiple data processing is required, or when data should be analyzed by many users.

The reason why metadata can be useful for multiple processing is simple: in this approach, a program reads only a small ("metadata") record. Usually, this is neither IO nor CPU consuming. Of course, the metadata records should be first constructed using a priory knowledge about which data characteristics are important for future analysis. We should also note that the process of construction of the metadata files can be rather CPU intensive.

In the next section, we will illustrate the metadata concept using a few short code snippets using Python/Jython scripting.

12.3.1 Using Built-In Metadata File

First, let us create data that will be used for our examples. Each event will consist of a string, representing the record number, and two arrays of the type POD. The size of both arrays is not fixed: it is determined by a Poisson distribution with the mean 500. One array is filled with a uniform random numbers, while the second array is filled with random numbers in the range between 0 and 1. We write 50,000 events into a serialized file "data.jser"

```
from jhplot.io import *
from jhplot import *

ps=math.Poisson(500)
def makeEvent(entry):
  p1,p2=POD("a"),POD("b")
  p1.randomUniform(ps.next(),0,1)
  p2.randomNormal(ps.next(),0,1)
  return [str(entry),p1,p2]

f=HFile("data.jser","w")
for i in range(50000):
  ev=makeEvent(i) # event with records
  if (i%1000 == 0):
     print "event=",ev[0]
  f.write(ev)        # write the event
f.close()
```

Listing 12.12 Creating a file with multiple events

Use this code with caution: it creates rather large file (about 300 MB). Now our task is to find a method that quickly identifies some interesting information without reading the entire file. For example, one can ask this question: How can we find events in which the sum of all elements inside both arrays is above some value? Of course, this can be any other data mining task. We have just picked up this task as it is simple to implement and, at the same time, captures the whole idea of using metadata information.

The code that reads the data file and counts all data records with the sum of all elements above `cut=320` inside both arrays can look like this

```
from jhplot    import *
from jhplot.io import *
import time

f=HFile("data.jser")
start = time.clock()
i=0; cut=320
while 1:
    event=f.read()
    if event == None:
        print "End of events"
        break
    if int(event[0])%1000 == 0:
        print "processed=",event[0]
    p1,p2=event[1],event[2]
    v=p1.getSum()+p2.getSum()
    if v>cut:  i+=1
t = time.clock()-start
print "Nr above "+str(cut)+" =",i,  " after time (s)=",t
f.close()
```

Listing 12.13 Analyzing complex data

The above code also performs very basic benchmarking. The output is listed below:

```
End of events
Nr above 320 = 170   after time (s)= 3.2
```

Of course, the number of the selected events and the execution time can be different for your test.

Now, let us assume that we need to process the created data file many times using different values of the variable `cut` in our event selection. This means that we should build a metadata by constructing an additional event record with the sum of all elements inside the arrays.

We are ready to design such a code using the Java class `HDataBase` discussed in Sect. 9.10.1. As the key value to be associated with each event record, we will use a string that consists of two parts: the event number and a value representing the sum of all elements inside both arrays. The event number and the value with the sum are separated by the underscore character.

```
from jhplot    import *
from jhplot.io import *
import time
```

```
f=HFile("data.jser")
start = time.clock()
db=HDataBase("data.db","w")
while 1:
    event=f.read()
    if event == None:
        print "End of events"
        break
    p1,p2=event[1],event[2]
    v=p1.getSum()+p2.getSum()
    db.insert(event[0]+"_"+str(v),event)
t = time.clock()-start
print "Entries=",f.getEntries(), " time (s)=",t
f.close()
db.close()
```

Listing 12.14 Creating metadata records

After execution of this script, our data will be converted into a flat-file database with the name "data.db". The time necessary to build such a database is a factor 5 larger than that used in the previous example. But this is fine—it will save our time when performing multiple processing.

Now we can read this database by iterating over all keys. We extract the event number and the sum value from the string. Then we access only events that have the sum of all elements above the cut value cut.

```
from jhplot  import *
from jhplot.io import *
import time

start = time.clock()
db=HDataBase("data.db")
keys=db.getKeys()
i=0; cut=320;
while keys.hasMoreElements():
    next = keys.nextElement();
    words = next.split("_")
    if int(words[0])%1000 == 0:
        print "processed=",words[0]
        if float(words[1])>cut:
        i +=1
        event=db.get(next)
t = time.clock()-start
print "Nr above "+str(cut)+" =",i, " after time (s)=",t
db.close()
```

Listing 12.15 Data analysis using metadata

The above code prints:

```
End of events
Nr above 320 = 170   after time (s)= 0.5
```

As you can see, we have obtained exactly the same answer as before, but this time the processing time is a factor 6 faster than when no metadata records were used. Thus, the advantage of using the metadata information is obvious.

12.3.2 External Metadata Files

There are many ways to create metadata records, and many ways to store such records. One can use, for example, the databases discussed in Sect. 9.10, or one can try other approaches. In any event, make some tests and learn what approach works the best for fast data access.

For example, the use of the metadata record as the key to access the needed data from a flat-file database is not the only approach. One can write metadata into an external file, thus completely decoupling the metadata from the main data.

The example below shows how to write a small external metadata file:

```
from jhplot import *
from jhplot.io import *
import time

f=HFile("data.jser")
start = time.clock()

meta=open("data.meta","w")
db=HDataBase("data1.db","w")
while 1:
    event=f.read()
    if event == None:
        print "End of events"
        break
    p1,p2=event[1],event[2]
    v=p1.getSum()+p2.getSum()
    db.insert(event[0],event)
    meta.write(event[0]+" "+str(v)+"\n")
t = time.clock()-start
print "Entries=",f.getEntries(), " time (s)=",t
meta.close()
f.close()
db.close()
```

Listing 12.16 Making an external file with metadata

The file "data.meta" contains rows with the event numbers and the sum of all elements inside the arrays in each event record. The database file contains the keys based on the event number only.

During data processing, first load the metadata file and identify interesting events. Then, read only selected event records using the keys as shown in this example:

```
from jhplot   import *
from jhplot.io import *
import time

start = time.clock()
db=HDataBase("data1.db")
i=0;  cut=320
for line in open ("data.meta", "rt"):
  key, sum = [x for x in line.split()]
  if int(key)%1000 == 0:
        print "processed=",key
  if float(sum)>cut:
        event=db.get(key)
        i=i+1
t = time.clock()-start
print "Nr above "+str(cut)+" =",i, " after time (s)=",t
db.close()
```

Listing 12.17 Using external metadata records

You will see the answer:

```
End of events
Nr above 320 = 170   after time (s)= 0.6
```

It shows a similar processing time as in the previous example which builds a single file-based database.

The approach in which we decouple the metadata file from the actual data can be rather convenient in some situations as it does not require the rebuilding the database file with new metadata entries every time we need to change our metadata definitions.

12.4 Multithreaded Programming

Modern computers typically have multiple number of CPUs or several cores within one CPU. If time-consuming tasks can be performed in parallel, this will increase program performance during data analysis or during numerical computations that typically do not require a lot of input data. In simple words, this is just breaking up a single programming task into pieces to be executed in parallel. The outputs will be combined upon completion of all pieces. The code segments should preferably be independent of each other. This can lead to the highest possible performance by taking advantage of modern multi-core systems.

We remind that the Java virtual machine provides the needed support for multi-threading. This significantly simplifies parallel programming when using Java and Java-based scripting languages. We have already discussed high-level Java methods for linear algebra that use multi-core processors in Sect. 5.1.6. Here, we will discuss how to use several threads and parallel processing for data-oriented tasks. We will create several examples to illustrate how to write analysis programs that contain two or more parts that can run concurrently.

12.4.1 Reading Data in Parallel

One can achieve better performance of a program using multiple threads when I/O bandwidth is not a bottleneck. Obviously, this depends on the hardware and on the actual calculations. Typically, numerical calculations can benefit from the use of parallel cores when the time necessary for processing events is larger than the time needed to access events from a file storage.

Let us illustrate the idea behind the parallel processing of data by looking at some concrete examples. As usual, first we will prepare data samples for our next example. We will make two data samples. Each file will have 500,000 events. Each event will have a record represented by a POD array with 20 random numbers distributed between 0 and 1:

```
from jhplot.io import *
from jhplot import *

f1=HFile("random1.jser","w")
f2=HFile("random2.jser","w")
for k in range(500000):
  p=POD()
  p.randomUniform(20,0,1.)
  f1.write(p)
  p.randomUniform(20,0,1.)
  f2.write(p)
  if k%1000==0: print "done=", k
f1.close()
f2.close()
```

Listing 12.18 Creating data files with arrays

Next, we will create a small class which reads the files "random1.jser" and "random2.jser". This class should accept a string with the file name, open this file and read the file record-by-record. We will refill the records, constructed from single POD array, with 200 random numbers distributed in accordance with a normal distribution. The Python threads discussed in Sect. 2.12 can be used to perform this task:

```
from jhplot.io import *
from jhplot import *
from threading import Thread

class testit(Thread):
    def __init__ (self,fin):
        Thread.__init__(self)
        self.fin = fin
        self.k =0
    def run(self):
        f1=HFile(self.fin,"r")
        self.k=0
        while 1:
            row=f1.read()
            if row == None:   break
            if self.k%10000==0:
                print self.fin+" done=",self.k
            row.sort()
            row.getStat()
            self.k=self.k+1
        f1.close()
```

Listing 12.19 Module "readthread.py"

Our new class inherits the `Thread` class, i.e., we sub-classing the `Thread` interface and overwrite its initialization and the method `run()`. In the method `run()`, we open the file to read its records. Then, we sort the 1D array and evaluate its statistical characteristics.

In the example below, we will read the two data files using two independent threads. Then, we perform a simple benchmarking by printing the total number of processed events and the time needed for our calculation:

```
from readthread import *
import time

print "Start at",time.ctime()
list=["random1.jser","random2.jser"]
# list=["random1.jser"] # uncomment it for second test
t1 = time.clock()
tlist = []
for f in list:
  cu=testit(f)
  tlist.append(cu)
  cu.start()

for t in tlist:
  t.join()
  print "From ",t.fin," processed: ",t.k
t2=time.clock()-t1
print "calculation takes",int(t2)," sec"
```

Listing 12.20 Reading data in two threads

The code reads the files in parallel. But you can always read a single file by removing one file name from the list (we have commented out this case). If your computer has several computational cores, then what you will likely see is that the total processing time for two files will be smaller than the processing time for a single file times a factor 2. For the author's computer with 8 cores, 3 s were spent to read two files in parallel, and the same time was spent to read one file (to do this test, simply uncomment the line shown in this example).

The advantages of this approach are obvious. If data are distributed between several files, one can read these files in parallel, thus one can process a larger volume of data for the same time period. We were able to parallelize our program without any effort since the Java virtual machine takes care of how to perform such parallelization.

We should note that it does not make much sense in using more threads than the number of available cores: if the number of cores is less than the number of threads, then you will end up in an illusion that you are still reading data in parallel; this will not be quite true, since threads will share the same computational cores and you will not gain higher performance in comparison with the case when the number of threads is equal to or less than the number of available processing cores.

While processing multiple files, you may face a bottleneck in the read–write speed, which is typically in the range of 50–100 MB/s for standard hard drives. The disk throughput is significantly larger for solid-state drives. To understand the hardware limitations, make the following calculation: run over a single data file, and estimate the read speed per second in the MB/s units. Then calculate how many threads you can use without reaching the disk speed limit. It is useful to make experiments with several files, and record what you see. At some point, you will observe that the increase in the number of threads does not change the performance of your program. This will be an indication of the disk I/O limitation, and that you have reached the allowed data throughput of your hardware.

12.4.2   Reading a Single File in Parallel

In the above example, we have considered the case when parallel processing of data is achieved by dividing data into several files. In principle, it is not too difficult to parallelize a program designed to read a single input file. As before, we should note that this approach makes sense for CPU intensive calculations which require input data.

Chapter 9 discussed the `PFile` class which can store data sequentially using the Protocol Buffers format. One important feature of this Java class is that one can access a particular event record using the method `read(inx)`, where `inx` specifies the position of an event record inside the file. This is important feature for parallel calculations, since one can jump to a necessary record within a file without uncompressing the entire file.

Let us use the class `PFile` for our next example. We will create a file with 1D arrays using the same approach as before:

```
from jhplot.io  import  *
from jhplot  import  *
import time

f=PFile("test.jpbu","w")
start  =  time.clock()
for i in range(90000):
  if (i%1000  ==  0):
     print "pocessed=",i
  p0= P0D("event"+str(i))
  p0.randomNormal(1000,0.0,1.0)
  f.write(p0)
print "PFile  time  (s)=",time.clock()-start
f.close()
```

Listing 12.21 Data writer. Module "pwrite.py"

Execution of this script creates the file "test.jpbu". Be careful since this file is rather large—typically, it has a size of 700 MB.

Next, we will prepare a module that takes the file name, initial and final position of records within the input file and a name of the thread which calls this class:

```
from jhplot.io import *
from jhplot import *
from threading import Thread

class testit(Thread):
   def __init__ (self,fin,process,i1,i2):
      Thread.__init__(self)
      self.fin = fin
      self.process=process
      self.k =0
      self.i1,self.i2 = i1,i2
   def run(self):
      f1=PFile(self.fin,"r")
      self.k=0
      for j in range(self.i1,self.i2):
         row=f1.read(j)
         row.sort()
         row.getStat()
         if row == None:  break
         if self.k%1000==0:
             print self.process+" done=",self.k
         self.k=self.k+1
      f1.close()
```

Listing 12.22 Module "readthread2.py"

The program reads the array, sorts it, and performs the calculation of all major statistical characteristics.

Now, let us check the created program. We will read all event records in a single thread as shown below:

```
from readthread2 import *
import time

print "Start at",time.ctime()
t1 = time.clock()
tlist = []

cu=testit("test.jpbu","cu",1,90000)
tlist.append(cu); cu.start()

for t in tlist:
    t.join()
    print "From ",t.fin," processed: ",t.k
t2=time.clock()-t1
print "calculation takes",int(t2)," sec"
```

Listing 12.23 Data reader using 1 thread

Record the time needed to run this script. For the calculation in parallel on a multi-core computer, we will create three threads. Each thread will read some fraction of data. The total number of the processed events will be exactly as in the single-threaded example:

```
from readthread2 import *
import time

print "Start at",time.ctime()
t1 = time.clock()
tlist = []

fi="test.jpbu"
cu = testit(fi,"cu1",1,30000)
tlist.append(cu); cu.start()

cu = testit(fi,"cu2",30001,60000)
tlist.append(cu); cu.start()

cu = testit(fi,"cu3",60001,90000)
tlist.append(cu); cu.start()

for t in tlist:
    t.join()
    print "From ",t.fin," processed: ",t.k
t2=time.clock()-t1
print "calculation takes",int(t2)," sec"
```

Listing 12.24 Reading data in parallel using three independent threads

Run this code and compare the answers. For a computer with more than three cores, the second program with three threads is typically 20–40 % faster. This is not due to a faster disk I/O (which is still the same!), but due to the fact that the program can perform certain tasks in parallel. One can easily check this by removing sorting and statistics calculations from the file "readthread2.py".

12.4.3 Numerical Computations Using Multiple Cores

Now, let us move on and consider CPU-intensive programs that do not require access to data. Thus, instead of reading and processing objects using extensive I/O, we will perform numerical calculations without any input data. This topic was discussed earlier in Sect. 5.1.6.

Again, we will prepare a class that inherits the properties of the Thread class. This time we pass a two-dimensional function F2D discussed in Sect. 3.4. In this example we will calculate the sum of all values in a double loop from 0 to 5000 inside the method run():

```
from threading import Thread

class functhread(Thread):
    def __init__ (self,func):
        Thread.__init__(self)
        self.func = func
        self.sum=0
    def run(self):
        for i in range(5000):
            for j in range(5000):
                self.sum += self.func.eval(i,j)
```

Listing 12.25 Module "functhread.py"

Next, we will perform numerical evaluations of four different functions in parallel using four threads, assuming that you have a computer with more than three processing cores:

```
from jhplot import *
from functhread import *
import time

print "Start at",time.ctime()
f1=F2D("cos(x*y)*x+y*0.5*cos(x)")
f2=F2D("cos(x*y)*y-x*2*cos(x)")
f3=F2D("cos(x*y)*x+y+sin(y)")
f4=F2D("cos(x*y)*y-x-sin(x*y)")
li=[f1,f2,f3,f4] # test with 4 functions
# li=[f1] # uncomment this line for single thread
t1 = time.clock()
```

```
tlist = []
for f in li:
  current = functhread(f)
  tlist.append(current)
  current.start()

for t in tlist:
  t.join()
  print "From ",t.func.getTitle()," processed: ",t.sum
t2=time.clock()-t1
print "calculation takes=",int(t2)," sec"
```

Listing 12.26 Evaluation of four functions in parallel

Run this script using the list of F1D functions as given in this example. Then, make a new test: uncomment the list with a single function f1. You will see that, for four functions, the calculation time is only insignificantly larger than the time necessary for the evaluation of a single function. For the author's computer, the time needed for the calculation of one function was 20 s, while 32 s required to process four functions. Again, this can be seen if one uses a computer with multiple cores.

The conclusion of this section is simple: take advantage of multiple cores that speedup calculations and make maximum use of your computer. First, think about how to split your program into independent segments to increase the efficiency of your calculations. Then, write a custom class based on the `Thread` class and use the `run()` method to perform the necessary calculation.

12.5   Data Consistency and Security

While your data analysis code grows, one may start to worry about integrity and consistency of data objects, which are either stored in the computer memory (non-persistent data) or saved in files (stored persistently). The integrity typically means that data are accurately defined by the data model used to create data, while consistency means that data do not contain wrong information, and satisfy possible constraints imposed on event records.

Surely, any data container can easily be modified inside an analysis code. In order to prevent deliberate (or accidental) modifications of data, one should find fingerprints of data objects, and compare them with those generated during the initialization. This should be done at runtime, i.e., during the code execution.

More importantly, it is often necessary to check consistency and integrity of data files on web sites, to make sure that the files have not been modified by someone. Again, what you will need is a fingerprint of your data, or some sort of unique signatures.

12.5.1 MD5 Fingerprint at Runtime

Maintaining data consistency is not too difficult for simple data containers: for example, you can use the well-known MD5 algorithm. The procedure is as follows: a publisher should provide a MD5 signature of an object or a file. Then a user confirms it by calculating the signature again, and comparing it to the one provided by the publisher. If the signatures do not match, then this will indicate that the object or the file was altered since it was created.

Below, we will generate an MD5 hash of a P1D array at runtime. This is simply a string with a hexadecimal number which reflects the counts of each byte of your object. Any recreation or modification of this object at runtime leads to a modification of this string.

We use the class MD5 which accepts any Java object as an argument. After generating an MD5 string, we regenerate it again after modifying the P1D container by adding a row of data:

```
>>> from jhplot.security import *
>>> from jhplot import *
>>> p1=P1D("test")
>>> p1.add(10,20)
>>> md5=MD5(p1)
>>> print md5.get()
da14f51d302dd6756457f16f989b8eb9
>>> p1.add(20,30) # now we will modify the array
>>> md5=MD5(p1)
>>> print md5.get()
c5671fff0834bcc09489798ec1f3f752
```

The method get() returns a hex-string representing the array of bytes of the input object. It should be noted that the printed strings in your case could be rather different from those shown here, since every new creation of objects leads to a different fingerprint. The important thing to remember is that they do not change at runtime.

One can perform a similar fingerprint using the Python module md5. This approach works for the DMelt objects that are convertible into strings. Usually, this means that such objects implement the method toString(). The syntax is somewhat different since we pass a string to the md5 instance:

```
>>> from jhplot import *
>>> import md5
>>> p1=P1D("test")
>>> p1.add(10,20)
>>> m=md5.new() # run MD5 fingerprint
>>> m.update(p1.toString())
>>> print m.hexdigest()
c5671fff0834bcc09489798ec1f3f752
```

As before, the string returned by the method `hexdigest()` can be rather different in your case.

There are a number of other algorithms that overperform the MD5 for integrity checks. For example, the SHA-256 method leads to a considerably stronger hash compared to the MD5 algorithm. You can call this algorithm as

```
>>> from jhplot import *
>>> import hashlib
>>> p1=P1D("test")
>>> p1.add(10,20)
>>> hash = hashlib.sha256(p1.toString())
>>> print(hash.hexdigest())
```

Note that the SHA-256 method is somewhat slower than the MD5 algorithm.

12.5.2 Fingerprinting Files

Analogously, the class `MD5` can calculate the MD5 checksum for file integration testing. In the case of files, a publisher should provide an MD5 signature, and a user must confirm it by calculating the signature and comparing it to the one provided by the publisher. If the signatures do not match then this will indicate that the file was altered.

This time, the fingerprint string is exactly the same as long as a file is not modified on the file system. Below, we calculate an MD5 checksum for the file `"file.txt"` and print it.

```
>>> from jhplot.security import *
>>> from java.io import *
>>> md5=MD5( File("file.txt") )
>>> print md5.get()
```

The string returned by the method `get()` is unique, and can be used for detecting file corruptions or deliberate modifications, since the chance of accidentally having two files with identical MD5 checksums is tiny: the MD5 algorithm creates a 128-bit (16-byte) hash value, therefore, the chance that two files have identical MD5 hex strings is 2^{-128} ($\simeq 2.9 \times 10^{-39}$).

In many cases, the MD5 checksum is not needed. If the file size is sufficiently large, the checksum algorithm is rather slow. In situations where you need to perform a fast check of data consistency, you can simply look at the file size in bytes

```
>>> import os
>>> print os.path.getsize("file.txt")
```

This command prints the file size in bytes. This is a perfectly acceptable approach for large files when requirements for data integrity and consistency are not very strong.

12.6 Real-Life Examples

Now, we have all needed information for performing a realistic data analysis on the Java platform. Unlike the previous chapters with programming recipes, this section contains extensive real-research examples based on Java scripting.

Here, we will learn how to simulate data and how to use data from the real world. We will show how to perform a full-scale data analysis to make a first step in extraction of knowledge about the underlying nature of data. We hope that the self-contained examples given in the following sections can provide the computational basis for conducting your own research projects using elements of statistics and numeric modeling.

12.6.1 Single-Particle Densities

In this section, we will analyze particle distributions, starting from a single-particle distribution, and then moving toward studies of multi-particle densities, correlations, and fluctuations.

As usual, before doing actual analysis, first we should prepare an event sample to be used for our research. We will create it in almost exactly the same way as scientists usually do when they model real-world data: we will simulate data samples using some mathematical principles. The simulated samples will help us understand and explain the underlying dynamics in the observed data, especially if there are missing data or data distorted by experimental apparatus.

We will consider the following experiment: particles created (or observed) in an experiment are counted in certain phase-space intervals. "Particles" could be real elementary particles produced by an accelerator machine or by cosmic rays, photons produced by a light source, and so on. The notion of "particles" can be used in a very generic way. For example, one can think about particles as people coming to a store during a certain time interval, cars entering a highway, telephone calls per time unit, and so on.

Each particle can be characterized by the same continuous characteristics. The measured number (N) of particles will depend on the time interval or/and the size of some spatial region where particles are detected.

12.6.1.1 Preparing a Data Sample

To be more specific about the distribution of particles, let us consider a Rutherford-like experiment, when a beam of particles from a radioactive source (or an accelerator) strikes a target (say, a thin foil). A special detector counts all particles after interactions with the target and measures the scattering angle θ. The distribution of particles is expected to be proportional to

$$\frac{1}{(1 - \cos(\theta))^2}. \tag{12.1}$$

This is the famous Rutherford formula for scattering of particles from a given nucleus. This form of the scattering angle is a signature for scattering off a point-like object without structure (all such point-like objects are inside the thin foil). A deviation of the measurement from this function is an indicator of structure inside the target.

The code to be shown below will be constructed from the code snippets that have already been discussed in the previous chapters. We create a Poisson distribution with the mean 10 (this can be any number!), and distribute particles in each event using our expected functional form in Eq. (12.1). In addition, we will need to store extra information about the incoming particle intensity, which was set to 1000 (using some arbitrary units). To do this, we create a dictionary at the very beginning of the file record.

```
from jhplot   import *
from jhplot.io import  *
from jhplot.math.StatisticSample   import   *

Xmin,Xmax=0.1,  1
f=F1D("0.01/(1-cos(x))^2",Xmin,Xmax)
p=f.getParse()
max=f.eval(Xmin)
f=HFile("events.jser","w");
pos=math.Poisson(10)
f.write(  {"Intensity":1000}  )
for  i  in  range(10000):
    if  (i%1000  ==  0):   print  "Event=",i
    a=randomRejection(pos.next(),p,max,Xmin,Xmax)
    f.write(a)
f.close()
```

Listing 12.27   Creating a data set using the Rutherford formula

The code writes events to the file "events.jser," where each event consists of a random number of particles with the scattering angles distributed as described above. The program is scalable to any arbitrary number of events, since we do not leave any objects in the computer memory, unlike the case when we create a list of events and then write it into a file.

12.6.1.2   Analyzing Data

Now, we come to the analysis of the event sample generated in the previous section. We remind that the file "events.jser" contains everything we need: the intensity of the incoming beam of particles and the event records for multiple observations. Each event contains a certain number of particles characterized by a some value (in our example, this is the cosine of the scattering angle).

Our task is to read all events from this file and plot what is called a differential cross section calculated as

$$\frac{d\sigma}{d\theta} = \frac{N}{\Delta \cdot L}$$

where L is the intensity of the incoming particles and Δ is the width of the histogram bin where particles are counted. As you can see, the distribution above is nothing but a probability density discussed in Sect. 7.1.1 scaled by the intensity value.

The distribution above conveys information about: (1) The shape of the particle distribution. Note that the distribution itself does not depend on the chosen bin size since we divide the counted number of particles in each bin by the bin size; (2) the intensity with which events are happening. This is due to the scaling of the density distribution by the intensity number, L.

Integration of the distribution must give a total cross section or a total rate of interactions, σ, which is simply $\sigma = N_{tot}/L$, where N_{tot} is the total number of counted particles after the interaction.

The code below shows how to extract the differential distribution and then visualize it:

```
from jhplot import *
from jhplot.io import *

f=HFile("events.jser")
a=f.read()
Intensity=a["Intensity"]
print "Intensity=",Intensity
h=H1D("Rutherford Scattering",40,0.1,1)
while(1):
    p=f.read()
    if p == None:
        print "End of events"
        break
    n=f.getEntries();
    ntot=len(p)
    if (n%1000 == 0):
        print "Event=",n, " particles=",ntot
    h.fill(p)
f.close()

c1 = HPlot()
c1.visible(); c1.setAutoRangeAll()
c1.setMarginLeft(100) # make more space for Y-label
c1.setLogScale(1,1)
c1.setNameX("cos &theta; [rad]")
c1.setNameY("d&sigma; / d cos &theta;")
h.scale(1/(h.getBinSize()*Intensity))
h.setFill(1)
c1.draw(h)
```

Listing 12.28 Analyzing a particle distribution

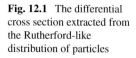

Fig. 12.1 The differential cross section extracted from the Rutherford-like distribution of particles

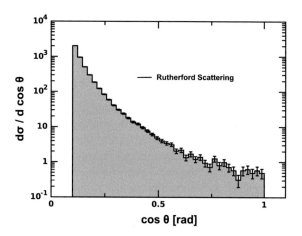

The code reads the data file and then extracts the intensity of the beam L using the known key. Then, we read the event records one by one, using the loop over all entries until the end of the file is reached. For each event, we extract the array of numbers representing the particle angles and then use this array to fill a histogram. Note that the method `fill(obj)` accepts arrays (not single values!). Then, we scale the histogram using the bin width and the intensity value. The resulting plot is shown in Fig. 12.1.

One can use the methods of the `H1D` class (see Sect. 7.1) to access a detailed information about this distribution, or even perform a fit as discussed in Sect. 11.2 to make sure that we indeed observe the famous scattering-angle cross section given in Eq. (12.1).

12.6.2 Fluctuations and Correlations

Now, we will discuss correlations between particles. The notion of correlations is directly related to the question of fluctuations inside small intervals [3]. Both phenomena ultimately imply the existence of forces between particles.

In the previous example, we distributed particles independently, ignoring the fact that this may not be true in real situations. This time, our plan is to model a data sample by introducing interdependence between particles. Then, we will measure the strength of correlations using several statistical tools included in DMelt.

12.6.2.1 Building a Data Sample for Analysis

We will assume that each particle can be characterized by a continues value y. For example, this can be a particle velocity, or a momentum, or the time at which a particle is detected. First, we will consider a sample without any correlations in the variable y, and then we will introduce correlations between particles.

In the simplest case, we assume that the number of particles falling into the measured interval $\Delta y = y_{max} - y_{min}$ is distributed in accordance with a Poisson distribution:

$$P_n = \frac{\lambda^n \exp(-\lambda)}{n!}$$

where λ defines the average number of particles in the counting interval, Δy. This distribution naturally arises when the number of occurrences of rare events is measured in a long series of trials. Inside of the interval Δy, particles are distributed independent of each other.

Let us prepare such an event sample, and save it in a file:

```
from jhplot   import *
from jhplot.io import *
from cern.jet.random.engine import *
from cern.jet.random import *

Ymin,Ymax=0,1
f=HFile("sample1.jser","w")
ps=Poisson(50,MersenneTwister() )
for i in range(10000):
    if (i%1000 == 0):  print "Event=",i
    p=P0D("Event="+str(i))
    ntot=ps.nextInt()
    p.randomUniform(ntot,Ymin,Ymax)
    p.sort()
    f.write(p)
print "Entries written =", f.getEntries()
f.close()
```

Listing 12.29 Creating data with independent particle production

This script assumes 10,000 generated events, with each event containing a certain number of particles arriving inside the interval [0, 1]. The Poisson distribution with the average 50 specifies the total number of particles per event. Then, we distribute particles uniformly inside the interval [0, 1] using the method randomUniform(). Each event is put into a P0D array which is then written into a serialized file.

To understand what is written, it is instructive to write the data into an XML file using the HFileXML() class, instead of HFile(). One can view such a file using any favorite editor, but to keep large data volumes in the XML files is not recommended since its size is a factor eight larger than in the case of the compressed serialized file.

So far, we have considered an idealized situation when particles were produced totally independent of each other. For each event, particles produced in certain regions do not know anything about the presence of other particles.

Actually, this almost never happens in reality: there are always interactions between particles which can be detected by looking at particle correlations at any

given event. For example, for particle or cosmic-ray physics, particles can interact
in accordance with some underlying dynamics.

Let us consider another example from real life. Assume that our "particles" are
shop customers counted in certain time intervals. Usually, people come to stores
in groups of relatives and friends, thus customers are never independent of each
other. Of course, such bunching of people in certain time intervals is due to a social
mechanism, which can be studied using various statistical tools.

There are two types of interactions. One type is attraction, which can be characterized by a positive correlation. In fact, this is the most frequent situation in our
example with the number of visitors per time interval. The second type of interaction
is a repulsion, which features anticorrelations. Probably, you have already guessed
what it means: this corresponds to a hypothetical town where everyone hates each
other and people avoid meeting other people in public places.

Now, let us introduce interactions between particles. Positive correlations will be
modeled by shifting particles closer to each other, while in the case of anticorrelation,
we will increase the distance between particles. Before doing this, let us determine
the distribution of distances between particles. This can be done with this modified
code:

```
from jhplot   import *
from cern.jet.random.engine import *
from cern.jet.random import *

Ymin,Ymax = 0,1
h=H1D("delta",100,0,0.2)
ps=Poisson(50,MersenneTwister() )
for i in range(10000):
    if (i%1000 == 0):  print "Event=",i
    p=P0D("Event="+str(i))
    ntot=ps.nextInt()
    p.randomUniform(ntot,Ymin,Ymax)
    p.sort()
    for j in range(p.size()-1):
        delta=p.get(j+1)-p.get(j)
        h.fill(delta)
c1 = HPlot()
c1.visible()
c1.setAutoRange()
c1.draw(h)
c1.drawStatBox(h)
```

Listing 12.30 Modelling distances between particles

The code above calculates the distances between particles in an event and plots the
corresponding histogram, see Fig. 12.2. The distribution has an exponentially falling
shape with the mean (and RMS) being around 0.02.

Try to fit this histogram using the method discussed in Sect. 11.4. This can be
done by appending the lines to the above code snippet:

Fig. 12.2 The distribution of distances between two particles in the created event samples

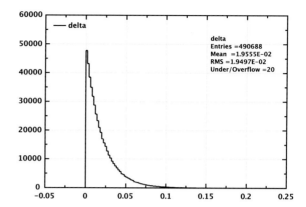

```
a=HFit(c1,"c1",h,"h")
a.addFunc("User1",  "Tooltip",  "b* exp(-a*x[0])","a,b")
```

and rerun the script. You will see a dialog window for interactive curve fitting. Select the "User1" function from the list on the left, and click the "Add" button. Then, push the "Fit" button. You will see that the histogram with the distances is well described by the function $1.8 \times 10^4 \times x^{31}$.

Here, we did not discover anything new: the exponential distributions are known to describe distances between events with uniform distribution in time or space. The distribution above is proportional to $\lambda \exp(-n\lambda)$, with the mean $1/\lambda$ and variance $1/\lambda^2$. It is obvious that λ depends on the full phase-space and on the average number of particles per event.

Our task is not to go into the details of this distribution, but try to introduce interactions between particles by either shifting them closer to each other (correlations), or moving them away from each other (anti-correlations).

Let us write a small function for this operation. We assume that the input for such a function is a POD object which keeps information about the original particle locations in an event. The output is a new POD with a modified distance between particles. We define R as a radius of interactions and will modify the distance between particles only if the size between any two particles is smaller than R. In this case, we will set a new interparticle distance as

```
from jhplot   import  *
from jhplot.io  import  *
from cern.jet.random.engine import  *
from cern.jet.random import  *

Ymin,Ymax=0,1
f=HFile("sample2.jser","w")
ps=Poisson(50,MersenneTwister()  )
R=0.01
for  i  in  range(10000):
```

```
      if (i%1000 == 0):   print "Event=",i
      p=P0D("Event="+str(i))
      ntot=ps.nextInt()
      p.randomUniform(ntot,Ymin,Ymax)
      p.sort()
      for j in range(p.size()-1):
          delta=p.get(j+1)-p.get(j)
          if delta<R:
              newP=p.get(j)+R/4
              p.set(j+1,newP)
      f.write(p)
print "Entries written =",f.getEntries()
f.close()
```

Listing 12.31 Modelling correlations between particles

The code above is sufficiently simple: the `newP` is used to substitute the original particle position if the distance `delta` is smaller than the interaction radius.

Let us prepare two samples with distinct interactions: in one case, we will use $R = 0.01$ and change the particle distances using the algorithm `newP=p.get(j)+R/4`, as given in the example above. This case corresponds to two-particle correlation, since particles will arrive in pairs, since the additional term $R/4$ is sufficiently small.

Finally, let us prepare another data sample using the previous code snippet. But now, we will substitute

```
newP=p.get(j)+R/4
```

with

```
newP=p.get(j)+R
```

The sample created after this code change contains anticorrelations since, normally, particles are separated by a smaller distance than R. Now we require that they will be separated by R. We will write our modified event records to the file "sample3.jser".

Let us emphasize again that the examples above are just toy models. We modeled simple interactions between two particles ignoring the fact that interactions can also happen between three or more particles. In the above case, we have introduced only two-particle correlations. Also, we did not change the functional shape of the distribution used to model distances between objects.

12.6.2.2 Analyzing the Data

Now we are equipped with three data samples in the files "sample1.jser," "sample2.jser," and "sample3.jser". The first file contains fully uncorrelated events and two other files contain data sets with toy-like interactions characterized by the radii

$R = 0.01$ and $R = 0.1$. How can we say what is inside such files, pretending that we do not know anything about their content?

In general terms, the main question is as follows: Assume we have a data sample which represents occurrence of some events at specific interval (time or some phase space), what tools should be used to understand interdependence of events? What is the strength of such interactions? Is it attraction ("correlations") or repulsion (anti-correlations)? What is the mechanism behind such an interaction?

The last question is difficult to answer, and, by no means do we offer a solution. Our task is rather modest: by looking at data, we will try to understand whether the events are independent or not. If we will find that events are heavily interdependent, then we will determine the type of such dependences (attraction or repulsion). Furthermore, we will try to estimate the strength of the interdependence.

We should also add that we are not going to offer a universal and comprehensive approach to deal with such kind of problems. Our task is to illustrate what can be done using Java scripting and the tools included in the DMelt package.

12.6.2.3 Plotting Multiplicities

First, let us read the prepared data files and plot a few basic distributions. We have the necessary software tools for doing this. The code below shows how to read the three data files prepared before and plot multiplicity distributions inside a small region [0–0.5]. We remind that our original particles are distributed inside the phase space window [0–1]:

```
from jhplot   import  *
from jhplot.io import  *
from java.awt import Color
import os

Ymax=0.5
h=[]
files=["sample1.jser","sample2.jser","sample3.jser"]

for f in range(len(files)):
    name=files[f]
    if os.path.exists(name):
        file=HFile(name)
        print "Open: "+name
        h.append(H1D(name,25,0,50))
        while(1):
            p=file.read()
            if p == None:
                print "End of events"; break
            i=file.getEntries();
            if (i%1000 == 0):  print "Event=",i
            n=0
            for j in range(p.size()):
                if p.get(j)<Ymax: n=n+1
```

```
            h[f].fill(n)
        file.close()

c1 = HPlot()
c1.visible();    c1.setRange(10,40,0,3000)
for i in range(len(h)):
    h[i].setFill(0)
    color=Color.getHSBColor(0.3*i,  1.0,  1.0);
    h[i].setColor(color)
h[1].setPenDash(2)
h[2].setPenDash(6)
c1.draw(h)
```

Listing 12.32 Plotting multiplicity distributions

Running this script will bring up a canvas with three distributions, each of them will
have a Poisson-like shape as shown in Fig. 12.3. They are all very similar, so there is
no way to say about the difference we have introduced for the two-particle distances.
Therefore, we should find a better solution to tackle the problem of correlations. One
solution would be to study fluctuations inside small phase space regions.

Since we have introduced two-particle correlations, we expect that the multiplicity
distributions inside small regions will deviate from the Poisson distribution: in the
case of interactions, it should be broader than the standard Poisson distribution (large
fluctuations). In the case of repulsion between particles, the distribution should be
narrower.

When fluctuations of separate particles are measured, it is convenient to transform
a multiplicity distribution $P_n(\delta)$ inside a region δ to the following observables:

$$\text{BP}: \quad \eta_q(\delta) = \frac{q}{q-1} \frac{P_q(\delta) P_{q-2}(\delta)}{P_{q-1}^2(\delta)}, \tag{12.2}$$

$$\text{NFM}: \quad F_q(\delta) = [n(\delta)]^{-q} \sum_{n=q}^{\infty} \frac{n!}{(n-q)!} P_n(\delta), \tag{12.3}$$

where the abbreviations denote the bunching parameters (BP) or the normalized
factorial moments (NFM). The parameter q defines the order of the BP or NFM.
Note that the experimental definitions of the BPs [4, 5] and the NFM [6, 7] are
different from those given above since, in reality, one should scan all phase space
regions to increase statistics when looking into ever smaller regions.

These two quantities measure deviations of a multiplicity distribution $P_n(\delta)$ from
a Poisson distribution for which $F_q(\delta) = \eta_q(\delta) = 1$. Note that such deviations are
measured differently by these three methods (for a review see [3]).

Uncorrelated particle production inside δ leads to the Poisson statistics, thus deviations of the NFM and BP from unity indicate correlations (interactions) between
particles leading to dynamical fluctuations (i.e., non-Poisson type of statistics). If
$\eta_q(\delta)$ or $F_q(\delta)$ are larger than one, one can say about positive correlations. This case
is expected for the first sample generated without including the toy interactions. In

Fig. 12.3 Distributions of particles prepared for three tests

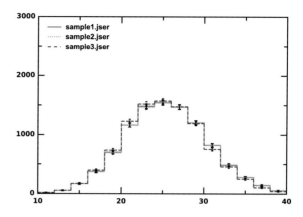

contrast, if both quantities are smaller than unity for ever smaller δ regions, one can say about anticorrelations (this, naturally, should be the case for our second sample with the repulsion between particles).

DMelt contains a Java library for calculations of both $\eta_q(\delta)$ and $F_q(\delta)$. We will consider the class called `BunchingParameters`, and also a similar class `FactorialMoments` from the package `jhplot.stat`.

We can initialize our calculations as

```
>>> bp = BunchingParameters (MaxOrder, Bins, Step, Min, Max)
```

where `MaxOrder` is the maximum order of BP, `Bins` is the number of bins used for the calculations, `Step` is any integer number to increase the binning step. The bins are defined as $1 + i * \text{Step}$, where i runs from 2 to `Bins`. This number of bins is used to divide the total phase space defined by `Min` and `Max`.

Once the object `"bp"` is created, we can use the method `run(array[])` inside the event loop, passing the array with all particles in the event. After the end of the event loop, we should call the method `eval()` to evaluate the BPs values. To access the values for BPs as a function of the number of phase-space divisions, one should call the method `getBP(order)`, which returns a `P1D` array with the BP values and their statistical errors for a given `order` of BP.

Now, we are ready. Let us read all three samples prepared in the previous section and calculate the BP up to the third order. The code is shown below:

```
from jhplot import *
from jhplot.stat import *
from jhplot.io import *
import os

Ymax = 0.5
bp2, bp3 = [], []
files = ["sample1.jser", "sample2.jser", "sample3.jser"]
```

```
for i in range(len(files)):
    name=files[i]
    if os.path.exists(name):
        f=HFile(name)
        print "Open: "+name
        bp=BunchingParameters(3,20,5,0,Ymax)
        while(1):
            p=f.read()
            if p == None:
                print "End of events"; break
            n=f.getEntries()
            if (n%1000 == 0):   print "Event=",n
            bp.run(p.getArray())
        bp.eval()
        res1=bp.getBP(2)
        res1.setTitle("BP_{2} for "+name)
        res1.setStyle("lp")
        res1.setSymbol(3+i)
        res1.setSymbolSize(8)
        bp2.append(res1)
        res2=bp.getBP(3)
        res2.setTitle("BP_{3} for "+name)
        res2.setSymbolSize(8)
        res2.setStyle("lp")
        res2.setSymbol(3+i)
        bp3.append(res2)
        f.close()
c1 = HPlot("BP",800,400,2,1)
c1.visible(); c1.setAutoRangeAll()
c1.cd(1,1)
c1.draw(bp2)
c1.cd(2,1)
c1.draw(bp3)
```

Listing 12.33 Studies of fluctuations

As you can see, we perform this rather complicated calculation using a few lines of the code. We build two lists that keep the results for η_2 and η_3 for all three processed samples, and then run over each of these samples. The result of this code is shown in Fig. 12.4. As expected, the BPs for the sample without interactions are indeed independent of $\delta = (Ymin - Ymax)/Bins$, and they are all equal to unity as expected for the Poisson statistics. The data sample with the interactions, when particles arrive in bunches, has a sharp increase for the calculated $\eta_2(\delta)$ with decreasing δ. The $\eta_3(\delta)$ does not show large variations with the number of bins, indicating that we are dealing with two-particle correlations. For the third sample, $\eta_2(\delta)$ decreases with δ (anticorrelations), and stays constant when δ reaches some δ_{min}. This roughly corresponds to anticorrelation radius in our code (see the parameter "R").

Let us calculate the NFM by replacing the class `BunchingParameters` with the name `FactorialMoments`. To retrieve the results, we should call `getNFM()` instead of `getBP()`. The resulting plot is shown in Fig. 12.5. One can see a similar result as that shown in Fig. 12.4. The only notable difference exists for F_3 for the

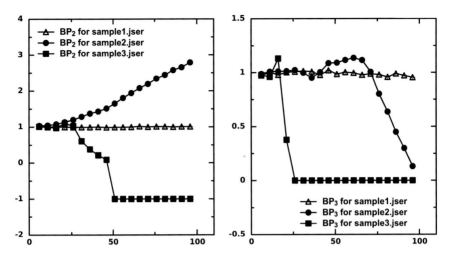

Fig. 12.4 Bunching parameters η_2 (*left*) and η_3 (*right*) as functions of the number of bins used to divide the total phase space [0–0.5]. The calculations were done for three data samples: with no correlations (the file "sample1.jser"), with correlations ("sample2.jser") and with anticorrelations ("sample3.jser")

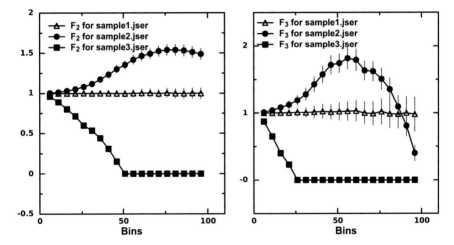

Fig. 12.5 The normalized factorial moments F_2 (*left*) and F_3 (*right*) as functions of the number of bins used to divide the total phase space. The calculations were done for three data samples: with no correlations (the file "sample1.jser"), with correlations ("sample2.jser") and with anticorrelations ("sample3.jser")

sample with interactions. This is not totally surprising since F_3 reflects not only three-particle correlations but also a contribution from two-particle correlations which (we know!) have been included when the sample was generated. One can find more details about the use and properties of BP or NFM in the review [3].

At the very end, you may still wonder why not just to look at the distribution of distances between two particles if we want to know how particles are distributed with respect to each other. We know that interparticle distances should be distributed in accordance with the exponential distribution in the case of the Poisson statistics. In principle, this can be done as well, but then we should find a method to compare such distributions with a reference distribution that is known to have exactly the same properties (multiplicities, shapes of single-particle distributions) but without any correlations between particles. This is not too easy and is outside the scope of our example: the use of the BPs, NFMs, and similar tools is very handy since they do not require the construction of such reference samples [3].

12.6.3 Analyzing Nearby Galaxies

In this section, we give another real-life example: this time we plan to analyze data on nearby galaxies. Our intention is not to be too scientific, nor do we plan to discuss detailed topics that require special knowledge. What we want to do is to illustrate the scientific programming using scripting on the Java platform. In a few lines of the code we will perform almost complete data analysis of macro-world structures.

Our analysis program will consist of reading a data file, fetching necessary data records, and plotting one type of values against another. Readers interested in science will immediately find areas where the discussed codes snippets can be reused for their own research, or they even can make a detailed study of the data file discussed in this section.

What we are going to do now is to analyze the catalog of Southern spirals galaxies (Mathewson+ 1996). This catalog is publicly accessible from the Astronomical Data server [8].

To simplify our example, we have prepared a file with the catalog to be used for this tutorial. Let us copy the file with the galaxy catalog from the Web:

```
from jhplot import *

http="http://jwork.org/dmelt/examples/data/"
xfile="Mathewson1996.tsv"
print Web.get(http+xfile)  # print download status
# view.open(xfile, 0  )    # uncomment if you run DMelt
   IDE
```

Listing 12.34 Getting data from the Web

The script fetches the file "Mathewson1996.tsv". Open this file and look at how the records are organized. If you run this script inside the DMelt IDE, uncomment the last line of this script. Now you can see that each entry in the row is separated by a space. Thus, if we want to read this file line by line, one should find a way to split each line of this file. Also, one can see that the comment lines always start with the symbol "#". Of course, these lines have to be ignored by our program. Finally, we split the strings into pieces and convert the substrings into float values.

We are dealing with a typical CSV file discussed in Sect. 2.15.4, thus it can be opened using the Java class CSVReader. But this class may lack flexibility to open files with a complex record structure. The good thing about Python is that to write a small code which parses a CSV file is very easy, perhaps as easy as when using a specialized library for CSV files.

Let us write a small Python code that reads the CSV file and then performs the necessary conversions. This code should split a string into n-size pieces using an arbitrary string as the delimiter, ignoring the lines which start from the symbols "#" or "*" (this will be necessary for our next example). Then, the code should convert a string into a float number. Normally, this can be done with the Python float() method (in the case if we expect a real number). But we should take care of situations when we need to convert strings such as "1/2," "2/3," etc., into float values.

Let us create a Python module file called "reader.py" with the following code:

```
def get(line, delim=None):
  s=line.strip()
  if s.startswith("*") or s.startswith("#"): return None
  if delim == None:
    return s.split()
  else:
    return s.split(delim)

def conv(s):
  try:
    return float(s)
  except ValueError:
    try:
      num, denom = s.split("/")
      return float(num) / float(denom)
    except ValueError:
      return None
```

Listing 12.35 "reader.py" module to read a CSV file

The function called get() takes any string and splits it into pieces using the delim as a delimiter string. If the delim is not used, we assume that the string should be split using a white space. In addition, we included the function convert() which accepts any string and converts it into a float value. For example, if we pass "1/2" to this function, it will return the float value 0.5. If no proper conversion is possible, the function will raise an exception (see Sect. 2.14) and return None.

Now, let us read this file and build a profile histogram discussed in Sect. 7.5. We would like to plot the average value of rotation velocities versus the face-on diameter of the spiral galaxies. By studying the file, one can easily see that we need to plot values in column 18 against the values in column 14. So, let us write the following simple code:

```
from jhplot  import *
from reader import *

c1 = HPlot()
c1.setGTitle("Southern Spiral Galaxies")
c1.setMarginLeft(100)
c1.visible(); c1.setRange(-1,4,100,200)

h=HProf1D("2447 galaxies (Mathewson+ 1996)",20,0.05,3.0)
for line in open("Mathewson1996.tsv"):
  tab = get(line)
  if tab == None or len(tab)<18: continue
  x=conv(tab[13])
  y=conv(tab[17])
  if x != None and y!=None: h.fill(x,y)

c1.setNameX("Face-on diameter [arcmin]")
c1.setNameY("< Rotation velocity [km/s] >")
h1=h.getH1D()
h1.setStyle("p")
c1.draw(h1)
```

Listing 12.36 Analyzing galaxies

The example shows how to open a CVS file and read it line by line. We split each line and make sure that we have the correct number of pieces (18). Then, we fill the profile histogram.

The resulting image is shown in Fig. 12.6. As we can see, there is rather obvious dependence of the average velocity on the galaxy radius. I am sure some readers will find this plot familiar, and will connect it to (in fact, known) physics phenomenon. If you are not one of them, then this would be your entry to the world of science.

For those who are impatient, here is a tip: there is a well-known discrepancy between the observed galaxy rotation curves and theoretical predictions (the so-called "galaxy rotation problem"). A possible solution to this problem is to hypothesize the existence of dark matter.

12.6.4 Analyzing Elementary Particles

In this section, we plan to study elementary particles from the Particle Data Book [9]. A typical size of the galaxies analyzed in the previous example is 10^{20} m, while the proton size is 10^{-15} m, so the examples in this book span more than 35 orders of magnitude in distances! Sure, the universe we leave in is a big place, and we have to find a way to study it.

Fig. 12.6 A dependence of
the average rotational
velocities on the galaxy size

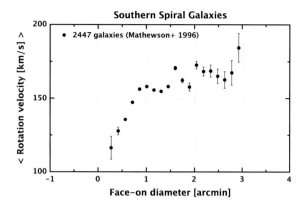

Let us first copy a file with particle characteristics from the Particle Data Group web page [9]. As before, to simplify this task, we have prepared an easy-to-copy file. As indicated in the comments for our next example, you can open this file in the DMelt editor for a visual examination:

```
from jhplot import *

http="http://jwork.org/dmelt/examples/data/"
xfile="mass_width_2008.csv"
print Web.get(http+xfile) # print download status
# view.open(xfile, 0  )   # uncomment if you run DMelt IDE
```

Listing 12.37 Reading a data file

The script fetches the data from the Web and copies to the local directory. If you run this script inside the DMelt IDE, uncomment the last line of this script which will open the file inside the editor.

Now, let us show how to read this file and how to understand correlations between different particle characteristics stored in this file. Let us plot the squared mass of the hadron on the J value, which is the total angular momentum of a hadron. We will use the same module "reader.py" developed in Sect. 12.6.3. But, in this example, we will read not only particle masses, but also errors on the masses and put all this information into the P1D container (see Sect. 4.2) for plotting. We will use a comma for value separation.

```
from jhplot   import *
from reader import *

p1=P1D("hadrons")
file=open("mass_width_2008.csv")
for line in file:
  tab = get(line,",")
  if tab == None : continue
  id=conv(tab[12])
```

```
  J=conv(tab[8])
  if id>100 and id<100000 and J != None:
    mass=conv(tab[0])
    er1=conv(tab[1])
    er2=conv(tab[2])
    if (mass<3E+3): p1.add(J,mass*mass,er1,er2)

c1 = HPlot()
c1.setGTitle("Hadrons")
c1.setNameY("M^{2} [MeV]")
c1.setNameX("J")
c1.visible()
c1.setMarginLeft(90)
c1.setRange(0,5,0,1E+7)
c1.draw(p1)
```

Listing 12.38 Analyzing elementary particles

Figure 12.7 shows the result.

As you can see, there is a correlation between the mass squared M^2 of hadrons and J. In fact, what we have discovered here using a few lines of the code is a reflection of the well-known observation that hadrons from the same family lie on special trajectories (the so-called Chew–Frautschi conjecture [10–12]). Such a relationship between J and M^2, also known as the principle of exchange degeneracy, is usually interpreted as a manifestation of the linear potential of the strong forces between constituent quarks inside the proton. More rigorous studies of such regularities using linear regression fits, and the most recent data, are discussed in the following articles [13, 14].

Here, we will stop the discussion of real-world data-analysis examples. We will come back to advanced techniques used for data analysis, data mining, and knowledge discovery in the subsequent chapters of this book.

Fig. 12.7 This figure shows a dependence of the mass squared of hadrons on angular momenta (J)

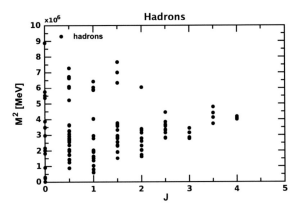

References

1. Cranshaw J et al (2008) A data skimming service for locally resident analysis data. J Phys (Conf Ser) 119:072011
2. Taylor T et al (2006) Human chromosome 11 DNA sequence and analysis including novel gene identification. Nature 440:495
3. De Wolf EA, Dremin IM, Kittel W (1996) Scaling laws for density correlations and fluctuations in multiparticle dynamics. Phys Rep 270:1–141
4. Chekanov SV, Kuvshinov VI (1994) Bunching parameter and intermittency in high-energy collisions. Acta Phys Pol B 25:1189
5. Chekanov SV, Kittel W, Kuvshinov VI (1997) Generalized bunching parameters and multiplicity fluctuations in restricted phase-space bins. Z Phys C 74:517
6. Bialas A, Peschanski RB (1986) Moments of Rapidity Distributions as a Measure of Short Range Fluctuations in High-Energy Collisions. Nucl Phys B 273:703
7. Bialas A, Peschanski RB (1988) Intermittency in multiparticle production at high-energy. Nucl Phys B 308:857
8. ADC, Astronimical data center, access to astronomy data and catalogs. http://adc.astro.umd.edu/
9. Eidelman S et al (2004) Review of particle physics. Phys Lett B 592:1
10. Chew GF, Frautschi SC (1961) Principle of equivalence for all strongly interacting particles within the s matrix framework. Phys Rev Lett 7:394–397
11. Chew GF, Frautschi SC (1962) Regge trajectories and the principle of maximum strength for strong interactions. Phys Rev Lett 8:41–44
12. Frautschi S (1968) Regge poles and S-matrix theory. Benjamin, New York
13. Anisovich V et al (2004) Quark models and high energy collisions. World Scientific Publishing Co., Pte. Ltd, Engelska
14. Chekanov SV, Levchenko BB (2007) Regularities in hadron systematics, Regge trajectories and a string quark model. Phys Rev D 75:014007

Chapter 13
Neural Networks

We will continue our discussion of data analysis, but this time we will concentrate on data mining, which is widely used for finding patterns for data classification and predictions. This topic was only briefly discussed in the previous sections.

In this chapter, we will discuss machine learning based on neural networks, computer programs inspired by artificial intelligence algorithms designed to remember complex patterns in data, and make decisions based on memorized information.

13.1 Introduction

Artificial neural networks are computer algorithms made up of a number of highly interconnected processing elements. They are processing devices that are loosely modeled after the neuronal structure.

A neural network is a powerful tool for data classification: if we know a source with input data, and know the output data, one can obtain a relationship between inputs and outputs. This can be done totally numerically, without using any predefined function as in the case of curve fitting discussed before. Once such a relationship is established, one can perform predictions based on the input, assuming that the data set has the same characteristics as those used for determination of the relationship during the so-called "training" procedure. Here, we recommend several books [1, 2] with detailed discussions about the neural networks. Below we will touch only the basics of this subject before showing the code design patterns using Java classes.

But what exactly should be used to perform such a classification of inputs and outputs? Generally, the concept of the neural network approach captures the essence of biological neural systems. In particular, it tries to simulate the brain's way of processing information by interconnecting billions of neural signals. Similarly, a neural network consists of many interconnected "neurons" in order to simulate biological neural systems.

© Springer International Publishing Switzerland 2016
S.V. Chekanov, *Numeric Computation and Statistical Data Analysis on the Java Platform*, Advanced Information and Knowledge Processing,
DOI 10.1007/978-3-319-28531-3_13

Let us take a look at a human brain more closely. It consists of a huge number of interconnected neurons. Each neuron has an input and a branching output structure, and it can be connected to up to 10,000 other neurons that pass signals to each other via synaptic connections. When a neuron is activated, it sends an electrochemical signal via the synapses to other neurons which may, in turn, fire signals to inputs of other neurons. The strength of the propagated signal depends on the efficiency of the synapses.

Mathematically, a neural network consists of a set of interconnected units, called nodes. Each node accepts a weighted set of inputs, and then it responds with an output. Each input has an associated weight representing the strength of that particular connection. A multilayer network with several layers of units is very popular. For such networks the output from one layer serves as input to the next.

The number of neurons in the input layer depends on the number of inputs, while the number of neurons in the output layer depends on the number of desired outputs. The number of hidden layers and the number of neurons in each such layer must be defined empirically.

Each neuron performs an accumulation of incoming pulses from its inputs. The accumulated information is stored as "weights." All weights are summed up and the total sum is passed into a nonlinear activation function. A hyperbolic tangent activation function, $S(x) = \tanh(x)$, is commonly used. Another popular choice is the sigmoid activation function, $S(x) = 1/(1 + e^{-x})$. The output of such neutron is formed by applying the sigmoid function as $S(d + \sum_{i=0}^{N} x_i w_i)$, where x_i are the input values from other neurons and w_i characterizes the weight for input connections i. A constant d is called the "threshold." It characterizes the neutron as a whole.

The neural net should be trained to adjust the parameters w_i and d. Once the input units have received the signal from outside and from the output units with the resulting signal, one can adjust the weights in the interconnected neurons such that one can perform predictions. During the learning process, the output vector of numbers is compared to the expected output. If the difference is not zero, the ("epoch") error on the prediction is calculated. The idea is to adjust the weights to minimize the difference between the neural-net output and the expected output. After the learning step, neural networks will memorize arbitrary associations between the input vector and the output.

13.2   A Basic Neural Network

Let us sketch a typical task that can be solved using the neural-net approach. Assume we have many events with the input vector of numbers. We also know some events in which the input vector produces a specific output. Now the main question is how we can make predictions based on the inputs.

In reality, we do not know an analytic relationship between inputs and outputs. In the case of machine learning based on neural networks, we are not interested in this at all, since the whole idea behind the neural network approach is to find a numerical relationship in terms of many weights associated with neurons. To translate a large number of weights into human understandable relationships between inputs and output(s) can be quite difficult, or even impossible.

For complex systems, we even do not know the contributions of input variables to the output. For example, some input contributions may have nothing to do with the actual outputs. The neural networks can deal with such situations as well.

13.2.1 Encog Approach

Our first example is based on the Encog Java project [3]. We will solve the well-known "XOR" problem in the dimension $n = 2$, consisting of binary data vectors (0 or 1 values). Let us create a CSV file where two first columns represent the inputs, and the last column defines the output. We will use Python to make this file:

```
data=" " "
0,0,0
0,1,1
1,0,1
1,1,0
" " "
fi=open("test.csv", "w")
fi.write(data)
fi.close()
```

Listing 13.1 Creating a CSV file with input and output

Our task is to create a multilayer feedforward neural network with two inputs and a single output. A feedforward neural network is built such that data from input to output units do not form a directed cycle. Thus, this type of network does not have feedback connections. This is the most popular network in many practical applications.

```
from org.encog import *
from org.encog.util.csv import *
from org.encog.util.simple import *
from jhplot import *

xf=CSVFormat.ENGLISH
f="test.csv"
tSet=TrainingSetUtil.loadCSVTOMemory(xf,f,False,2,1)
nn=EncogUtility.simpleFeedForward(2,3,0,1,True)
EncogUtility.trainToError(nn,tSet,0.01)  # train to 1%
EncogUtility.evaluate(nn, tSet)          # print test
IO.writeXML(nn,"network.xml")            # save to XML file
```

Listing 13.2 Creating a neural net and train using a CSV file

The class `loadCSVTOMemory` reads the data where "2" specifies the number of inputs. Then, the method `simpleFeedForward` builds a neural network using the resilient propagation (RPROP) algorithm, which is one of the best general-purpose supervised training methods. The arguments of the method `simpleFeedForward` tells that we have 2 inputs, one hidden layer with 3 neurons, no second hidden layer (0), and `True` means that we use a hyperbolic tangent activation function. If `False` is set, this will mean that the network uses the sigmoid activation function. Finally, we save the trained neural network to the XML file "network.xml", so you can read the weights using your favorite editor and inspect the stored values. The output of this code is

```
Input=0.0, 0.0, Actual=-0.0791, Ideal=0.0
Input=0.0, 1.0, Actual= 0.8675, Ideal=1.0
Input=1.0, 0.0, Actual= 0.8684, Ideal=1.0
Input=1.0, 1.0, Actual= 0.0710, Ideal=0.0
```

As you can see, we have rather close prediction since the output values ("actual") are close the input ("ideal"). If you expect integer numbers, one can round the output numbers to the closest integer values, and will discover that the predictions match exactly the expectation.

Let us restore the trained neural net and visualize it. We also perform a basic analysis of the neural net by printing the stored weights and other characteristics.

```
from org.encog.visualize import *
from org.encog.neural.networks.structure import *
from jhplot import *

nn=IO.readXML("network.xml")  # restore from XML file
f=NetworkVisualizeFrame(nn)
f.setDefaultCloseOperation(1)
f.setVisible(True)                # show image of NN

print "Analyze network:"
a=AnalyzeNetwork(nn)
print a.toString()
print "Values=",a.getAllValues().tolist()
print "Weights=",a.getWeightValues().tolist()
print "Nr of connections=",a.getTotalConnections()
```

Listing 13.3 Restoring the neural network

Let us give another example. This time, instead of reading CSV files, we will use a list of values, and build the neural network in a more programmable fashion. This time we will use the sigmoid activation function, plus we will control the training process while we run the code:

```
from org.encog.ml.data.basic import *
from org.encog.neural.networks import *
```

```
from org.encog.neural.networks.layers import *
from org.encog.neural.networks.training.propagation.
    resilient import *
from org.encog.engine.network.activation import *

INPUT= [ [0,0], [1,0], [0,1], [1,1]]
OUT=    [   [0],   [1],   [1],   [0]]
nn=BasicNetwork()
nn.addLayer(BasicLayer(None,True,2))
nn.addLayer(BasicLayer(ActivationSigmoid(),True,3))
nn.addLayer(BasicLayer(ActivationSigmoid(),False,1))
nn.getStructure().finalizeStructure()
nn.reset()   # reset neural net

tSet=BasicMLDataSet(INPUT,OUT)
train=ResilientPropagation(nn,tSet)

epoch,error=1,1
while (error > 0.01): # training iterations
  train.iteration()
  error=train.getError()
  if (epoch%10==0):
      print "Epoch #",epoch," Error:",error
  epoch+=1

for pair in tSet: # show results
  out=nn.compute(pair.getInput())
  x=pair.getInput().getData(0)
  y=pair.getInput().getData(1)
  z=pair.getIdeal().getData(0)
  print x,",",x,", predict=",out.getData(0),", ideal=",z
```

Listing 13.4 Programming a neural network using Encog

The output of this code will be similar to the previous example, but this time it will print the training error for each epoch. One can easily visualize epoch errors using the HPlot canvas and the P1D array as discussed in the previous sections. This example can easily be extended to meet your needs.

13.2.2 Using Neuroth

Alternatively, one can use the Neuroph Java package [4] designed to construct neural networks with different architectures. We will be quick in the description of this package, since the basic idea behind the neural network concept has been discussed in the previous section.

We will show again how to solve the "XOR" problem using the same steps as before, i.e., we will train the neural net on the input data, save it to a binary file and then restore the neural net for further examination. We will create a neural net with 2 inputs, 3 hidden neurons, and 1 output. As indicated in the code, we use a hyperbolic tangent activation function.

```
from jhplot import *
from org.neuroph.core.data import *
from org.neuroph.nnet import *
from org.neuroph.util import *

data=DataSet(2,1)   # create a dataset
data.addRow(DataSetRow([0,0],[0]))
data.addRow(DataSetRow([0,1],[1]))
data.addRow(DataSetRow([1,0],[1]))
data.addRow(DataSetRow([1,1],[0]))

func=TransferFunctionType.TANH
nn=MultiLayerPerceptron(func,[2, 3, 1])
learningRule = nn.getLearningRule()
learningRule.setBatchMode(True)
nn.learn(data) # Training neural network
IO.write(nn,"neuroph.jser")

print "Read network and test:"
nn=IO.read("neuroph.jser") # restore from XML file
for dataRow in data.getRows():
    nn.setInput(dataRow.getInput().tolist())
    nn.calculate()
    networkOutput = nn.getOutput();
    print "Input: ", dataRow.getInput().tolist()
    print "Output: ", networkOutput.tolist()
```

Listing 13.5   Solving the XOR problem with the Neuroph package

The output of this code is similar to that of the previous example. Please look at the API of the imported Java classes.

13.3   Backpropagation with Multiple Outputs

Now we will show an example that illustrates how to build a backpropagation neural network to memorize relationships between words. This can be considered as a simple vocabulary. Unlike many standard dictionaries, our neural network can generate a probability for wrongly associated word using ASCII characters for input and outputs. Thus, to build a word, we will need several neurons in the output.

This example is based on the Java classes [5] for the backpropagation algorithm that uses a computed output error to change the weight values in backward direction. Unlike the previous examples with numeric (integer) values, our next neural network will deal with ASCII characters. The input pattern consists of 10 ASCII characters while the output consists of seven characters. Here are two example lines of the input file:

```
London0000 England
Washington USA0000
```

where 0 is used to fill the missing characters for the 7- and 10-character input and outputs. We will train the network to predict the country name from the input town name. Our input layer will contain 10 neurons, while the output layer will have 7 neurons, i.e., one neuron for each character in input and output. Obviously, these characters need to be converted to a machine representation.

Let us create such a network:

```
from jhpro.nnet import *
from jhplot import *

http="http://jwork.org/dmelt/examples/data/"
print Web.get(http+"ascii2bin.cnv")
print Web.get(http+"towns.pat")

bpn=BackpropagationNet()
bpn.readConversionFile( "ascii2bin.cnv" ) # ascii->binary
bpn.addNeuronLayer( 10 )   # input layer (must be 10)
bpn.addNeuronLayer( 4 )    # hidden layer (vary)
bpn.addNeuronLayer( 7 )    # output layer (must be 7)
bpn.connectLayers()
for i in range (bpn.getNumberOfLayers()):
  print "layer ", i,":",bpn.getNumberOfNeurons(i)," neurons"
  print "weights: ", bpn.getNumberOfWeights()
bpn.readPatternFile( "towns.pat" ) # pattern file
bpn.setLearningRate( 0.25 )
bpn.setMinimumError( 1E-06 )   # minimum error to stop
bpn.setAccuracy( 0.5 )         # training accuracy
bpn.setMaxLearningCycles( -1 ) # -1 = no maximum cycle
bpn.setDisplayStep(500)        # show every 500 steps
bpn.resetTime()
while ( bpn.finishedLearning()==False ):
  bpn.learn()
  if bpn.displayNow(): print bpn.getError()
print "err:",bpn.getError()
IO.write(bpn,"backpropagation.jser")
```

Listing 13.6 Training backpropagation neural network using words

Here is an explanation of a few features that we did not discuss before. This example downloads the file "ascii2bin.cnv" used to convert the ASCII characters to a binary representation. The number of binary values that represent the ASCII characters can be changed freely. The second file "towns.pat" contains multiple lines of "town-country" associations. It will be used for training the neural network. The learning process will continue until the error between the input and outputs reaches $\times 10^{-6}$. In this example, the trained accuracy was set to 0.5. We will discuss the reason for this in the following paragraphs. Finally, we will save the neural network with the trained weights in a file.

Now it is time to run this network. We will use 3 inputs. Two inputs were taken from the trained set, and one input contains a misspelled input "Moskow" instead of "Moscow". The neural net will attempt to reproduce the country name from the town names. Here is the code:

```
from jhpro.nnet import *
from jhplot import *

bpn=IO.read("backpropagation.jser")

def compare(recall, target):
 acc=0;   xl=len(recall)
 rp,tp = list(recall), list(target)
 for i in range(xl):
     if ( rp[i]==tp[i] ): acc=acc+1
 if ( acc == xl): print "correct!"
 else:
    e="("+str(acc)+"/"+str(xl)+":"+str(((acc*100)/xl))+"%) "
    print "out=",recall," expected=",target,"err=",e

a=["Madrid0000","Spain00"]   # town -> country
compare(bpn.recall(a[0]), a[1] )
b=["Moscow0000", "Russia0"]  # town -> country
compare(bpn.recall(b[0]) , b[1] )
c=["Moskow0000", "Russia0"]  # misspelled "Moscow"
compare(bpn.recall(c[0]), c[1] )
```

Listing 13.7 Backpropagation neural network using words

The function output(x,y) compares characters of the predicted word with the expectations. The method list(s) in this example creates arrays of the characters from the strings. The output of the above code is three lines with the word "correct". Thus, the misspelled word was correctly associated with the expected country name. The fact that the neural network did not notice the misspelled word "Moskow" is due to the trained accuracy of 0.5 which was set in the previous code example. If we will train the network with the accuracy set to a smaller value, say to 0.05, the mistake in the input will be detected and the output will be

```
correct!
correct!
out= Russiq2  expected = Russia0 err= (5/7:71%)
```

So, the neural network correctly reproduced the country name in the first two cases, but the last case failed with the probability 85 % for the correct character assignments (only 6 characters were predicted correctly out of 7).

One possible practical application of this example is to build a dictionary that ignores small typos, assuming that the accuracy value for the neural network is not very small.

13.4 Numeric Predictions

Now we will consider more sophisticated neural network which involves several analysis stages, such as data generation, preparation of data for training, neural network validation, and final analysis. This time we will learn how to predict floating point numbers using several numeric inputs. This type of neural networks has a broad range of applications in natural sciences, medical science, market predictions, engineering, and other fields.

13.4.1 Generating a Data Sample

We will start with a preparation of the data sample which will be used later for a classification task of our numeric data. Let us assume that we have four variables generated in accordance with some random distributions. Then, we generate an output value from the inputs using some functional form. The vector with inputs and outputs will be stored persistently in a serialized file as shown below:

```
from java.util import Random
from jhplot  import *

inp=PND("Data")
out=PND("Output")
r= Random(2L)
for i in range(2000):
    x1=10*r.nextDouble()
    x2=1+2*r.nextDouble()
    x3=1/(1+2*r.nextDouble())
    x4=1+10*r.nextDouble()
    inp.add([x1,x2,x3,x4])  # array with inputs
    out.add([x1/x4-x2+x3])  # output
d={"input":inp,"out":out}   # map with input and output
IO.write(d,"data.jser")
print "file is ready"
```

Listing 13.8 Creating a data set for neural network studies

The code generates four random variables and calculates the output value y using the relationship $y = x_1/x_4 - x_2 + x_3$. This can be any function, of course.

So, the task is seemingly clear. We have the input array and the output (which, in general, could be also an array). The question is how to establish the relationship between these variables pretending that, in reality, we know nothing about it.

13.4.2 Data Preparation

Generally speaking, it is advisable to standardize the input and to rescale the output to the range [0, 1] or [−1, 1], depending on the output activation function.

Standardization means that we scale the data such that all inputs have the same ranges. Without this procedure, a variable with the largest scale may dominate, thus producing a bias toward this variable. To circumvent this problem, each column should be standardized, i.e., transformed to:

$$S_i = (X_i - \bar{X})/\sigma,$$

where X_i is the original value, $\bar{X}$ is the mean and σ is the standard deviation of the input data.

In order to standardize a PND, apply the following command:

```
>>>    pnd.standardize()
```

where "pnd" is an input PND data container. The example below shows how to use this method:

```
from jhplot import *
pnd = PND("pnd")
pnd.add([0.02,10,3])
pnd.add([0.01,6,1])
pnd.add([0.03,12,5])
print pnd.standardize()
```

Listing 13.9 Standardizing input data for neural networks

This shows the data after the standardization procedure:

```
   0.0     0.21    0.0
  -1.0    -1.09   -1.0
   0.99    0.87    1.0
```

In contrast, a normalization or rescaling means that each column is transformed such that all column values are either between [0, 1] or [−1, 1], depending on the range of the activation function. To normalize a PND, use the rescale() method.

One should pass "0" as an argument if rescaling should be done for the range [0, 1]:

$$S_i = (X_i - X_{min})/(X_{max} - X_{min})$$

In the case of the range [−1, 1], one should apply this transformation:

$$S_i = 2 * (X_i - midrange)/(X_{max} - X_{min}),$$

$$midrange = 0.5 \cdot (X_{max} + X_{min}),$$

which is implemented by the method `rescale(1)` of the PND Java class. This method returns an array which will be used to transform the PND array back to its original form. In this case, the output of the rescale method should be passed as input to the `rescale(obj)` method. Consider the example:

```
from jhplot import  *
pnd=PND("pnd")
pnd.add([0.02,10,3])
pnd.add([0.01,6,1])
pnd.add([0.03,12,5])
a=pnd.rescale(0)
print pnd
pnd.rescale(a)
print pnd
```

Listing 13.10 Rescaling the data for neural networks

The code prints the rescaled and restored 2D arrays:

```
 0.5   0.66   0.5
 0.0    0.0   0.0
 1.0    1.0   1.0
 # second array:
 0.02  10.0   3.0
 0.01   6.0   1.0
 0.03  12.0   5.0
```

One can see that the rescale method accurately restores the original array once you have passed the output from the first `rescale(0)` call, which converts each column to the range [0, 1]. Analogously, if the `rescale(i)` method is called with argument "1", all values will be moved to the range [−1, 1], and the conversion back will be done in exactly the same way as before.

In the next example we read the saved containers, standardize the input, and normalize the output. Then we save the resulting container into a dictionary with the keys "input" and "output". We also save the array keeping the scaling factors, which can be used to restore the output data later.

```
from jhplot import *

d=IO.read("data.jser")
inp=d["input"]
out=d["out"]
inp.standardize()
scale=out.rescale(0)  # save scale factor!

d={"input":inp,"out":out,"scale":scale}
IO.write(d,"data_scaled.jser")
print "file is ready"
```

Listing 13.11 Preparing data for neural network studies

It should be noted that the standardization should be approached with caution because it discards certain information. Generally, there are no "rules of thumb" to be applied to all scenarios.

Now we are ready to use a fraction of the rescaled data for training of our neural network. In the next section, we will briefly discuss the machinery behind the construction of a neural network, which comes with the DMelt package.

13.4.3   Building a Neural Net

First, let us discuss how to build and visualize an abstract neural network, before continuing with our next example. In DMelt, a neural network can be constructed using the Java class HNeuralNet which is based on the Encog project [3]. In order to illustrate its usage, we will build a feedforward neural network with an input layer, two hidden layers, and one output layer. The input layer will contain four neurons, two hidden layers will have five neurons each, and the output layer will have a single output:

```
from jhplot import *
net = HNeuralNet()
net.addFeedForwardLayer(4)  # 4 input neurons
net.addFeedForwardLayer(5)  # hidden layer 1 with 5
    neurons
net.addFeedForwardLayer(5)  # hidden layer 2 with 5
    neurons
net.addFeedForwardLayer(1)  # output (single value)
net.reset()                 # randomize the weights
net.showNetwork()
```

Listing 13.12 Building a feedforward neural network

The above code adds layers, staring from input to the output, passing integer arguments, and specifying the number of interconnected neutrons in each layer. The next step is to reset all neuron thresholds to some (random) values. This is done with the method reset(). The constructed network can be visualized with the method

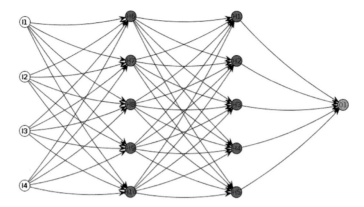

Fig. 13.1 A representation of a neural network with two hidden layers. The threshold values for the activation matrix is shown for each neutron. In this example, the initialization was done with the `reset()` method

`showNetwork()`. The result of this method is a pop-up GUI window displaying the structure of the network. This window is shown in Fig. 13.1. In this diagram, each layer has the following notation: "I" is the input layer, "H" is the hidden layer, and "O" is the output layer.

After designing the network structure, we should set the input and the output data for training. This can be done with the method `setData(in,out)`, where `in` and `out` are data arrays of the type PND. One can look at the data by calling the `editData()` method which opens an editor with the neural-net input and output.

Next, the created network should learn how to adjust the weights and the neuron thresholds. We will use the backpropagation training algorithm that feeds values forward and then calculate the error by propagating it back to the earlier layers. This is done by applying the method:

```
>>> net.trainBackpopogation(b,max,learnRate,mom,err)
```

where:

b	is `true` if a dialog window is required to show the learning rate and its errors. The smaller the error on Y, the higher the chances that the learning is successful. Make sure that the error value does not change with the iteration (or epoch) number. If b is Java `false` ("0" in Jython), the monitoring of the learning rate and its error cannot be done.
max	is the maximum number of epochs for learning, i.e., the learning continues until `max` is reached, but only if the epoch error is larger than `errEpoch`. If we reach a maximum count of iterations or epochs, this will mean that the training was not successful.

learnRate   is the rate at which the weight matrix will be adjusted based on learn-
             ing (the so-called learning rate). This number is usually between 0.01
             and 0.2.
mom   is the influence that previous iteration's training δ will have on the
             current iteration. Usually the range 0.01–0.4 is assumed for this
             value.
err   is the "epoch" error for training at which the learning should be
             stopped. If the specified error is not reached during the learning, the
             program stops the learning after reaching a certain number of events
             given by "max".

It should be noted that the epoch error can also be obtained with the method
getEpochError() without opening the control window.

It is always a good practice to call the showNetwork() method to look at the
new threshold values. Once the network was trained, one can save the network in a
file, so one can restore it later. This can be done as

```
>>> net.save("test.eg")
```

One can restore the saved network at any time by calling the method:

```
>>> net.read("test.eg")
```

where "test.eg" is an input file with the neural network.

13.4.4 Training and Verifying

The data after standardization and rescaling are ready for the next step-training. We
will reset all network weights and load the first half of the prepared data (1000 rows).
The training will be done using 10,000 epoch iterations. During training, we will open
a frame displaying the value of the epoch error. Once the training is finished, we will
save the results into the file "test.eg", attributing trainedNN string to the name of
our network. We also display the neural network after the training in order to verify
that all thresholds indeed have been changed after the training. The code below shows
the entire training process:

```
from jhplot import *

d=IO.read("data_scaled.jser")
inp=d["input"].getRows("input",0,1000)
out=d["out"].getRows("result",0,1000)

net=HNeuralNet()
net.addFeedForwardLayer(4)
net.addFeedForwardLayer(4)
net.addFeedForwardLayer(1)
net.reset()
```

```
mn =10000
net.setData(inp,  out)
print net.trainBackpropagation(True,mn,0.01,0.02,0.001)
net.showNetwork()

print  "Epoch  error=",net.getEpochError()
net.save("test.eg")
```

Listing 13.13 Training a neural net

Figure 13.2 shows training error versus epoch. As expected, the error gets smaller with increasing number of epochs.

Next, we will verify the performance of the network training using the second half of our input sample. We will read the data and the network saved in the previous example, and generate predictions from the input. Then we will restore the input (remember, previously we had to scale the input data to the range [0, 1]) using the "scale" arrays, and apply the recovery procedure to the predicted array. Then, we will save the restored output and network predictions in a separate file for the next analysis step. The code that verifies the training procedure is given below:

```
from  jhplot  import  *

d=IO.read("data_scaled.jser")

min,max  =1000,2000
inp=d["input"].getRows("input",min,max)
out=d["out"].getRows("result",min,max)
scale=d["scale"]

net=HNeuralNet()
net.read("test.eg");

pred=net.predict(inp)
pred.rescale(scale)
out.rescale(scale)

d={"predicted":pred,"expected":out}
IO.write(d,"data_verify.jser")
```

Listing 13.14 Neural network verification

Note it does not have any graphical output, since it creates a new file that will be used in a program to show the result.

Finally, we will visualize the outputs and the predicted values. We will print the predicted and the expected values and calculate the ratio of the predicted outputs to the original outputs. Our hope is that this ratio should have values close to one, and have reasonably small spread. Let us create a histogram with this ratio as shown below:

Fig. 13.2 Training error versus epoch for the neural network discussed in the text. Note that the actual graph from the reader's program can differ since it depends on random initial conditions

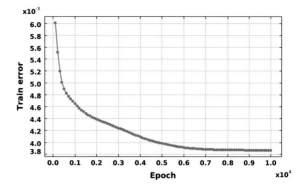

```
from jhplot import *

d=IO.read("data_verify.jser")
predicted=d["predicted"]
expected=d["expected"]

c1=HPlot()
c1.visible()
c1.setNameX("predicted/expected")
c1.setRangeX(-2,5)

h=H1D("ratio",50,-2,5.)
for i in range(predicted.size()):
    p=predicted.getRow(i)
    x=expected.getRow(i)
    d1=p.get(0)
    d2=x.get(0)
    if (d2 !=0):
            h.fill(d1/d2)
    if (i<20): print "predicted=",d1," expected=",d2

c1.draw(h)
c1.drawStatBox(h)
```

Listing 13.15 Using the neural network for predictions

The script brings up a canvas with the filled histogram shown in Fig. 13.3. One can see that the histogram has a peak at 1 as expected. Thus, we have confidence that our neural network can make reasonable predictions. However, the distribution for the ratio has long tails, with RMS of the distribution close to 0.9. The mean value is also slightly shifted from 1. All of these features indicate that the predictions are not perfect and that, for some rare events, the predicted values can be far away from the original values.

Fig. 13.3 The distribution of the predicted to the expected values using a feedforward neural network after training

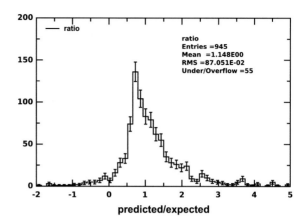

13.5 Bayesian Networks

A Bayesian network (or "belief" network) is a graphical model for manipulating probabilistic relationships between variables of interest, and building decision scenarios. The network is based on graphs with random variables and their conditional dependencies. Such networks are widely used in managing uncertainty in science, engineering, business, and medicine [6, 7].

13.5.1 Creating Bayesian Network Using Scripts

We illustrate Java scripting of a Bayesian network using the open-source Jayes package [8]. We learn how to build such a network using this package and, at the same time, we try to understand the basic concept of the idea behind such networks. We will closely follow the Wikipedia definition and Java examples included in the Jayes package.

First, we will consider a situation borrowed from the Wikipedia article on Bayesian networks. Imagine a garden where the condition of the grass (wet or dry) depends on two factors: weather (rain or sunny) and sprinkler (switched on or off). Naturally, the use of the sprinkler depends on weather. This can be illustrated as a graph shown in Fig. 8.5 of Sect. 8.3. Assume we know the direct influence relationships among variables. For example, the conditional probability that the sprinkler is on or off depends on the weather condition (rain or sunny). We know the probability of rains and the conditional probabilities of each building block of our graph. Now we can ask complex questions that are not too obvious from a naive analysis of this probabilistic graph, such as "what is the probability of rain given that we observe wet grass?"

The key building block of our network is the `BayesNode` Java class. It is designed to define outcomes, parents, and a table with conditional probabilities,

$P(x_i|x_j \ of$ parents). Let us show how to solve the question posed above by creating a three of `BayesNodes` blocks with known conditional probabilities:

```
from bayesnet.jayes import *
from bayesnet.jayes.inference.jtree import *
from java.util import *

net=BayesNet()
a=net.createNode("weather")     # node=1
a.addOutcomes("rain", "suny")
a.setProbabilities(0.2, 0.8)    # P(rain)=0.2

b=net.createNode("sprinkler")   # node=2
b.addOutcomes("on", "off")
b.setParents(Arrays.asList(a))
b.setProbabilities(
#   on  | off
  0.01, 0.99,  # weather=rain
  0.4,   0.6)  # weather=suny
c=net.createNode("grass")         # node=3
c.addOutcomes("wet", "dry")
c.setParents(Arrays.asList(a, b))
c.setProbabilities(
  # wet  |  dry
  # weather=rain
  0.99,     0.01,  # sprinkler=on
  0.8,       0.2,  # sprinkler=off
  # weather=suny
  0.9,       0.1,  # sprinkler=on
  0.0,       1.0)  # sprinkler=off
inferer=JunctionTreeAlgorithm()
inferer.setNetwork(net)
evidence=HashMap()
evidence.put(c, "wet")                  # grass is wet
inferer.setEvidence(evidence)
beliefsC = inferer.getBeliefs(a)
print "P(weather | grass is wet)=",beliefsC.tolist()
```

Listing 13.16 Building a simple Bayesian network (I)

Our code constructs the Bayesian network with the nodes defined by the conditional probabilities. The comments included in this example explain each step. The output of this code is the probability for the rain when wet grass is observed:

```
P(weather | grass=wet)= [0.35768, 0.64231]
```

It shows that the probability for rain ("weather $=$ rain") is 35.8 %.

We can also ask more complicated questions, such as what is the probability for sprinkler being switched on for rainy weather and when grass is wet? Simply append this code to the above program:

```
evidence=HashMap()
evidence.put(a, "rain")
evidence.put(c, "wet")
inferer.setEvidence(evidence)
beliefsC = inferer.getBeliefs(b)
ans=beliefsC.tolist()
print "P(sprinkler | weather=rain, grass=wet)=",ans
```

Listing 13.17 Working with Bayesian networks (II)

You will see a smaller value for the conditional probability (1.2 %).

Let us extend the above example by creating a Bayesian network for visitors of a hospital. A visitor can be a smoker or a nonsmoker (with certain probability). The race of the visitor is also known with some probability. We also know the state of the health (which directly depends on the race, and on whether the visitor is a smoker or a nonsmoker).

Let us calculate the probability for smoking white visitors who are entering the hospital with symptoms of illness.

```
from bayesnet.jayes import *
from bayesnet.jayes.inference.jtree import *
from java.util import *

net=BayesNet()
a=net.createNode("visitor") # node=1
a.addOutcomes("smoking", "non-smoking")
a.setProbabilities(0.2, 0.8)

b=net.createNode("race")      # node=2
b.addOutcomes("asian", "hispanic", "white")
b.setParents(Arrays.asList(a))
b.setProbabilities(
# asian | hispanic | white
  0.2,       0.3,        0.5, # visitor=smoking
  0.3,       0.4,        0.3) # visitor=non-smoking

c=net.createNode("condition")   # node=3
c.addOutcomes("ill", "healthy") # can be "ill" or "
    healthy"
c.setParents(Arrays.asList(a, b))
c.setProbabilities(
  # ill | healthy
  #  visitor == smoking
     0.1,   0.9,   # race=asian
     0.0,   1.0,   # race=hispanic
     0.5,   0.5,   # race=white
  # visitor == non-smoking
     0.2,   0.8,   # race=asian
     0.0,   1.0,   # race=hispanic
     0.7,   0.3)   # race=white
inferer=JunctionTreeAlgorithm()
```

```
inferer.setNetwork(net)
evidence=HashMap()
evidence.put(c, "ill")
evidence.put(b, "white")
inferer.setEvidence(evidence)
beliefsC = inferer.getBeliefs(a)
print "P(visitor | ill, white)=",beliefsC.tolist()
```

Listing 13.18 Building a Bayesian network for hospital visitors

This prints [0.229, 0.77], where the first number corresponds to smoking white visitors. The comments in this code should be sufficient to read conditional probabilities used to create this network.

We hope that the examples shown above are sufficient to design complex Bayesian networks with multiple number of requested outcomes.

13.5.2 HBayes Method

The Bayesian network is also included into the DMelt library using the package called JavaBayes [9]. It calculates marginal probabilities and expectations, and performs robustness analysis. In addition, it allows a user to import, create, modify, and export networks.

The network editor can be called using the HBayes class:

```
>>> from jhplot import *
>>> HBayes()
```

This brings up a GUI editor to be used to create a Bayesian network, together with a console window. A user can follow the step-by-step instruction given in the console. The JavaBayes network is well documented and will not be discussed here further.

13.6 Kohonen Self-organizing Map

In addition to the neural network algorithms which require a "learning" procedure, there is another class of unsupervised learning algorithms. Such algorithms attempt to find the most appropriate topological description of input data.

Self-organizing feature maps (SOFM) belong to a popular class of unsupervised learning algorithms. These maps are useful for classification and visualizing high-dimensional data in low-dimensional views. A well-known type of SOFM is a Kohonen network proposed by the Finnish professor Kohonen [10]. The goal of the Kohonen network is to map input data of arbitrary dimension to a discrete map of low dimensions in order to discover some underlying structure in data.

The Kohonen SOFM does not require a target output, and there is no supervision whatsoever as for the traditional neural networks. The network has nodes (or units) at specific topological positions and vectors of weights of the same dimension as the input vectors. Instead, where the node weights match the input vector, that area of the lattice is selectively optimized to more closely resemble the data. From an initial distribution of random weights, and over many iterations, the SOFM eventually settles into a map of stable zones. Each zone is effectively a feature classifier, so you can think of the graphical output as a type of feature map of the input space. Training the Kohonen network requires many iterations following this procedure:

1. Each node weight is initialized with a random value.
2. A vector is chosen randomly from the set of training data.
3. Every node is examined to calculate which weights are similar to the input vector. The winning node that has the smallest Euclidean distance to the input pattern is commonly known as the "best-matching unit."
4. Next, the radius of the neighborhood of the best-matching unit is calculated. Initially, this value is set to the radius of the lattice, but diminishes after each step. For any nodes found inside the radius of best-matching unit, the node weights are adjusted to make them more like the input vector. The closer a node is to the best-matching unit, the more its weights get altered. Weights of the nodes that are far away from best-matching units do not change significantly. Then, we repeat Step 2 for N iterations.

Below we will consider a number of examples of how to construct and run a Kohonen SOFM using an open-source package [5] developed in Java.

13.6.1   Kohonen SOFM in 2D

First, we will prepare the data and graphical canvas for our example. We will create a random data in $X-Y$ that follows a wave pattern.

```
from jhplot    import   *
from java.util import   *
import math

c1 =HPlot()
c1.visible(); c1.setLegend(0)
c1.setRange(0,100,0,150)
c1.setMarginLeft(70)
c1.setNameX("X"); c1.setNameY("Y")

p1= P1D("input data")
rand = Random(1L) # reproducible random numbers
inputSize=100      # number of random points
for i in range(inputSize):
   x=i+10*rand.nextGaussian()
```

```
      y=50+30*math.cos(0.2*x)+10*rand.nextGaussian()
      if (x>0 and x<100): p1.add(x,y)
c1.draw(p1)
```

Listing 13.19 Preparing data and graphical canvas for a Kohonen SOFM

Next, we will use the data from the previous example for the input of the Kohonen SOFM of size 4 × 4. The training will stop when the difference between the original data and the network nodes is less than 0.01. Then we will run the Kohonen SOFM in which neurons are organizing themselves according to certain input values, and show the resulting map on top of our original data. Simply append the code shown above to the previous example.

```
from jhplot import *
from jhpro.nnet import *
from java.awt import Color

kfm=KohonenFeatureMap()
mapSizeX=4             # map size is 4x4 neurons
maxCycle=100000000
im = InputMatrix(inputSize, 2)
kfm.setMaxLearningCycles(maxCycle)
kfm.createMapLayer(mapSizeX,mapSizeX ) # create a XY map
kfm.setStopArea(0.01)                       # stop learning here
kfm.setInitActivationArea(1)
kfm.setInitLearningRate(0.6)
im.setInputXY(p1)    # set X and Y data
kfm.connectLayers(im)

i=0
p2=P1D("weights")
p2.setColor(Color.red); p2.setStyle("l")
while kfm.finishedLearning() == False:
    kfm.learn()
    i=i+1
    if (i%50 ==0):
        weights=kfm.getWeightValues()
        p2.clear(); c1.clearData()
        print "rate=",kfm.getLearningRate()
        print "Activation area=",kfm.getActivationArea()
        print "Elapsed time=",kfm.getElapsedTime()
        for j in range(mapSizeX*mapSizeX):
                p2.add(weights[0][j], weights[1][j])
        p2.sort(0) # draw line after sorting in X
        c1.draw([p1,p2])
print "Final activation area=",kfm.getStopArea()
print "SOM=",p2
```

Listing 13.20 Building a Kohonen SOFM with self-organizing neurons

Fig. 13.4 An illustration of a Kohonen SOFM in 2D. The *symbols* show the original input data, while the *line* shows the resulting map from the Kohonen SOFM

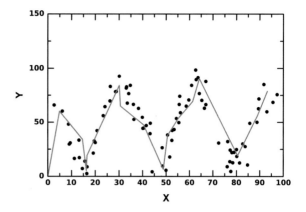

The resulting plot is shown in Fig. 13.4. As you can see, the 4 × 4 map shown with red lines resembles the original data.

13.6.2 Kohonen SOFM in 3D

Now we can easily extend the above 2D example to the 3D case. The only change compared to the previous code is that we will create 3D data with the components X, Y, Z. We should also use the appropriate 3D canvas and the P2D container to visualize the data and the trained SOFM.

```
from jhplot   import   *
from java.util import   *

c1=HPlot3D("Canvas",600,500)
c1.setRange(-100, 100, -100, 100, -100,100)
c1.visible()
pc=P2D("Input Data")
rand = Random(1L)   # reproducible random numbers
inputSize=200        # number of random points
for i in range(inputSize):
     x=5*rand.nextGaussian()
     y=5*rand.nextGaussian()
     z=20*rand.nextGaussian()
     if (i>100):
             x=10+10*rand.nextGaussian()
             y=-20-30*rand.nextGaussian()
             z=50-10*rand.nextGaussian()
     pc.add(x,y,z)
c1.draw(pc)              # draw data
```

Listing 13.21 Preparing data for a Kohonen SOFM in 3D

Now, append the following code to the previous code example:

```
from jhplot   import   *
from jhpro.nnet import *
from java.awt import *
from java.util import Random
import math

kfm=KohonenFeatureMap()
mapSizeX,mapSizeY=4,4        # 4x4 map size
maxCycle=1000000             # max Nr of iterations
im=InputMatrix(inputSize, 3);
kfm.setMaxLearningCycles(maxCycle);
kfm.createMapLayer(mapSizeX, mapSizeY) # create a map
kfm.setStopArea(0.02)    # value of error to stop learning
kfm.setInitActivationArea(1)
kfm.setInitLearningRate(0.6)
im.setInputXYZ(pc)                        # insert X-Y-Z data
kfm.connectLayers(im)
p2=P2D("SOM weights")
p2.setSymbolSize(10)
p2.setSymbolColor(Color.red)
i=0
while kfm.finishedLearning() == False:
    kfm.learn()
    if (i%50 ==0):
      p2.clear()
      print "print   rate=",kfm.getLearningRate()
      print "Activation area=",kfm.getActivationArea()
      print "Elapsed  time=",kfm.getElapsedTime()
      weights=kfm.getWeightValues()
      for j in range(mapSizeX*mapSizeY):
          p2.add(weights[0][j], weights[1][j],weights[2][j])
      c1.draw(p2)
      c1.updateData()
    i=i+1
c1.update()
```

Listing 13.22 Running a Kohonen SOFM in 3D

The resulting figure is shown in Fig. 13.5. The large red circles show the network nodes after training. As expected, the Kohonen algorithm follows the trends of the original 3D data (small black dots), and captures the clusters in the input data.

13.7   Bayesian Self-organizing Map

Now we will consider a Bayesian self-organizing map (BSOM) which represents an algorithm for estimating a probability distribution from input data on the basis of a Bayesian stochastic model. The data-recognition algorithm of the BSOM tries

Fig. 13.5 An illustration of a Kohonen SOFM in 3D. The *large red symbols* show the resulting Kohonen SOFM that captures the topology of the input data shown with the *small black dots* (color figure online)

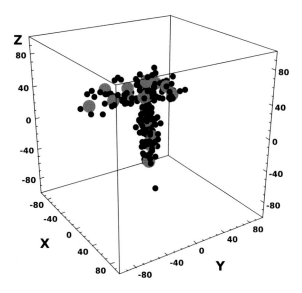

to find relationships in high-dimensional space and, if successful, it converts this knowledge into a simple relationship using low-dimensions.

We will consider an example that helps us make the BSOM explanation more illustrative. To do this, we will use the Java class HBsom based on the BSOM program [11].

The code below fills a histogram, converts it into a P1D container which is then used for the algorithm input. We use 30 interconnected units to analyze the topological shape in 2D, see the method setNPoints. The last line of the code creates a GUI window with the visualized data points and interconnected units which are the BSOM result. Also, we save the created input data to the ASCII file "data.txt" (see the method toFile()). This file will be used for the example to be shown in the next subsection.

```
from java.util import *
from jhplot  import *

h1 = H1D("Data",20,-100.0, 300.0)
r = Random()
for i in range(2000):
    h1.fill(100+r.nextGaussian()*100)

p1d=P1D(h1,0,0)
p1d.toFile("data.txt")

bs=HBsom()
bs.setNPoints(30)
bs.setData(p1d)
bs.visible()
```

Listing 13.23 Illustrating a Bayesian self-organizing map (BSOM)

To start the BSOM algorithm, one should set up the values α and β. Then press the button "learn". Initially, the BSOM units are positioned randomly. During the execution of the program the algorithm calculates the most optimal positions for interconnected units. The resulting plot is shown in Fig. 13.6.

The parameter α represents the strength of topological constraints, while β is the noise level in the data. The parameters α and $1/\beta$ can be related to the temperature of a physical system. For $\alpha \simeq 0$, topological constraints on the units are not imposed and the BSOM can be viewed as a data clustering based on a spherical Gaussian mixture model. When β is infinitely large, BSOM is similar to the K-means clustering algorithm.

Running this algorithm can help find an optimal configuration for data description. Usually, one should start learning from a high value of α and a low value of β. We note that BSOM can automatically search for the optimal values of the parameters by pressing the "auto" button. The "density" button shows the estimated data density.

13.7.1 Noninteractive BSOM

It is more practical to run the BSOM program in a noninteractive mode. This can be done by removing the line with the `visible()` method and inserting the line `run()`. The result of the algorithm can be retrieved with the method `getResult()` which returns the output in the form of `P1D` array. Then we plot the results as in the code below:

Fig. 13.6 The BSOM algorithm in action using an interactive mode

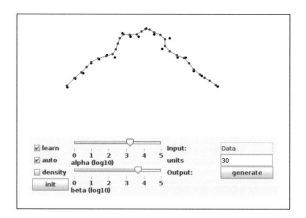

```
from jhplot  import *
from java.awt import Color

c1 = HPlot("Canvas")
c1.setGTitle("Bayesian Self-Organizing Map")
c1.visible()
c1.setAutoRange()

p1d=P1D("data","data.txt")
p1d.setErrToZero(1)
bs=HBsom()
bs.setNPoints(30)
bs.setData(p1d)
bs.run()
result=bs.getResult()
result.setStyle("pl")
result.setColor(Color.blue)
c1.draw(p1d)
c1.draw(result)
```

Listing 13.24 Running a noninteractive BSOM

One should note that we set all statistical errors of the input P1D to zero. The resulting P1D is shown with the symbols connected with the lines, see Fig. 13.7.

Let us tell more about the use of the HBsom Java class in the noninteractive mode. The method run() executes the learning algorithm. The learning stops if topological changes between each step are smaller than some parameter, which can be set via the method setDelta(). One can get the total number of iterations using the method getNiterations().

For more complicated topologies of data in 2D, it is important to adjust the initial α and β values using the method setAlphaBeta(a,b).

Fig. 13.7 Running the BSOM algorithm in a noninteractive mode

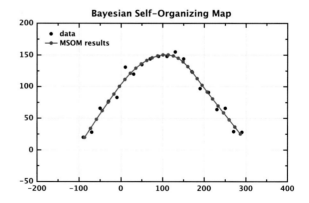

13.8   Neural Network Using Python Libraries

DMelt packages several libraries for neural network studies implemented in Python. In particular, one can use the PyANN program [12] which was initially written in Python (to be more exact, in CPython). As you already know, numerical programs with large number of loop iterations implemented in Python are slow. Therefore, we will not gain much in terms of execution performance when running this package using Jython, compare to the dedicated Java or C++ libraries. But there are several other advantages in using the Jython approach: (1) The code is fully "transportable" and can be run using CPython; (2) There is a direct access to the original code if one needs to understand it and make necessary modifications; (3) Finally, you will get all benefits of the Python programming language, including its interactivity for easy debugging.

In this section we will illustrate how to use the PyANN library using DMelt. The corresponding library is located inside the directory "pyann" in "python/packages". If you are using the DMelt IDE, this directory is imported automatically. If you do not use this IDE, you should import this Python module manually inside the code examples to be shown later.

Now we will learn how to run a feedforward neural network implemented in Python. As in the previous section, we will divide our work in several steps: (1) We will prepare a common module to make our codding compact; (2) we will create files with the input data for training and verification. We will use the `shelve` module from Python which is discussed in Sect. 2.15.6; (3) then we will build a neural network and train it. Finally, we will verify the performance of this algorithm.

Let us first make a common module that imports the necessary libraries from the PyANN package. Here, we will also define several global parameters, as well as functions for input and output. The example code is given below:

```python
import sys,math,random,shelve

import pyann.mlp
from pyann.mlp.layer import *
from pyann.mlp.monitoring import *
from pyann.mlp.training import *
from pyann.mlp.training.backprop import *

WINDOW, MIN_VALUE, MAX_VALUE = 10,-1,1

def writeData(data,valid,pattern):
    sh=shelve.open("data.shelf")
    sh["data"]=data
    sh["validate"]=valid
    sh["pattern"]=pattern
    sh.close()

def readData():
    return shelve.open("data.shelf")
```

Listing 13.25 A Python module "NNpython.py"

Using this module, we will create a file with the input data. We will use the function sin(i) as a template to generate data patterns. The data will have 73 rows, 63 will be used for training and the rest rows are inputs for validation. Each row of the data will be in the form of tuples with 10 inputs. The neural network will have one output (also in the form of tuple but with one value).

```
from NNpython  import *

data = [] # Build data set
for i in range(-360, 361, 10):
    data.append(math.sin(i))
print len(data)

start = len(data)-WINDOW-1
pattern=[] #  Build pattern
while len( data[start : start+WINDOW] ) == WINDOW:
  input = tuple(data[start : start+WINDOW])
  output = (data[start+WINDOW],)
  pattern.append(pyann.mlp.Pattern(input, output))
  start -= 1

valid = [] # validation pattern.  20% of data
for i in range(int(len(pattern)*0.2)):
 valid.append(pattern.pop(random.randrange(len(pattern))))

writeData(data,valid,pattern)
```

Listing 13.26 Creating a data set

In the above example, use the method type() for debugging. It helps to understand what this code is doing. All data will be saved into a file.

Now we come to the training. First, we build a neural net with 10 inputs, 5 nodes for the hidden layer and one output node. Then we read the data and train the net using rather self-explanatory methods. The script prints the training ("epoch") errors. Finally, we will verify the network using our verification sample.

```
from NNpython  import *

layers = (SigmoidInputLayer(WINDOW),\
 SigmoidLayer(5),\
 SigmoidOutputLayer(1,minValue=MIN_VALUE,
 maxValue=MAX_VALUE))
net = pyann.mlp.Network( layers )

sh=readData()
data=sh["data"]
validation=sh["validate"]
patterns=sh["pattern"]

monitor = VerboseMonitor()
stopOnMaxIter =MaxIterationsStopCondition(500)
stopOnMinError = MinErrorStopCondition(0.01)
```

```
stopOnVMinError =MinValidationErrorStopCondition(0.005)

trainer =BackpropagationTrainer(net, monitor)
trainer.setLearningRate(0.7)
trainer.setMomentum(0.5)
trainer.setRandomize(True)
trainer.setStopConditions((stopOnMaxIter,\
        stopOnMinError,stopOnVMinError),\
        joiner = "or")

trainInfo = trainer.train(patterns, validation)
print """
Epochs: %(epoch)s
Final error: %(error)s
Validation set error: %(validationError)s
""" % { "epoch": trainInfo.getIterationNumber(),
  "error": trainInfo.getError(),
  "validationError": trainInfo.getValidationSetError()}

for i in range(370, 725, 10):
    start = len(data) - WINDOW
    predicted = net.classify( tuple(data[start:]) )[0]
    data.append(predicted)
    print "sin(%s) = %s" % (i, predicted)
```

Listing 13.27 Neural network training and predictions

Run this code to see the result of the training. The code snippet prints out the predicted values, which are close to the true values of the input $\sin(i)$ function.

References

1. Beale R, Jackson T (1990) Neural computing: an introduction. Institute of Physics Publishing, Bristol
2. Bharath R, Drosen J (1994) Neural network computing. McGraw-Hill, New York
3. Heaton J, contributions, The encog project. http://code.google.com/p/encog-java/
4. Sevarac Z et al, Neuroph, java neural network framework. http://neuroph.sourceforge.net/
5. Frhlich J (2004) Neural Networks with Java, diploma thesis, Fachhochschule Regensburg http://www.nnwj.de/
6. Neapolitan RE (2003) Learning bayesian networks. Prentice Hall, Upper Saddle River
7. Jensen FV, Nielsen TD (2007) Bayesian networks and decision graphs (information science and statistics), 2nd edn. Springer, Berlin
8. Kutschke M, Neuroph, Java neural network framework. https://github.com/kutschkem/Jayes
9. Cozman FG, Javabayes: Bayesian networks in Java. http://www.cs.cmu.edu/javabayes/
10. Kohonen T (2001) Self-organizing maps, 3rd edn., Springer Series in Information Sciences, vol 30. Springer, Berlin
11. Utsugi A (1997) Neural Computation 9:623
12. PyANN, a Python framework to build artificial neural networks. http://sourceforge.net/projects/pyann

Chapter 14
Finding Regularities and Data Classification

Regularities and distinct patterns observed in data can be used to better understand the underlying process that creates such data. Thus, an important part of any realistic data analysis is to search for any unusual features that can be used later for data comprehension. Observation of structures in data can lead to discoveries of new laws of nature, and also be used for more pragmatic tasks, such as prediction of future dynamic changes.

This chapter will continue the discussion of data mining and knowledge discovery algorithms on the Java platform. As the reader may already have noticed, this book has already discussed several topics relevant to finding regularities in multidimensional data. For example, Sects. 13.6 and 13.7 have discussed self-organizing maps, which are excellent tools in exploratory phase of data mining.

In the following sections we will continue the discussion of algorithms designed to find unusual features in data. Such algorithms go much beyond the simple statistical characteristics discussed in Chap. 10. At the same time, unlike the neural networks discussed in Chap. 13 , they do not require the learning step to memorize input data.

14.1 Cluster Analysis

Clustering algorithms are important tools for unsupervised classification of data. The main idea is to classify a given data set through a certain number of clusters by minimizing distances between objects inside each cluster. In general terms, data clustering is a method to group objects such that they belong to the same group depending on the similarity of objects. A detailed discussion of the clustering analysis can be found, for example, in several books [1–3].

The data clustering in DMelt is based the jMinHEP Java library [4]. This library includes:

© Springer International Publishing Switzerland 2016   505
S.V. Chekanov, *Numeric Computation and Statistical Data Analysis
on the Java Platform*, Advanced Information and Knowledge Processing,
DOI 10.1007/978-3-319-28531-3_14

- The K-means algorithm (single and multi-pass) which classifies data by minimizing the sum-of-square distances between data points and the so-called geometric centroids, assuming that the number of clusters is fixed a priory. Clustering is done in a single pass starting from seed centroids positioned at random locations. Initially, the algorithm assigns groups of objects to "clusters" which have the closest centroid, and then it recalculates the positions of new centroids. This step is repeated until the centroids no longer move.

  This algorithm has one disadvantage. It depends on the initial conditions, thus jMinHEP also has the option to rerun the algorithm multiple number of times until a stable solution is found. Typically, you will need to define a number of expected clusters, a maximum number of iterations, and/or a precision with which the clustering is done. To get started, you may rely on the default values for precision and for the maximum number of iterations.
- The C-means (fuzzy) algorithm. Unlike the K-means approach, this method of clustering allows data points to belong to two or more clusters. The algorithm usually uses a "fuzziness" coefficient (typically, it is set to 2) and accuracy of the calculations.

  When clusters overlap, it is not easy task to plot such clusters. In this algorithm, each point has a membership probability to belong to a particular cluster. However, one could supply some probability (typically above 0.6) to illustrate data points which have a membership probability above the specified value.
- The agglomerative hierarchical clustering algorithm. For this clustering method, there is no need to supply a predetermined number of clusters. The algorithm tries to determine the so-called "natural grouping". The clustering continues until all points are clustered into a single cluster.

Clustering can be done with the package `jminhep.algorithms`. One should follow these steps to perform clustering:

- Create a data container using the class `DataHolder`. This object keeps data in a multidimensional phase space, i.e. each data point can be characterized by an arbitrary number of values.
- Each point is represented by the `DataPoint` class from the package `jminhep.cluster`. Data points can be defined in a multidimensional phase space. One should populate this data holder from an external source, or fill it on-the-fly by simulating events. We will show how to do this in the example below.
- Create and initialize a clustering algorithm class and pass it the constructor of the class `DataHolder`.
- Run the algorithm in a loop and retrieve the output.

Finally, one can print or plot the retrieved centroids of clusters by projecting multidimensional space into $X-Y$ plane and, optionally, one can plot the original points used for clustering.

14.1.1 *Preparing a Data Sample*

First, let us learn how to fill data into the `DataHolder` container. Once this data holder is ready, we will save the data in a file, and then we will read this file in order to illustrate clustering analysis.

One single point should be represented by a `DataPoint` object. For example, a point with five components can be filled as

```
>>> from jminhep.cluster import *
>>> p=DataPoint([1,2,3,4,5])
>>> print p.getDimension()
5
>>> print p.toString()
(1.0,2.0,3.0,4.0,5.0)[0]
```

Then, one should add each point represented by the `DataPoint` class to the `DataHolder` container. Below, we add two data points:

```
>>> from jminhep.cluster import *
>>> d=DataHolder("data")
>>> p=DataPoint([1,2,3,4,5])  # 1st point
>>> d.add(p)
>>> p=DataPoint([5,6,7,8,9])  # 2nd point
>>> d.add(p)
```

One can apply the method `add(obj)` multiple number of times to fill in the `DataHolder` container. One can get the data back to the usual arrays with the `getArrayList()` or `getArray()` methods.

Here are a few useful methods of how to retrieve information with the input data:

```
>>> from jminhep.cluster import *
>>> size=d.getSize()     # get size
>>> p=d.getRow(n)        # get DataPoint at index "n"
>>> min=d.getMin()       # get min
>>> max=d.getMax()       # get max
>>> print d.toString()   # print
>>> d.clear()            # clear data
```

Now let us create data with three clusters. The data points will be generated in 3D. We assume that the data points will be clustered around three cluster centers, at the positions $(-8, 5, 10)$, $(10, 20, 5)$, and $(10, 1, 20)$. This can be done using a Gaussian random-number generator as

```
from java.util import Random
from jminhep.cluster import *
from jhplot import *

data=DataHolder("Clusters in 3D")
```

```
r=Random(1L)
for i in range(20): # 1st cluster
   a =[]
   a.append(    r.nextGaussian()-  8)
   a.append( 2*r.nextGaussian()+  5)
   a.append( 3*r.nextGaussian()+10)
   data.add( DataPoint(a) )
for i in range(30): # 2nd cluster
   a =[]
   a.append( 10*r.nextGaussian()+10)
   a.append(  2*r.nextGaussian()+20)
   a.append(  5*r.nextGaussian()+  5)
   data.add( DataPoint(a) )
for i in range(10): # 3rd cluster
   a =[]
   a.append( 3*r.nextGaussian()+10)
   a.append( 2*r.nextGaussian()  +1)
   a.append( 4*r.nextGaussian()+20)
   data.add( DataPoint(a) )
IO.write(data,"data.jser")
```

Listing 14.1   Creating data for a clustering analysis

We should note that the sizes of the clusters (in terms of the Gaussian standard deviation, σ) are different—this is given by the scaled factors used to multiply the random values created by the method nextGaussian(). The second cluster is broad since its radius spans 10σ (arbitrary) units.

We have written the data in a compressed serialized file. One can also use the writeXML() method to write data into an XML file for easy viewing. We should remind: if you need to write a lot of data, the best way to do this is to use the HFile or HFileXML Java classes, see Sect. 9.3. We can write each DataPoint object persistently in a loop. This can be done straightforwardly, since these Java objects can be serialized. As explained before, this approach allows creation of rather large files since we do not store all data in the computer memory before writing them to the hard drive.

The previous example shows how to fill the data in three dimensions. But what method to use to display such data, especially when there are more than three dimensions? One possible option is to project the DataHolder into a P1D array, and plot it using the standard HPlot class. For example, if our data are represented by the DataHolder object, one can view the 1st and 2nd components as

```
from jhplot import *

data=IO.read("data.jser")
c1=HPlot()
c1.visible()
c1.setAutoRange()
p1=P1D(data,0,1) # fill 1st and 2nd component
```

```
HTable(p1)          # show data in a table
c1.draw(p1)         # draw 1st and 2nd component
```

Listing 14.2 Visualizing data for used for clustring

This code brings up a frame with the table and a 2D canvas with the data shown as $X-Y$ (or scatter) plot.

14.1.2   Clustering Analysis

We will assume that the created data from the previous example were successfully saved in the file "data.jser", and they can be extracted and viewed. Now one can perform a realistic clustering analysis. Generally, to run any clustering algorithm, one needs to follow these steps:

```
>>> from jminhep.algorithms import *
>>> alg = [Algorithm]Alg(data)
>>> alg.setClusters(NumberOfClusters)
>>> alg.setOptions(some options)
>>> .. more options ..
>>> alg.run() # run over the data
>>> print "algorithm: " + pat.getName()
>>> .. get results
```

In the above example, "[Algorithm]" is the name of a selected clustering algorithm. Below we list several choices:

KMeansAlg(data)	the standard K-Means cluster algorithm;
KMeansExchangeAlg(data)	K-Means cluster algorithm using the exchange mode;
FuzzyCMeansAlg(data)	Fuzzy (C-means) cluster algorithm;
HierarchicalAlg(data)	an agglomerative hierarchical clustering algorithm.

Then, one should set the expected number of clusters and other options. For example, for the C-means algorithm, one should set the number of expected clusters, the numerical precision of clustering and the so-called fuzziness parameter. It is possible to set the maximum number of iterations to stop clustering if no appropriate solution can be found. It is also useful to set the probability association if you need to know which points belong to which cluster (this is not necessary for the K-Means algorithm). The description of each mode can be found with the method getName().

To run the algorithm over the data, call the method run(). This method should be invoked for a fixed number of clusters. One may also consider the *runBest()* method. In this case, the program will attempt to determine the number of clusters by rerunning the algorithm many times over the same data set. Then it calculates the most optimal solution by minimizing the so-called "compactness" of the cluster configuration. The smaller the compactness, the higher the chance that a particular cluster solution is the most optimal.

After the clustering procedure (i.e., after running the method `run()`), one can find the "compactness" of the cluster configuration by calling the method `getCompactness()`. The centroid positions are retrieved with `getCenters()`, and the data-point association can be obtained using several built-in methods. Please look at the API of this library.

Now we have all the machinery needed to perform a clustering analysis. Below, we show a detailed example for the K-means algorithm. First, we prepare a function `printAnswer()` designed to print all relevant information about the cluster configuration. Then, we will run the K-means algorithm for: (1) a single pass using three clusters; (2) 10 passes with different random seed locations. The algorithm returns the cluster configuration with the smallest compactness; (3) Running the K-Means algorithm using multiple number of passes, and determining the best possible configuration with the smallest compactness.

```
from jminhep.algorithms import *
from jhplot import *

data=IO.read("data.jser")
def printAnswer(alg):
  print "Name="+alg.getName()
  print "No of final clusters:" +str(alg.getClusters())
  print "No of points: " +str(alg.getNumberPoints())
  print "Compactness: " +str(alg.getCompactness())
  centers = alg.getCenters()
  print centers.toString()

alg=KMeansAlg(data)  # initialize K-means
alg.setClusters(3)
alg.setOptions(1000,0.001)
alg.run()                # run the algorithm
printAnswer(alg)
alg.run(10) # run 10 times with different seeds
printAnswer(alg)

alg=KMeansExchangeAlg(data)
alg.setEpochMax(200)
alg.runBest()   # find smallest compactness
printAnswer(alg)
```

Listing 14.3 Running the K-means algorithm

The clustering results will be printed by the `printAnswer()` function. The shortened output from the execution of the above example is given below:

```
kmeans algorithm fixed cluster mode
No of final clusters:3
No of points: array('i', [24, 26, 10])
Compactness: 0.978823657656
0   (-8.475, 6.985, 8.286)[0]
1   (13.213, 20.17, 5.398)[0]
2   ( 8.9301,0.416, 19.00)[0]

kmeans algorithm for multiple iterations
No of final clusters:3
No of points: array('i', [22, 30, 8])
Compactness: 0.937913317119
0   (15.3939, 20.0829, 6.292)[0]
1   (-2.2662, 2.94504, 12.46)[0]
2   (-5.15491, 20.783, 2.120)[0]

kmeans using exchange method for best estimate
No of final clusters:2
No of points: array('i', [28, 32])
Compactness: 0.971777186195
0   (11.72, 20.337, 5.0967)[0]
1   (-3.090, 3.968, 12.080)[0]
```

This output shows the positions of the found clusters and the values of compactness. One may immediately see that the answers from all these algorithms are different. This is not too surprising—the clustering analysis is an inherently ambiguous task, especially when dealing with overlapping clusters.

Try to make some modifications to the input data to reduce the cluster overlaps. One can simply set all multiplicative factors for the method nextGaussian() to a smaller value (say, 1) in the code shown in Sect. 14.1.1. This reduces the cluster size, and thus overlaps between the clusters. After rerunning the above code with the K-means clustering, you will see that all three algorithms give similar answers.

Now, let us visualize our clustering analysis. We will read the data prepared in the previous subsection and plot the first and second components of data in the $X-Y$ plane. This time, however, we will run the C-means algorithm with a fixed number of expected clusters. We will also explicitly set the strength of the association between data points and a particular cluster to 0.7.

```
from java.awt import Color
from jhplot    import *
from jminhep.algorithms import *

data=IO.read("data.jser")
c1=HPlot()
c1.visible(); c1.setAutoRange()
p1=P1D(data,0,1) # 1st and 2nd component

alg = FuzzyCMeansAlg(data)
```

```
alg.setClusters(3)  # expect 3 clusters
# 10,000 iterations, 0.001 (precision), 1.8 (fuzziness)
alg.setOptions(10000, 0.01, 1.8)
alg.setProb(0.7)  # association probability
alg.run()          # run the algorithm
print "Algorithm: " + alg.getName()
print "Compactness: " + str(alg.getCompactness())
print "No of final clusters:" + str(alg.getClusters())
centers=alg.getCenters()
print "No of points in cluster=",alg.getNumberPoints()
p2=P1D(centers,"Centroids",0,1)
p2.setColor(Color.red)
p2.setErrAll(0); p2.setSymbol(9); p2.setSymbolSize(15)
c1.draw([p1,p2])  # show original data and cluster centers
```

Listing 14.4 Running the C-means algorithm with visualization

The resulting plot is shown in Fig. 14.1. The figure displays the input data projected in 2D (black dotes), and the red crosses show the centers of three clusters. You may wonder why the cluster centers are shifted from the visually expected positions. The answer is simple: the clustering was done in 3D, while Fig. 14.1 shows only a 2D projection of the data.

14.1.3 Interactive Clustering Analysis

One can also perform the clustering analysis with the JMinHEP program [4]. This program can be used to select the type of the cluster algorithm and the initial conditions using a handy interactive menu. You can start the jMinHEP cluster program using the HCluster Java class, which executes the main GUI of the program. Do not forget to append the data in the form of DataHolder object:

Fig. 14.1 The result of the C-means clustering algorithm shown in 2D. The clustering analysis is performed using data points in 3D. The algorithm is required to identify exactly three clusters

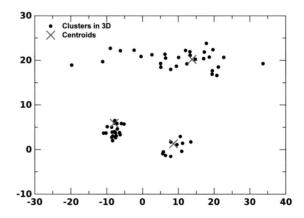

```
from jhplot   import *
data=IO.read("data.jser")  # load data
c1=HCluster(data)           # start GUI
```

Listing 14.5 Running an interactive clustering analysis

Note that data can also be loaded using the menu of this frame. This package uses different colors to visualize the membership of data points to certain clusters.

14.2 Clustering Particles into Jets. Real-Life Example

Let us discuss a real-life situation where the clustering analysis is a typical procedure to identify regularities in data. Collisions of subatomic particles at high energy create many other particles which can be recorded in detectors. Figure 14.2 shows a result of a typical collision of two particles at the CERN laboratory. Scientists working in this laboratory are searching for new discoveries in the head-on collisions of protons, similar to those discussed in Sect. 10.5.6.2. Particles from collisions can form "jets"—cones of many high-energetic particles. They are considered as experimental signatures of quarks and gluons, the basic building blocks that make up matter. Scientists are hunting for such clusters of particles, since they carry the information about the mechanism by which quarks and gluons are produced. Particles and jets are characterized by three numbers: the transverse energy, p_T (reflecting energies of produced particles), azimuthal angle (ϕ) in the detector frame and the so-called

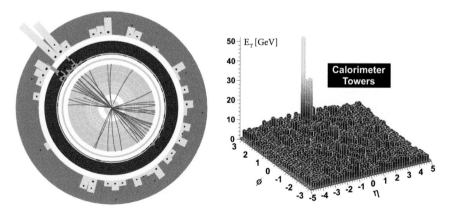

Fig. 14.2 The result of a collision of two accelerated particles at the LHC experiments at CERN. The *left plot* shows many particles produced as a result of such collisions in the perpendicular plane of the recording detector, while the *right plot* shows the particle density in this collision as a function of η and the azimuthal angle ϕ variables. The spikes shown on this image are "jets"—collimated spray of stable particle recorded in detectors. All credits for this image goes to CERN. For clarity of presentation in this book, the colors were inverted from the original image [5]

pseudo-rapidity (η) describing the angle of a particle relative to the direction of the original colliding particles (the "beam" axis). Please look at Sect. 4.4.5 which discussed how to characterize a typical particle with the Lorentz vector.

Let us make a numeric mini-experiment. We will create a typical result of particle collisions with random particles distributed in the (p_T, ϕ and η) phase space. Then we will make clusters of collimated high-energetic particles representing jets. We will use the Gaussian random numbers to build such groups of particles. First, we will create 200 random particles in the plane ($-\pi - \pi$) in ϕ, and in the similar range of η. The transverse energies will also be described by a Gaussian distribution (but using positive values since energies cannot be negative), Then we will add two jets of particle, or clusters, of the size (diameter) of $R = 1$ in the $\eta - \phi$ plane. The center of one cluster with 100 particles will be positioned at (1, 1), while the second cluster with 80 particles is at (-1, -2). We will keep all particles in an array, and save this array in the file "particles.jser".

```
from jhplot import *
from java.util import *
from hephysics.jet import *

plist=ArrayList()
def buildCluster(events,R,c_eta,c_phi,energy):
    r = Random(1L)   # reproducible random numbers
    for i in range(events):
        pt=energy*abs(r.nextGaussian())
        phi=c_phi+2*R*(r.nextDouble()-0.5)
        eta=c_eta+2*R*(r.nextDouble()-0.5)
        p=ParticleD(); p.setPtEtaPhiM(pt,eta,phi,0)
        plist.add(p)
R=1.0
buildCluster(200,3.14,0,0,1)    # 200 random particles
buildCluster(100,R/2,1,1,10)    # 50 at eta=1 and phi=1
buildCluster(80,R/2,-1,-2,20)   # 80 at eta=-1 and phi=-2
IO.write(plist,"particles.jser")
```

Listing 14.6 Building a list of particles

Note that we use the Java class `ParticleD` which characterizes a typical particle, and then we fill the `ArrayList` with 380 particles. 200 particles have relatively low energy (simulated with using a Gaussian distribution with $\sigma = 1$), while the clusters of particles will have large energy ($\sigma = 10$ and 20).

Now we will read the file with our data and process the array of particles with the so-called anti-k_T clustering algorithm [6] using the Java class `JetN2` [7]. This jet algorithm behaves very much like a cone that groups particles together using a distance measure defined in the $\eta - \phi$ plane and the transverse energy. This clustering algorithm is exactly what is used[1] by the LHC experiments.

---

[1]Note that this algorithm is implemented in Java for the usage by DMelt and other projects, such as the HepSim project [8], while the LHC experiments typically use a C++ version of this algorithm.

The code shown below reads the data and reconstructs jets using the expected jet size $R = 1$ and the minimum transverse momentum 20 gigaelectronvolt (GeV). Then we extract jet characteristics and plot the original data together with the found jets using the $\eta - \phi$ plane.

```
from java.awt import Color
from java.util import *
from jhplot    import *
from hephysics.jet import *
from jhplot.shapes import *

D=1.0
plist=IO.read("particles.jser")
ktjet =JetN2(D,"antikt",20) # min pT=20 GeV
ktjet.buildJets(plist)
print ktjet.toString()
jets=ktjet.getJetsSorted()

c1 = HPlot("jets",500,500)
c1.visible(); c1.setRange(-4,4,-5,5)
c1.setNameX("&eta;"); c1.setNameY("&phi; [rad]")

pp=P1D("particles")
for i in range(plist.size()):
      p=plist.get(i)
      eta=p.eta(); phi=p.phi()
      pp.add(eta,phi)

for k in range(jets.size()):
    jj=jets.get(k)
    cic= Circle(jj.eta(), jj.phi(), D/2.)
    cic.setFill(1); cic.setColor(Color.red)
    c1.add(cic)
c1.draw(pp)
```

Listing 14.7 Clustering particles into jets

Figure 14.3 shows the result of particle clustering into jets. The red circles indicate the position of jets, while the black points show particles stored in the array with the simulated data. As you can see, the code correctly identifies two jets, including their positions.

14.3 Smoothing and Interpolation

As follows from the title of this section, now we will talk about the smoothing algorithms that attempt to capture the main characteristics of data (or patterns) by reducing inessential information, such as statistical noise, which is usually present in real measurements. Such a procedure is used for image processing, data reduction,

Fig. 14.3 An illustration of clustering of particles into energetic jets. The input data for the clustering procedure in the η–φ space are shown as *black dots*. The *red circles* show the reconstructed jets as the result of the clustering procedure described in the text (color figure online)

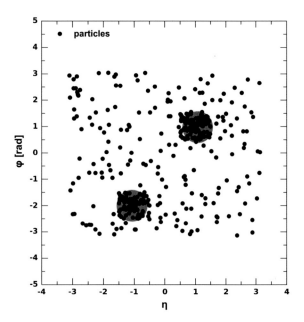

noise filtering, signal processing, and in many other areas. A smoothing can be achieved using several algorithms, such as the moving average [9], Lowess [10], SPlines [11] and other methods (see the book [12] for a review of the smoothing algorithms). Note that the smoothing procedure, generally speaking, does not require analytic functions that are essential for linear and nonlinear regressions discussed in Chap. 11.

Interpolation is closely related to the concept of data smoothing. The goal of interpolation is to find values at positions between known data points. For example, in situations with missing data (or measurements), it is important to interpolate data using observed patterns or trends by estimating values that lie between recorded data points.

We will follow the main strategy of this book by creating the data to be used to illustrate the main idea behind the smoothing algorithms. Let us create a complex histogram by using random numbers. The histogram will contain a broad peak and some background.

```
from java.awt   import *
from java.util import *
from jhplot import *

h1=H1D("Complex data",50,-2.0,3.0)
h1.setFill(1)
h1.setColor(Color.blue)
r=Random(1L)    # 1L for reproducibility
for i in range(1000):
   h1.fill(r.nextGaussian()-2)   # broad shape
   # a peak with mean=1 and sigma=0.4
```

```
   if  (i<500):  h1.fill(r.nextGaussian()*0.4+1)
IO.write(h1,"histo.jser")
```

Listing 14.8 Creating a histogram for smoothing

Now, let us read the histogram from this file and apply a smoothing algorithm. We will use the Lowess (or Loess) [10] smoothing algorithm based on a locally weighted polynomial regression.

```
from jhplot import *
from jhplot.stat import *

c1 = HPlot()
c1.visible()
c1.setRange(-2,3,0,100)
h1=IO.read("histo.jser")
c1.draw(h1)
k=Interpolator(h1)
p1=k.smoothLoess(0.2,0,0.4)   # runs Loess algorithm
p1.setStyle("l")              # shows as line
c1.draw(p1)                   # shows smoothed data
```

Listing 14.9 Applying the Lowess smoothing to a histogram

According to the description of the `Interpolator` Java class, the method `smoothLoess` takes three arguments: the bandwidth, robustness, and accuracy. In our example, we set their values to 0.2, 0, 0.4, respectively. Please read the description of these parameters using the API specification of this class. Here, we only mention that the smoothness parameter is strongly affected by the first value, 0.2, which defines the bandwidth in the Lowess/Loess fit. If you would decrease its value, the fit will closer follow the input data.

Figure 14.4 shows the original histogram and a line representing the result of the Lowess smoothing procedure. As you can see, we have achieved our goal of data smoothing using this technique: the figure shows a smooth curve running through a set of data points (histogram heights).

Note that one can also apply the Lowess interpolation algorithm to scattered $X-Y$ data. Even more: scattered data can contain uncertainties on X and Y values. As we already know from the previous chapters, such $X-Y$ data with uncertainties can be represented by the P1D Java class. Let us make a small example which will illustrate how apply the Lowess/Loess interpolation to such data:

```
from java.util import *
from jhplot import *
from jhplot.stat import *

c1 = HPlot()
c1.visible()
c1.setRange(0,100,0,150)

p1=P1D("Data") # X-Y data points with errors
```

```
r=Random(1L)      # 1L for reproducibility
for i in range(100):
    p1.add(i,i+r.nextGaussian()*0.2*i,0.1,0.1,2,2)
c1.draw(p1)

k=Interpolator(p1)
p1=k.smoothLoess(0.2,0,0.4)
p1.setStyle("l")
c1.draw(p1)
```

Listing 14.10 Interpolating scattered data

The result of this code is shown in Fig. 14.5. As before, the smoothness of the interpolation can be regulated using the argument of the method smoothLoess.

Let us consider another example where we will extract the smoothed polynomial curve which can be later used for an interpolation. This time we will use the Cubic smoothing spline algorithm, a method that can conveniently extract the fitted curve,

Fig. 14.4 An illustration of the Lowess/Loess histogram smoothing. The histogram is shown with error bars indicating statistical uncertainties, while the smooth *black line* shows the result of the smoothing procedure (color figure online)

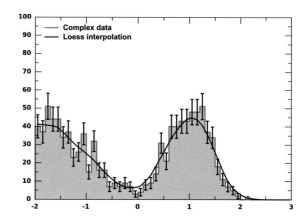

Fig. 14.5 An illustration of the Lowess/Loess interpolation of scattered data

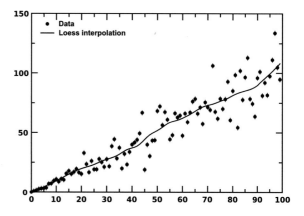

which can be used for interpolation to new regions with missing data. Let us come back to the example with the histogram and show how to do this:

```
from jhplot import *
from jhplot.stat import *

c1 = HPlot()
c1.visible()
c1.setRange(-2,3,0,100)
h1=IO.read("histo.jser")
c1.draw(h1)

k=Interpolator(h1)
sp=k.interpolateCubicSpline(0.95, 2)
fit=sp.getSplinePolynomials()
print "Length=",len(fit)

p1=P1D("CubicSpline")   # visualize smoothing
nbins=100
xmin,xmax=-2,3
h=(xmax-xmin)/float(nbins) # step size
print "Step=",h
for i in range(nbins): # print interpolated results
    z1=xmin+i*h
    z2=sp.evaluate(z1) # evaluate using spline
    print i,z1, z2     # print the result
    p1.add(z1,z2)      # fill smoothed X-Y
p1.setStyle("l")
p1.setSymbolSize(1)
c1.draw(p1)
```

Listing 14.11 Smoothing using the Cubic smoothing spline

This algorithm applies the smoothing factor 0.95 which is used as an argument for the method `interpolateCubicSpline`. The closer this value is to 1, the closer the curve follows the original data. The second argument of this method specifies the treatment of uncertainties in the curve fitting. In this example, the value 0 means that all errors are ignored for the smoothing procedure.

14.4 Peak Identification

Searches for unusual features and patterns in observed data lie at the heart of data mining and knowledge discovery. For example, data may contain quasi-periodic signals, such as peaks. The human eye, a wonderful instrument built by nature, can catch tiny peaks on a smooth background relatively easily. But what if one needs to process large data samples in order to identify peaks, and then to measure their positions, heights, and widths? To help with this, one can run the peak finder algorithms which are designed to detect peaks, report about their shape characteristics, calculate their significance, positions, and widths. These algorithms are useful to identify peaks in time-series of financial markets, in chromatographic data, in mass spectroscopy, gamma-ray spectroscopy, particle colliding experiments, image processing, and many other areas. It is worth mentioning that the Higgs discovery in particle collisions was possible by identifying peaks above the predicted backgrounds (see Sect. 10.5.6).

DMelt has a number of Java classes to identify peaks in data. Let us consider a simplest peak finder [13] proposed for gamma-spectrum analysis. We will illustrate this algorithm by preparing an array with values "10", assuming arbitrary units. We will call this array "background". Then, we will insert "peaks" at positions 50 and 70. They will have the heights 50 and 100.

Now, our goal is to run a peak finder algorithm which will attempt to identify the peaks. The algorithm requires a "sensitivity" parameter and the value of the full width at half maximum (FWHM) for the expected peaks. In our example, the sensitivity will be set to 2, which gives a 3% chance for any observed peak to be false. The value of FWHM will be set to 1. Let us run the algorithm using the code given below:

```
from jhplot import *
from jhpro.tseries import *

spec=[10]*100    # background
spec[50]=50      # peak 1
spec[70]=100     # peak 2
sensitivity=2
width=1
p=PeakFinder("Peaks",spec,sensitivity,width)
mu=p.getPeaks()
print mu.tolist()
```

Listing 14.12 Example of a peak finder for arrays

The output of this code is a list with the detected peaks:

```
[[Peak
  Position = 49.99 +/- 0.0
  Area = 60.0 +/- 0.0
  FWHM = 1.0 +/- 0.0
], [Peak
  Position = 69.99 +/- 0.0
  Area = 110.0 +/- 0.0
  FWHM = 1.0 +/- 0.0
]]
```

You can change the sensitivity parameter to some other value, and you will see a different answer. This shows that the algorithm needs to be trained before using it, i.e., you cannot blindly rely on the algorithm if you do not know what sizes of peaks are expected. Therefore, first run tests on data with known peak characteristics to adjust the parameters of this algorithm.

Peak finding is related to the task of determining a correct background shape using theoretical or known distributions. For example, one can smooth data and then extract any deviations from the smoothed background spectrum. The topics of data smoothing and interpolation were discussed in Sect. 14.3. Here we will remind again that smoothing can be achieved using the moving average [9], Lowess [10], SPlines [11] and other algorithms.

Now, we will consider another peak-detection algorithm [14] used for the extraction of statistically significant peaks in event-counting distributions taking into account statistical or other uncertainties. This algorithm does not perform the usual smoothing, since it simply walks over the data points and checks that the first derivative has a downward-going zero-crossing at the peak maximum. Then, it makes the decision about possible peak features. This algorithm is best suited for high-statistics histograms when the presence of random noise is minimal.

The code shown below fills a histogram with a broad Gaussian distribution with mean 0 and standard deviation 300. This will represent a background distribution. Then, we add two Gaussian peaks on top of this distribution, at the position 400 and 700, and at width 20 and 10, respectively. The numbers of events for each peak are 1000 and 500, see the first arguments in the method `fillGauss`. Then, we run the peak identification algorithm using the input histogram with data. Note that, unlike the previous example, we use realistic data similar to any counting experiments, since each point of the background and peaks has statistical uncertainty. The algorithm assumes sensitivity "0.7" and reports peaks which have heights with the significance "3." This value is closely related to the 3σ criterion for statistical significance discussed in the previous chapters.

```
from jhplot import *
from java.util import *
from sys import path
path.append(SystemDir+"/python/packages/npfinder/")
from npfinder import *

h=H1D("data",100,0,1000)
h.fillGauss(30000,0,300)  # background
h.fillGauss(1000,400,20)  # peak 1
h.fillGauss(500,700,10)   # peak 2
data=GetData(h)
peaks=FindPeaks(data,0.7,3)
for peak in peaks:
    print peak.toString()
```

Listing 14.13 Running a peak finder using a histogram

This code requires importing the Python package "npfinder" from the DMelt package. If you run it outside the DMelt IDE, you should define the variable `SystemDir` so it will point to the DMelt installation directory.

The above algorithm correctly identifies the peaks at the positions around 400 and 700. In addition, the example code reports the start and end positions of the peaks. Figure 14.6 shows the image of the histogram with the identified peaks. This image

Fig. 14.6 An illustration of
a peak finder algorithm with
the identified peaks above
background in a counting
experiment. The *red line*
indicates the expected
background under the peak,
while the *red dots* show the
identified peaks (color figure
online)

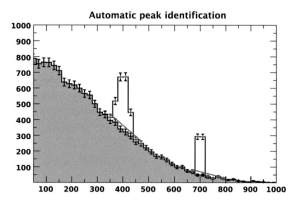

can easily be obtained by adding the plotting canvas `HPlot` to the above code, and
drawing the histogram as shown in the previous sections. After some experimentation
with the above example, it is not too difficult to see that one can miss the peaks after
decreasing the number of events contributing to the first peak from 1000 to 500. To
find smaller peaks, one needs to reduce the sensitivity parameter "0.7" to a smaller
value, and to decrease the significance value.

Generally speaking, it is difficult or even impossible to design a completely automatic and reliable peak finder if you do not know exactly what peaks you are looking
for. To make a reliable peak finder, one should make a priory assumption about the
expected peaks, and to develop a technique to "train" the algorithm on known peaks
to assess the uncertainties on the measured positions and other peak characteristics.

14.5 Principal Component Analysis

This chapter continues our discussion of data mining algorithms to be applied to data.
This time we will discuss principal component analysis (PCA), a popular method to
reveal relationships among variables by rotating the original data to new coordinates,
reducing complexity of data. After the transformation, data can be described with a
few uncorrelated variables.

Data points distributed in multiple dimensions can have certain patterns or shapes
that are usually indicative of some underlying dynamics. The PCA can help identify
geometrical features or patterns in data. For example, data in $X-Y$ space can be
distributed in elongated form. One can try to fit the data with an ellipsoid, such that
the ellipsoid axes represent "principal components".

To obtain more intuition, let us show an example of the so-called eccentricity
[15, 16] which is calculated as $ECC = 1 - v_{min}/v_{max}$, where v_{max} (v_{min}) maximal
(minimal) values of variances of data along the principle (minor) axis, assuming
that data points are distributed on the $X-Y$ plane. To find the principle axis, we
first need to calculate the center of data in $X-Y$, and then transform the data to a
new coordinate system such that the largest variance comes to lie on the "principal"

axis. Then, the minor axis will be orthogonal to the principle axis. The calculated eccentricity value ECC ranges from 0 (for perfectly circular data in $X-Y$) to 1 (for infinitely elongated shape).

Now it is time to do some experimentation. Let us generate an asymmetrical random distribution using Gaussian random numbers, assuming that the standard deviations σ along X and Y axis have different values. We assume the Gaussian in the X direction has $\sigma = 1$, while along the Y-coordinate it has $\sigma = 0.3$. Then, we will run a PCA algorithm and calculate the eccentricity and the rotation angle of a possible ellipse. Here is the code:

```
from jhplot   import  *
from java.util import  *
from jhpro.stat import  *

sd=0.3
c1  = HPlot("Data",450,300)
c1.visible();  c1.setRange(-3,3,-3,3)
c1.setNameX("X");  c1.setNameY("Y")
p1  = P1D("Data")
r=Random(1L)   # reproducible random numbers
for i in range(100):
    p1.add(r.nextGaussian(),sd*r.nextGaussian())
pca=EEcentricity(p1)
h=HMLabel(["ECC  =%0.3f"%pca.getEccentricity(),
           "Ang  =%0.3f"%pca.getAngle()],-1,2)
c1.add(h)
c1.draw(p1)
```

Listing 14.14 Eccentricity analysis in the $X-Y$ plane

This code creates a canvas with the random data shown in Fig. 14.7, together with the values of the eccentricity and rotation angles. Now make a new test: create a symmetric distribution of data points, i.e., replace 0.3 with 1. You will see that the eccentricity value will be close to 0.

Fig. 14.7 An illustration of a PCA analysis used to calculate the eccentricity. The input was generated using random data points elongated in the X direction. This image indicates the eccentricity value and the rotation angle

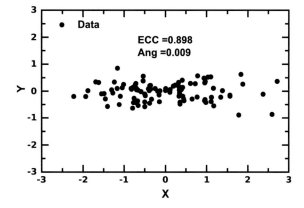

Let us consider a full-scale example that shows an internal mechanism behind the PCA analysis. This time we use the PCA Java class, and fill the data array P1D without using the random numbers. Then, we calculate coordinates of the projection vectors (eigenvectors), means, and standard deviations for each component in X and Y. One can also access the covariance matrix of the PCA transform:

```
from jhplot    import  *
from jhplot.stat import  *

c1=HPlot()
c1.visible(); c1.setAutoRange()
p1 = P1D("Data")
p1.add(3,4); p1.add(4,4); p1.add(5,5)
p1.add(6,4); p1.add(8,7); p1.add(9,6)
c1.draw(p1)

p=PCA(p1)
print p.getEigenvalue(0)," ",p.getMean(0)," ",p.getStd(0)
print p.getEigenvalue(1)," ",p.getMean(1)," ",p.getStd(1)
print p.getSummary()
```

Listing 14.15 Running a PCA analysis in 2D space

As a bonus, one can calculate the eccentricity $1 - v_{min}/v_{max}$ discussed above by using the values given by the getEigenvalue(i) ($i = 0, 1$) method.

The PCA is a popular technique in many areas of science. For example, the eccentricity can characterize the geometrical positions of objects in space. For example, the orbital eccentricity of Pluto is quite large ($\simeq 0.249$), while Earth's orbital eccentricity value is tiny (0.016). The eccentricity and PCA can also be used to characterize statistical distributions of data. For example, it can be used to study the shapes of the distributions of particles inside jets [16], which were briefly described in Sect. 14.2.

14.6   Decision Trees

We will continue with the topic of data classification by discussing how to organize data into certain categories. We will illustrate this concept using the so-called decision trees (for a recent review, see [17]).

Unlike the neural networks discussed in the previous chapter, the decision tree is a graph with branching, where every branching can have possible decision outcome.

Let us build a simple decision tree using the DMelt Java libraries. Our goal is to design a code that finds the outcome by answering a few questions with possible choices "Yes" or "No". Our next example illustrates a typical decision tree with 7 nodes organized into several branches:

```
from jhpro.dtree import *

dt = DecisionTree()
dt.createRoot(1,"Does animal eat meat?")
dt.addYesNode(1,2,"Does animal have stripes?")
dt.addNoNode(1,3,"Does animal have stripes?")
dt.addYesNode(2,4,"Animal is a Tiger")
dt.addNoNode(2,5,"Animal is a Leopard")
dt.addYesNode(3,6,"Animal is a Zebra")
dt.addNoNode(3,7,"Animal is a Horse")
print dt.outputBinTree()

c=dt.queryBinTree(dt.rootNode)
print c.toString()                        # current question
c=dt.queryBinTree(c.yesBranch)   # set Yes
print c.toString()                        # current question
c=dt.queryBinTree(c.yesBranch)   # set Yes
print c.getNodeAnswer()                   # get node answer
print c.toString()                        # print answer
```

Listing 14.16   Creating a simple decision tree

The script creates a decision tree from the answers "Yes" and "No", and then tests it using two questions/answers. We intentionally did not make a figure with the graph showing the structure of this decision tree. Please analyze the code. You can see that there are several possible answers to the questions, and we program these answers using the nodes addYesNode for "Yes" and addNoNode for "No". The structure of this decision tree is printed as:

```
[1]       nodeID = 1     Does animal eat meat?
[1.1]     nodeID = 2     Does animal have stripes?
[1.1.1] nodeID = 4     Animal is a Tiger
[1.1.2] nodeID = 5     Animal is a Leopard
[1.2]     nodeID = 3     Does animal have stripes?
[1.2.1] nodeID = 6     Animal is a Zebra
[1.2.2] nodeID = 7     Animal is a Horse
```

After the tree is created, we can ask two questions. The program walks through the nodes with possible choices, such that the output of the previous query is the input for the next query of the decision tree. The output of the code is

```
1 Does animal eat meat?      (Yes)
2 Does animal have stripes? (Yes)
Answer=4 Animal is a Tiger
```

Here, the answer "(Yes)" was added to the output to make the example clear. The last line shows the final answer to the above questions, and prints the node ID $= 4$ that corresponds to the answer. If the node is not final and other questions are to follow, the method `c.getNodeAnswer()` returns −1, instead a positive value for the node ID. As a test, replace `c.yesBranch` (Yes answer) with `c.noBranch` (No answer). It will print

```
1 Does animal eat meat? (No)
2 Does animal have stripes? (No)
Answer=7 Animal is a Horse
```

As you can see, data can easily be classified from the given input. One can build complex decision trees using additional possible scenarios.

Note that the DMelt libraries include several other algorithms to construct decision trees. Some of them can be initialized from ASCII text files.

References

1. Spath H (1980) Cluster analysis algorithms. Wiley, New York
2. Kaufman L, Rousseeuw P (2005) Finding groups in data: an introduction to cluster analysis. Wiley, New York
3. Jain A, Dubes R (1988) Algorithms for clustering data. Prentice Hall, Upper Saddle River
4. Chekanov S The jminhep package. http://hepforge.cedar.ac.uk/jminhep/
5. ATLAS, An ATLAS event with asymmetric jets (2014). http://cds.cern.ch/record/1696935
6. Cacciari M, Salam GP, Soyez G (2008) The anti-kT jet clustering algorithm. JHEP 04:063
7. Chekanov S, Pogrebnyak I, Wilbern D (2015) Cross-platform validation and analysis environment for particle physics, arXiv:1510.06638
8. Chekanov S (2015) HepSim: a repository with predictions for high-energy physics experiments. Adv High Energy Phys 2015:136093
9. Kenney JF, Keeping ES (1962) Mathematics of statistics, Pt. 1. Van Nostrand, Princeton
10. Cleveland WS (1979) Robust locally weighted regression and smoothing scatterplots. J Am Stat Assoc 74:829
11. Schoenberg J (1946) Contribution to the problem of approximation. Q Appl Math 4:45 and 112
12. Simonoff J (1998) Smoothing methods in statistics, vol 2. Springer, New York
13. Hnatowicz V, Ilyushchenko V, Kozma P (1990) Peak identification. Comp Phys Comm 60:111
14. Chekanov S, Erickson M (2013) A nonparametric peak finder algorithm and its application in searches for new physics. Adv High Energy Phys, vol 2013, Article ID 162986
15. Ayoub B (2003) The eccentricity of a conic section. Coll Math J 34:116
16. Chekanov S, Proudfoot J (2010) Searches for TeV-scale particles at the LHC using jet shapes. Phys Rev D 81:114038
17. Rokach L, Maimon O (2014) Data mining with decision trees., Series in machine perception and artificial intelligence, theory and applications, WorldScientific, Singapore

Chapter 15
Miscellaneous Topics

This chapter discusses several Java libraries and classes which, besides being educational, are helpful for scientific, technical and financial computing. Their use directly translates to increased productivity for numerical computation and data analysis on the Java platform.

15.1 Working with Data

15.1.1 Downloading Files from the Web

Several examples discussed in the previous chapters used the Java class Web to download files from the Internet. Let us show again how to call class:

```
>>> from jhplot import *
>>> Web.get("http://www.jython.org/Project/news.html")
```

In the case of success, the code will return the download status "Success" and will copy the file "news.html" to the local directory. Since the Web class is implemented in Java, you can use the same approach for all scripting languages supported by Java, or call it within Java programs.

Alternatively, the same can be done using the Python module "web.py" located in the directory "macros/system." This module is similar to the well-known "wget" program available on the Linux and Unix platforms, but it is somewhat simplified and reimplemented in Python. In the case of the HTTP protocol, a file can be downloaded using the JythonShell as

© Springer International Publishing Switzerland 2016
S.V. Chekanov, *Numeric Computation and Statistical Data Analysis on the Java Platform*, Advanced Information and Knowledge Processing, DOI 10.1007/978-3-319-28531-3_15

```
>>> from web import *
>>> wget("http://www.jython.org/Project/news.html")
```

This command retrieves the file "news.html" from the server to the current directory, and shows the progress status during download.

15.1.2   Extracting Data from Figures

DMelt includes a remastered `Dexter` package for extraction of data points from figures in raster formats (GIF, JPEG or PNG). To start this program, import the Java package `debuxter`, and run this command:

```
>>> from debuxter import JDebux
>>> a=JDebux("file")
```

where `"file"` is the file name of the image in raster format. If you have a PDF, EPS or PS figure, enlarge this figure as much as possible before converting it to the image file in raster formats with the file extensions "png", "jpeg" or "gif". This can simplify the data extraction process. There are plenty of tools around that can be used for such transformation. For example, for Linux/Unix, use the command `convert`.

After execution of the above lines, a frame with the `Dexter` program will pop up. The frame shows the inserted figure, and surrounds it with the necessary menus for data calibration. Here is a complete example code that downloads the image shown in Fig. 11.3 from the Web, and prepares it for image digitization:

```
from jhplot import *
from debuxter import JDebux
fig="signal_bkg.png"
a=Web.get("http://jwork.org/dmelt/examples/data/"+fig)
a=JDebux(fig)
```

Listing 15.1   Extracting data from images

Figure 15.1 shows the application window to extract the data. You should draw the X and Y axes with the mouse, and specify the minimum and maximum values for the axis ranges. After calibration of the X and Y axes, one can locate data points with the mouse, print their (x, y) values. and save the resulting coordinates. Read the description of this tool by clicking on the help menu.

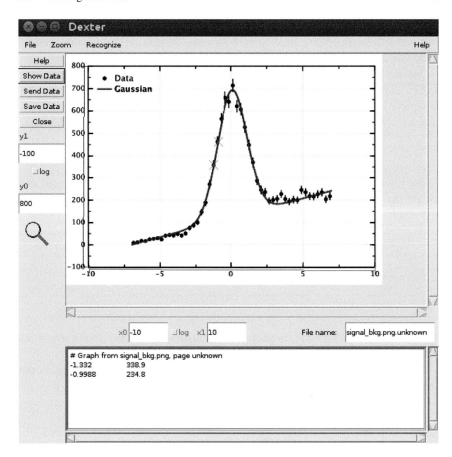

Fig. 15.1 A program for extraction of data points from an image ("image digitization")

15.1.3   Tables and Spreadsheets

This section discusses a few useful methods to display data in tables and spreadsheets.

In some cases, instead of plotting on canvases, it is convenient to show data as tables. DMelt provides excellent support for this. Use the HTable class as in the example shown below:

```
>>> from jhplot import *
>>> HTable(obj)
```

where obj is an object created by classes from the package jhplot, such as H1D, P1D or F1D. After executing this command, you will see a frame with a table populated with the numbers representing one of such objects (by default, the numbers are formatted using the Java scientific format). Using this table, one can sort data by clicking the top header of each column, or search for a particular value or string. Note that it is impossible to modify the data since the class HTable is designed for examining data containers.

One can pass a Boolean value telling how you expect the data to be formatted. If the data have to be shown as formatted strings in the Java scientific format, use this constructor:

```
>>> HTable(obj,1)
```

If data values should not be formatted, use "0" (Java false) instead of "1."

One can call to the `HTable` object differently: if a data container is created, one can use the `toTable()` method. For example, to display an `H1D` histogram, call this constructor:

```
>>> HTable(h1)   # where h1 is H1D histogram
```

or, alternatively,

```
>>> h1.toTable()
```

As before, if `toTable()` is used with the argument "0", the numbers will not be formatted in the scientific format.

The `HTable` viewer can be filled by the Java arrays (or by Python lists). In this case, it accepts three arguments:

```
>>> HTable("description", names, arr2d)
```

where `names` is a list with the column names (string type), and `arr2d` is the input data in the form of a 2D array. Let us give an example illustrating this:

```
from jhplot import *

a=[[10,20],[300,4000]]
b=["X","Y"]
HTable("table",b,a)
```

Listing 15.2 Table filled with numeric values

This brings up a table filled with numeric values. The number of columns will be determined automatically by the `HTable` class.

For more flexibility, one can export data to a full-featured spreadsheet using the `SPsheet` Java class. It is designed to build a spreadsheet-like table in which the data can be modified and edited in the usual way as for any Excel-like application.

To create an empty spreadsheet, use

```
>>> from jhplot import *
>>> SPsheet()
```

One can open a CSV file by passing `CSVReader` object. This class was discussed in Sect. 9.6.2.1 where we also have shown a small script used to open a CSV file in the spreadsheet.

The functionality of the `SPsheet` class goes beyond simple manipulations with the CSV files: if one needs to build a spreadsheet using data stored in other formats, call the constructor:

```
>>> SPsheet(obj)
```

where `obj` represents an object of the type `H1D`, `P1D` or `F1D`. Now the data can be changed and saved to a CSV file.

15.2   Measurements with Units

If uncertainties on measurements are known, you are probably dealing with the topic of uncertainty propagation described in Sect. 10.6. Here, we will discuss calculations involving quantities with specific units and, if available, a propagation of uncertainties using such quantities. Such situations are covered by the JScience project [1], which is one of the Java packages contributed to the DMelt libraries.

First, we will show how to work with the International System of Units (SI), a modern form of the metric system which is widely used for measurements in science, medicine, and technology. Let us show how to import basic international units into your program:

```
>>> from javax.measure.unit.SI import *
>>> dir() # list available attributes
```

You will see the imported units:

```
["AMPERE", "ATTO", "BECQUEREL", "BIT", "CELSIUS",
 "CENTI", "CENTIMETER", "CENTIMETRE", "COULOMB",
 "DECI",  "FARAD", "FEMTO", "GIGA", "GRAM", "GRAY",
 "HECTO", "HERTZ", "JOULE", "KATAL", "KELVIN",
 "KILO", "KILOGRAM", "KILOMETER", "KILOMETRE", ...]
```

Here, we show only a few SI units to save page space.

Similarly, one can import units that do not belong to the International System of Units, but are still often used due to their cultural or historical importance, or due to their usage in certain areas. See the example below:

```
>>> from javax.measure.unit.NonSI import *
>>> dir() # list available attributes
```

which prints a long list of supported units (we show a few only):

```
["ANGSTROM", "BYTE", "INCH", "OUNCE", "POUND" ..]
```

The library can be used to perform calculations with the SI and non-SI units, as well as conversions between different units. Let as show a simple conversion between units using the class `Amount`. The next code converts 100 ± 5 kg to pounds assuming uncertainties:

```
from org.jscience.physics.amount import *
from javax.measure.unit.SI import *
from javax.measure.unit.NonSI import *
m=Amount.valueOf(100, 5,  KILOGRAM) # 100 +- 5  kg
print m.to(POUND)
```

Listing 15.3 Conversion from kg to lb

The answer of this conversion is 220 ± 11.

Now, let us perform calculations with kilograms taking into account uncertainties. Again, the central Java class for this calculation is `Amount`. We will calculate $5 * m_1/2 + m_2$, where m_1 and m_2 are masses expressed in kilograms. These masses are assumed to have uncertainties. The code is shown below:

```
from org.jscience.physics.amount import *
from javax.measure.unit.SI import *
m1=Amount.valueOf(100, 5,  KILOGRAM) # 100 +- 5  kg
m2=Amount.valueOf(200, 20, KILOGRAM) # 200 +- 20 kg
m=m1.divide(2).times(5).plus(m2)     # m1/2 * 5 + m2
print m.getEstimatedValue()," +- ",m.getAbsoluteError()
print m.to(KILOGRAM)
```

Listing 15.4 Calculations with units and uncertainties

One can easily convert the final answer to pounds using the previous example.

Now, you are equipped to do more complex calculations with units, mixing them as you need. For example, let us calculate the velocity of a falling object. We will assume that there is an uncertainty associated with the time measurement in our experiment. After importing the necessary JScience classes, our calculation can be programmed as shown in this code:

```
from org.jscience.physics.amount import *
from org.jscience.physics.amount import *
from javax.measure.unit.SI import *
TIME=Amount.valueOf(100,5,SECOND) # 100 +- 5 seconds
v=TIME.times(Constants.g)     # t*gravitation constant
print  v
```

Listing 15.5 Calculation of the velocity of a falling object

The output of this code is 980 ± 49 m/s.

The JScience package supports exact or arbitrary precision measurements, coordinate systems, and conversions between currencies after setting appropriate currency exchange rates. Look at the Java API of this package to learn more.

15.3 Cellular Automaton

A cellular automaton was introduced by Von Neumann and Ulam as a simple model for self-reproduction [2]. A comprehensive discussion of this topic can be found in [3].

On an abstract level, the automaton is the model of a spatially distributed process. It consists of an array of cells, each of which is allowed to be in one of a few states. During simulation, each cell looks to its neighbors and determines what state it should change to using a predefined simple rule. Such steps are repeated over the whole array of cells, again and again. The evolution of such a system with time can be very complex. The most popular two-dimensional cellular automaton is the "Game Of Life" invented by mathematician J. Conway in the 1960s.

We will consider the cellular automaton in two dimensions when each cell is characterized by x and y values. To start the cellular automaton, one should first create an instance of the HCellular class.[1] Once the instance is created, it should be easy to check the available methods of this Java class as

```
>>>from jhplot  import *
>> c=HCellular()
>> print c.getRules()
  [Aggregation, Aqua, AquaP2,
   BlockVN, Check24, Check29,
   Check35, Check25ByGA, CyclicCA8,
   CyclicCA14, VN, Life, Life2,
   Generation, GMBrain, Hodge,
   Ising, Stripe]
```

Each rule is defined by a string that specifies the Java class used for the initialization. The rules can be applied to the instance HCellular using the method setRule(str). Learn about the rules using the DMelt documentation system. The code shown below illustrates how to initiate the well-known "Game Of Life":

```
>>> c.setRule("Aggregation")
>>> c.visible()
```

The first method sets the rule "Aggregation" from the list of available rules. As for any canvas, the method visible() brings up a window with the 2D cellular automaton. Run the algorithm by pressing the button "Start" and you will see how this cellular automaton evolves for the specified rule.

---

[1] This instance is based on the Cambria [4] package.

The initial configuration can be accessed using the methods shown below:

```
>>> print c.getRule()
>>> print c.getInitString()
```

One can change the initial configuration for a cellular automation with the method
`setInitString`. A convenient way to write configuration files and read them
back using Python or Java classes was discussed in Sect. 2.15.

15.4 Image Processing

For most of us, an image is a visual representation of something. From the point
of view a data analyst, an image is an array with numbers. A typical image is built
from the rectangular grid of pixels, thus it has a height and a width counted in pixels.
Each pixel has a color in the form of a 32-bit integer. Most images have pixels in
the so-called RGB color space described using 3 variables: Red, Green, and Blue.
This is usually defined as a list of three values (red, green, and blue), where "red",
"green", and "blue" are integer values. Each value runs from 0 (no color) and to 255
(the highest level of the color). For instance, (0, 0, 0) means the darkest black, while
the list (255, 0, 0) corresponds to the brightest red color. This short excursion to the
internals of digital images should be enough to get started with image manipulation
using DMelt libraries.

To explore the topics of image processing and manipulation, we will need to
download a test image. For example, one can download the DMelt logo image using
the `Web` class as described in Sect. 15.1.1. But here we will use the `IJ` class from
the ImageJ package [5] instead, since it reads the image directly from a given URL
to the computer memory, and prepares it for image processing:

```
from ij import *
from ij.process import *

imp=IJ.openImage("http://jwork.org/dmelt/logo.png")
imp.show()          # show the image in a frame
print type(imp)     # print class name
print dir(imp)      # supported manipulation methods
```

Listing 15.6 Loading an image using the ImageJ package

This script reads the image from the Web, shows it in a frame, and prints the list of
methods associated with the class `IJ`.

We should also remind that you can open this image in the `IEditor` class as

```
from jhplot import *
Web.get("http://jwork.org/dmelt/logo.png")
IEditor("logo.png") # open inside ImageJ
```

Listing 15.7 Openning an image in ImageJ

which prepares this image for editing in the ImageJ GUI.

The goal of this section is to learn how to program with images using Java scripting, instead of the GUI approach. The scripting concept will provide you with a tool to work with many images at the same time, or performing repetitive tasks on images that are hard to do using graphical user interfaces.

First, let us check the image size in pixels, and extract the array with the pixel data:

```
from ij import *
imp = IJ.openImage("http://jwork.org/dmelt/logo.png")
print imp.getWidth(), "x", imp.getHeight()
pix = imp.getProcessor().convertToFloat().getPixels()
print pix #  print array with pixels
```

Listing 15.8 Extracting the pixel array from an image

The script prints the array on the screen for visual examination.

One can obtain statistical characteristics of image colors by reading the array with pixels using the class `ColorStatistics` from the `ij.process` package:

```
from ij import *
from ij.process import *
from ij.measure import *

imp = IJ.openImage("http://jwork.org/dmelt/logo.png")
ip = imp.getProcessor()
stats = imp.getStatistics()   # get ColorStatistics
print "Number of pixels=",stats.pixelCount
print "Mean=",stats.mean
print "Standard deviation=",stats.stdDev
print "Kurtosis = ",stats.kurtosis
print "Skewness=",stats.skewness
print ip.getHistogram()   # print histogram
```

Listing 15.9 Statistical analysis of image colors

This example can be used for a simple image identification program. For example, one can build a script that finds duplicate or modified copies of images in your image collection.

Now let us create and show histograms with frequencies of color components. We remind that there are three colors that define the actual pixel color, so we should fill three histograms for each. This can be achieved by using the DMelt histogram class `H1D` and the ImageJ classes:

```
from ij import *
from ij.process import *
from jhplot import *
from jhplot.utils import *
from java.awt import *

imp = IJ.openImage("http://jwork.org/dmelt/logo.png")
```

```
pix=imp.getProcessor().convertToRGB().getPixels()
rgb=ColorPixel.getPOI(pix) # get pixels in the usual R,G,B

c1 = HPlot(); c1.visible();
red = H1D("Red",17, 0, 255)
green = H1D("Green",17, 0, 255)
blue = H1D("blue", 17, 0, 255)

red.setColor(Color.red)
green.setColor(Color.green)
blue.setColor(Color.blue)

red.fill(rgb[0])
green.fill(rgb[1])
blue.fill(rgb[2])

c1.setRange(0,255,1,10000)
c1.setLogScale(1,1)
c1.draw([red,green,blue])
```

Listing 15.10 Building the histogram with image colors

The output of this code is shown in Fig. 15.2. Now you can extend this code as you need. For example, one can make a script that compares the histograms of these images using different statistical tests discussed in Sect. 10.5.5. This simple code snippet can be extended to a more sophisticated program for the detection of differences between similar images.

Fig. 15.2 Tree histograms for pixel colors (*red, green, blue*) extracted from the image shown in Listing (15.10) (color figure online)

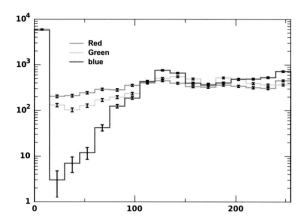

15.4.1   Image Modification

The extracted array with the image pixels can be modified, and then a new image can be created from the modified pixel array. Let us show an example in which we find a minimum color value, and then subtract it from the pixel array.

```
from ij import *
from ij.process import *
from java.lang import Math

imp = IJ.openImage("http://jwork.org/dmelt/logo.png")
pix = imp.getProcessor().convertToFloat().getPixels()
min = reduce(Math.min, pix) # minimium pixel value
# new array with the minimal value subtracted
pix2=map(lambda x: x-min,pix)
proc=FloatProcessor(imp.width, imp.height, pix2, None)
img1=ImagePlus("modified",proc)
img1.show() # show the modified image
```

Listing 15.11   Image creation after changing the pixel array

The example shows a frame with a new black-and-white image.

Let us show another example. The next code illustrates the so-called "edge detection"—a procedure of identifying points at which the image brightness sharply changes.

```
from ij import *
from ij.process import *
from ij.measure import *
from javax.imageio import *
from java.io  import *

imp = IJ.openImage("http://jwork.org/dmelt/logo.png")
ip=imp.getProcessor()
ip.findEdges()
bimg=ip.getBufferedImage()

outputfile=File("new.png")  # save the image
ImageIO.write(bimg,"png", outputfile)
imp=IJ.openImage("new.png") # shows the image
imp.show()
```

Listing 15.12   Edge detection of images

Figure 15.3 shows the original image and the images after subtraction of the minimum value.

As usual, we invite the reader to explore the methods of the ImageJ package. One can simply use the dir() approach to list the available methods, or look at the Java API of the classes imported in these examples.

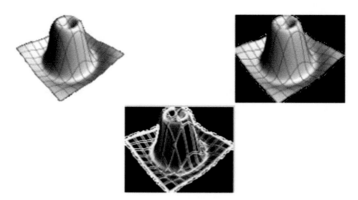

Fig. 15.3 The original image (*left*) and two images after the transformations discussed in Listings (15.11) and (15.12)

15.4.2 Transforms Using Multiple Cores

How to takes advantage of multiple processing cores when performing computations with matrices was discussed in Sect. 5.1.6. It was mentioned that Parallel Colt includes common for image processing trigonometric transforms, such as the Discrete Fourier Transform (DFT), Discrete Hartley Transform (DHT), Discrete Cosine Transform (DCT), and Discrete Sine Transform (DST).

Here, we will continue this discussion in the context of images. In particular, we will show how to use multiple CPU cores for a fast Fourier transform (FFT), computing the discrete Fourier transform (DFT) of double precision data. The FFT allows the transformation of images from the spatial domain to the "frequency" domain, which is commonly used to reduce background noise (see [6] for a review).

As a simple exercise that shows how to use DFT, let us consider transforms of random matrices. The code shown below uses the Parallel Colt library optimized for multiple processing cores:

```
from cern.colt.matrix.tdcomplex import *
from cern.colt.matrix.tdouble   import *

M=DComplexFactory2D.dense.random(100,100)
M.fft2()        # compute 2D DFT in-place
M=DComplexFactory2D.dense.random(100,100)
M.fftColumns()  # compute 1D DFT of each column of M
M=DoubleFactory2D.dense.random(100, 100)
M.dct2(1)  # compute 2D Discrete Cosine Transform (DCT)
print M
```

Listing 15.13 Discrete Fourier transforms for random matrices

The example can give you some ideas about how to apply the same approach for images: simply create matrices with image pixels, and use them as the input for the Java classes shown in this example.

15.5 Market and Financial Analysis

There is nothing special about analysing financial data when it comes to the DMelt Java classes. You can use the same graphic canvases and data containers as in the examples discussed in the previous chapters. But you should be aware of a few specialized charts and data containers commonly used in analysis of financial data. We will discuss such topics in this section.

Let us show how to use the familiar DMelt Java classes to build financial plots. We will start from a simple example that shows how much your money can grow using the power of compound interest. The formula for the value after t periods can be written as $P(1+i/n)^{t \cdot n}$, where P is the initial investment (principal amount), i is annual nominal interest rate and n is the number of times the interest is compounded per year. We simplify the code assuming yearly interest ($n = 1$), and then calculate the compound interest for every year using 5 % interest rate. The code is shown below:

```
from jhplot import *

principal=10000 # starting principal (initial investment, %)
rate=0.05       # percentage of interest rate (5%)
maxyear=20      # maximum years to consider

c1=HPlot()
c1.visible(); c1.setAutoRange()
c1.setMarginLeft(100)
p1=P1D("Value $$")
p1.setDrawLine(1)    # show as connected lines
for year in range( 1, maxyear ):
   amount = principal*(1.0+rate)**year
   print "%4d%16.2f" % ( year, amount )
   p1.add(year,amount)
c1.setNameX("year")
c1.setNameY("balance $$")
c1.draw(p1)
```

Listing 15.14 Compound interest calculations

The code creates the image shown in Fig. 15.4.

Now, we will consider a few other topics that can be interesting for people doing financial calculations. As before, we will pay less attention to the theoretical background of our examples. Instead, we will discuss in more detail the formulation of

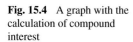

Fig. 15.4 A graph with the calculation of compound interest

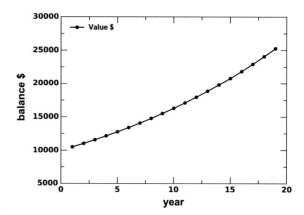

algorithmic ideas and concrete Java packages designed for financial calculations and data visualization.

15.5.1   Time Series

A sequence of data points consisting of successive measurements made over a time interval is called "time series." In fact, any list or any data container considered so far can be used for this type of data. But, in finance and other related areas, there is a specially designed data structure that can be used to keep such data in files. Popular formats for the time series are ASCII text files, MATLAB, and Microsoft Excel formats. DMelt uses the JStatCom library [7] which can help integrate numerical computations with these formats.

   Let us write a typical time series to an ASCII file, and then read it back using the class `HStatData` designed to keep statistical data with time series.

```
data="""
/*
Description goes here.
10 observations starting from 1970 (1st quarter),
and finishing in 1972 (2nd quarter)
*/
  3   1970.1 4
 inve    income   profit
0.00551 0.01057 0.00435
0.02297 0.02211 0.00041
0.03783 0.01636 0.01178
0.09436 0.03193 0.01015
0.01359 0.02138 0.00218
0.02445 0.00192 0.00283
0.03325 0.03581 0.00656
0.07644 0.01471 0.00032
```

```
0.00086 0.01808 0.00512
0.02159 0.02827 0.08169
"""
file = open("tseries.dat", "w")
file.write(data)
file.close()

from jhpro.tseries import *
js=HStatData("Data","tseries.dat")
js.toTable()
```

Listing 15.15 Creating a time series file

The code creates the ASCII file in the format discussed in [8]. There are 10 rows in this file, called observations. Each row has 3 values "investment", "income" and "profit". After file creation, we read the time series from the file and show it in a table. Note that we can also pass an URL to stream the data over the Internet.

Let us comment on the file format used to create the data series. Missing values can be codded with NaN. Values can use the exponential notations, i.e., $1.1e-10$. The comments are included using the $/*.. \text{ text } ..*/$ method. The files have a header and metadata describing the time series. The first number defines the number of variables (3 in this example), while the last number defines "periodicity". For example, "4" means quarterly data, "1" means yearly data, "12" means monthly data. The start date (1970.1) means the first quarter of 1970, if the next number is "4". If the next number is "12," then 1970.2 means "February." Here is another example for the header:

```
3  1960.0  365 /* 1 day periodicity */
```

In this case, we start from the 1st day of 1960, and the last observation will be after 10 days, assuming the total number of observations is 10. Here is another example:

```
3  1960.0  8760 /* 1 hour periodicity */
```

which corresponds to 10 h of observations. Generally, the time of observations can also be specified as an extra column. This, however, is not the most disk-saving approach when dealing with large data.

The data file can be saved in a number of formats as shown in this example:

```
from jhpro.tseries import *
js=HStatData("Data","tseries.dat")
js.saveData("data.mat","matlab")
js.saveData("data.xls","Excel")
js.saveData("data.gauss","GaussDat")
js.saveData("data.dat","txt")
```

Listing 15.16 Exporting time series into external files

The code indicates which file format is used to store the data.

What can be done with the time series data? First and foremost, we should try
to plot one numeric variable against another. For example, we can plot the second
column against the third one:

```
from jhplot   import *
from jhpro.tseries import *
js=HStatData("Data","tseries.dat")
ind1=1 # 2nd column
ind2=2 # 3rd column
p1=js.getP1D(ind1, ind2)
print p1.toString()
c1=HPlot()
c1.visible(); c1.setAutoRange()
names=js.getVarNames().vec()
c1.setNameX(names[ind1])
c1.setNameY(names[ind2])
c1.draw(p1)
```

Listing 15.17 Exporting time series data to external files

Now, how about plotting the observations against the time intervals on the x-axis?
First, we should look at the internals of the several classes that return information
about the time. This code illustrates how to print several important time-interval
characteristics.

```
from jhplot   import *
from jhpro.tseries import *

js=HStatData("Data","tseries.dat")
print js.getDescription()
ind1=1 # 2nd column
y=js.getColumn(ind1)
ts=js.getTS()
t=ts[0]
print t.numOfObs() # number of observations (10)
print t.values()    # values
print "Start=",t.start()          # start time
rr=t.range()                      # time ranges
print rr.timeAxisStringArray() # periodicity
print rr.subPeriodicity()
print js.getVarNames().vec()    # variable names

for j in range(t.numOfObs()):
  dd=rr.dateForIndex(j)
  print dd, " ", dd.doubleValue(), " values=",y[j]
```

Listing 15.18 Accessing time ranges

It prints the time ranges in the human-readable and the values of the second
column:

Fig. 15.5   Showing a time
series using connected lines

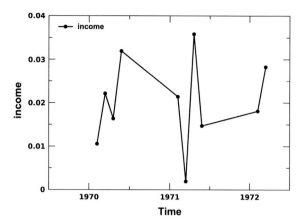

```
1970 Q1    1970.1   values= 0.01057
1970 Q2    1970.2   values= 0.02211
1970 Q3    1970.3   values= 0.01636
1970 Q4    1970.4   values= 0.03193
1971 Q1    1971.1   values= 0.02138
1971 Q2    1971.2   values= 0.00192
1971 Q3    1971.3   values= 0.03581
1971 Q4    1971.4   values= 0.01471
1972 Q1    1972.1   values= 0.01808
1972 Q2    1972.2   values= 0.02827
```

The Java class used to access such values is called TSDate. It is included to the
com.jstatcom.ts package. Now, we can easily display the time series using
the P1D class discussed in the previous sections. The result of the code from Listing 15.19 is shown in Fig. 15.5.

```
from jhplot   import *
from jhpro.tseries import *

js=HStatData("Data","tseries.dat")
ind1=1               # 2nd column vs time
y=js.getColumn(ind1)
t=js.getTS()[0]
rr=t.range()          # time ranges
names=js.getVarNames().vec()
pp=P1D(names[ind1])
pp.setDrawLine(1)    # show as connected lines
for j in range(t.numOfObs()):
        dd=rr.dateForIndex(j)
        pp.add(dd.doubleValue(),y[j])
c1=HPlot()
c1.setMarginLeft(90)
c1.visible(); c1.setAutoRange()
```

```
c1.setNameX("Time")
c1.setNameY(names[ind1])
c1.draw(pp)
```

Listing 15.19 Plotting time series

15.5.2 *Financial Charts*

Since we have entered the subject of financial charts, we will discuss several other approaches to display charts used in market analysis and finance.

Let us come back to the time series topic. One can draw the time series charts using the JFreeChart library and the class `HPlotChart` as discussed in Sect. 8.4. In our next example, we will create the data points that represent a rising trend of some value (profit, merchandise value, etc.) versus time, over the period between 1900 and 2015. We will introduce fluctuations, since they are usually present in real-life situations. For this purpose, we will use the Gaussian random numbers. The standard deviation of the normal distribution will be set to 1.2. Then, we will build two time series. The original time series will include yearly fluctuations. The second time series will be created after applying a smoothing procedure, which will help us see the trends in the data. Our final code to produce Fig. 15.6 is shown below:

```
from org.jfree.chart import *
from org.jfree.data.time import *
from java.util import Random
from java.awt import BasicStroke
from jhplot import *

r = Random(1) # reproducible numbers
eur = TimeSeries("Profit")
for i in range(0,115): # data for 115 years
  t=Day(1,1, 1900+i)
  eur.add(t,1.2*r.nextGaussian()+0.1*i)
mav=MovingAverage.createMovingAverage(eur,"10 year moving
    average",3650,0)
data=TimeSeriesCollection()
data.addSeries(eur)
data.addSeries(mav)

chart=ChartFactory.createTimeSeriesChart("Timeseries","Year"
    ,"Profit $",data,1,1,1)
rend =chart.getPlot().getRenderer();
rend.setSeriesStroke(0, BasicStroke( 2.0 ) )
c1=HPlotChart(chart); c1.visible()
```

Listing 15.20 Time series with a moving average

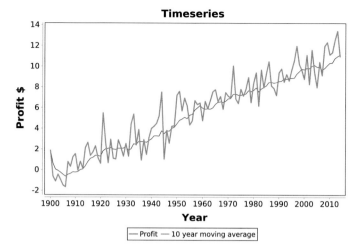

Fig. 15.6   A chart with a time series (*the red line*) and the calculated moving average (*the blue line*) (color figure online)

Fig. 15.7   Showing a candlestick chart used to describe price movements of a security, derivative, or currency

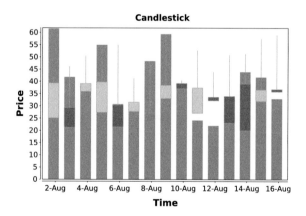

Now, let us consider a second example. This time we will make a candlestick chart using the JFreeChart library. This chart is used in finance to display the high, low, opening and closing prices, and is often considered to be an effective way to interpret price, and to study emotions of traders. An example of such a chart is shown below:

```
from java.util import *
from org.jfree.chart import *
from org.jfree.data.xy import *
from java.lang import Math
from jhplot import *

def createData(year,  month, date):
```

```
  cal = Calendar.getInstance()
  cal.set(year, month - 1, date)
  return cal.getTime()

date=[]; high=[]; low=[];
open=[]; close=[]; volume=[]
cal=Calendar.getInstance();
cal.set(2008, 5, 1)

for i in range(15): # fill data
  date.append(createData(2008, 8, i+1))
  high.append(30+Math.round(10) + (Math.random()*20.0))
  low.append(30+Math.round(10) + (Math.random()*20.0))
  open.append(10+Math.round(10) + (Math.random()*20.0))
  close.append(10+Math.round(10) + (Math.random()*20.0))
  volume.append(10.0+(Math.random() * 20.0))
data =  DefaultHighLowDataset("",date,high,low,open,close,
    volume)
chart=ChartFactory.createCandlestickChart("Candlestick","Time"
    ,"Price",data,False)
c1=HPlotChart(chart); c1.visible()
```

Listing 15.21 Creating a candlestick chart

The result of this code is shown in Fig. 15.7.

JFreeChart is a complete Java library, and its full description is outside the scope of this book. We should mention that this package is well suited for many custom plots. Please look at the Java API of this library.

References

1. JScience, Java Tools and Libraries for the Advancement of Science. http://www.jscience.org/
2. von Neumann J (1963) The general and logical theory of automata. Wiley, New York
3. Wolfram S (2002) A new kind of science. Wolfram Media, Champaign
4. Suzudo T Cambria Java package. http://www001.upp.so-net.ne.jp/suzudo/
5. ImageJ Java library. http://rsb.info.nih.gov/ij/
6. Brigham E (1988) Fast fourier transform and its applications. Prentice Hall, Upper Saddle River
7. JStatCom. Release 2.7 (2009). http://www.jstatcom.com/
8. Lütkepohl H, Krätzig M (eds) (2004) Applied time series econometrics. Cambridge University Press, Cambridge

Chapter 16
Using Other Languages on the Java Platform

One prominent feature of the DMelt software environment for numerical and statistical computations is that it can be used with different programming languages on different operating systems. Unlike many statistical programs, DMelt is not limited by a single programming language. A developer can use this program with several scripting languages, such as Python/Jython, BeanShell, Groovy, and Ruby. DMelt also includes support for an Octive-like interpreted language with a syntax very similar to MATLAB.

Most examples listed in this book are written in the Python high-level programming language using the Jython implementation. It allows a dramatic simplification of programming codes by taking advantage of the elegant features of Python. This leads to a substantial productivity gain for scientific computations. But Java is the heart of DMelt; therefore, a developer can benefit from the large number of numerical and graphics libraries developed by the Java community, in addition to the standard Python modules.

This chapter is geared to help you convert the Jython examples of this book into other scripting languages, as well to the Java language itself in order to embed numerical and statistical calculations into Java applications.

16.1 Python Scripting with DMelt

Throughout the book, we have considered a programming style with direct calls to Java classes using the Python syntax. Jython, as an implementation of the Python programming language, is well designed for such tasks.

You can program all the examples discussed in this book using Java. But you will find several limitations while doing this. For example, Java does not support operator overloading, which is one of the important features of modern high-level programming languages. We will remind that "operator overloading" means that one

© Springer International Publishing Switzerland 2016
S.V. Chekanov, *Numeric Computation and Statistical Data Analysis
on the Java Platform*, Advanced Information and Knowledge Processing,
DOI 10.1007/978-3-319-28531-3_16

can define custom types and operations. For example, one can redefine operators like
"+" or "−" depending on the types for which these operators are used.

Operator overloading is useful since this feature can significantly simplify programs, making them more concise and readable. One can easily see a problem if one
wants to do a lot of elementary mathematical operations with objects. In this case,
one should call Java methods that usually have long names. Of course, for simple
mathematical operations, such as addition and subtraction, it is more convenient to
use "+" and "−".

DMelt allows the use of operator overloading by calling Jython classes that are
directly mapped to the corresponding Java classes. Such Jython classes inherit the
methods of the Java classes from the `jhplot` package. At the same time, the most
common arithmetical operators are overloaded. The package that can do this is called
`shplot` (where the first letter "s" means "scripted" or "simple" hplot), in contrast
to name `jhplot` ("Java j-hplot").

Below is an example of how to build Jython histograms to perform some common
operations:

```
from java.util import *
import sys
from shplot import *

c1=hplot("scripting",1,2)
c1.visible()
c1.setAutoRange()

h1=h1d("histogram1",200,-5,5)
h2=h1d("histogram2",200,-5,5)

r=Random()
for i in range(500):
    h1.fill(r.nextGaussian())
    h2.fill(0.2*r.nextGaussian())
h1=h1+h2                         # add 2 histograms
c1.draw(h1)                      # draw
h1=h1*2                          # scale by a factor 2
c1.draw(h2)

c1.cd(1,2)                       # go to the next plot
c1.setAutoRange()
h1=h1-h2                         # subtract 2 histograms
h1=h1/h2                         # divide  2 histograms
h1=h1*10                         # scale by factor 10
h1=h1/100                        # divide by 100
c1.draw(h1)
```

Listing 16.1 Jython scripting with operator overloading

This example works with the DMelt IDE. If you run this code using a batch mode,
do not forget to import the Python modules located in the directory "macros/shplot"
(relative to your installation directory). The above example illustrates a mapping

between Python class h1d (1D histogram) and the Java class H1D. Analogously, the class hplot corresponds to the Java HPlot class. One should note that all Jython classes that inherit Java JHPlot classes have exactly the same names as the corresponding Java classes from the jhplot package. The only difference is that the class names have to be typed using lowercase letters. This means that the statement such as

```
>>> from   jhplot import  *
>>> c1=HPlot("scripting")
```

creates an instance of the Java class HPlot. However, when one uses lowercase letters for the same class name, the corresponding Python instance of this object is created:

```
>>> from   shplot import  *
>>> c1=hplot("scripting")
```

In this case, the major arithmetical operators are overloaded since "c1" is a pure Jython object.

Analogously, one can use the Python equivalent of the Java H1D class:

```
>>> from   shplot import  *
>>> h1=h1d("test",200,-5,5)
>>> h1.   # press [F4] to see all methods
```

One may notice that all the methods of H1D Java class are inherited by the Jython class h1d. Thus, even if you write a code using Jython shplot classes, one can easily access the Java methods:

```
>>> from   shplot import  *
>>> h1=h1d("test",200,-5,5)  # build a Jython histogram
>>> h1.setFill(1)            # accesses a Java method
```

In the above example, setFill(1) was applied directly to the Java class H1D. If one needs to subtract, add, divide, scale the data, use "−", "+", "/", "*" or "*scale" (where "scale" is a numeric constant), instead of the long oper(obj) or operScale(obj) statements of the Java H1D class.

One can go even further in redesigning our codding using Jython. When plotting an object on a canvas, such as a histogram, one can say that a histogram was added to the canvas. So, why not, instead of the usual draw(obj) method, use the operator "+"? Yes, this is also possible: Here is a code that shows how to draw three histograms on the same canvas using a single line:

```
>>> c1+h1+h2+h3
```

Of course, the same operation can be done with any other DMelt class, such as H1D or P1D from the `jhplot` package, which can also be drawn on the canvas. Note: subtraction of objects (which may correspond to the removal of objects from the canvas) is not implemented.

In the same spirit, data can be shown in a table. Below, we display histogram values in a pop-up table:

```
>>> htable()+h1
```

where the Jython `htable` class is a direct mapping of the `HTable` Java class.

As before, one should use the DMelt code assist to learn more about Jython/Java methods. This time, the code assist will be applied to Jython objects, rather than to Java class instances.

Note that not all Java classes have been mapped to the corresponding Jython classes. Look at the directory "macros/shplot" to find out which Jython modules are available. In the same directory, one can find several useful examples (their names contain the string "_test").

16.1.1 Operations with Data Holders

In the case of Jython scripting, all major operations for P0D, P1D, and PND are overloaded as well:

```
>>> from   shplot import *
>>> p1=p1d("test1")   # a Jython p1d based on P1D
>>> p2=p1d("test2")
>>> p1=p1+p2          # all values are added
```

Note that, for the latter operation, all y-values are added (and errors are propagated, respectively), while x values remain the same. This operation cannot append arrays. Instead, use the `merge()` method.

Here are a few more examples:

```
>>> p1=p1+p2     # all values are added
>>> p1=p1-p2     # subtraction
>>> p1=p1*p2     # multiplication
>>> p1=p1/p2     # division
>>> p1=2*p1      # scale by 2
>>> p1=p1/10     # scale by 0.1
```

To draw objects on a canvas, use "+" as before:

```
>>> c1+p1    # add p1d object to a  canvas
```

Similarly, one can work with one-dimensional (P0D) or multidimensional (PND) arrays.

16.1.2 Adding Python Modules

Unlike Java jar libraries, which are typically located in the directory "lib" inside the DMelt installation, external third-party Jython modules should be put into the directory "macros/user". This directory is imported automatically when a custom script is executed using the DMelt IDE.

For the Python packages that do not use Jython calls to Java classes, one should use the directory "packages" inside the directory "python". This directory is always scanned by the DMelt IDE, and it is included into the Python's list "os.path".

One can add other directories visible for the Python codding, but you should always import the directory path in the initialization file "jehepsys.py". This file is located in the directory "macros/system". It is called every time you run a custom Python script using the DMelt editor.

16.2 Using Java Programming

DMelt consists of Java libraries deployed in the form of jar files. Thus you may choose to write a Java code instead of Jython scripts using the Python language. Many examples of this book use a mix of Python and Java classes. Usually, we warn the reader about the usage of pure Python modules which will prevent from converting the examples of this book into the corresponding Java code.

Similar to the Jython code examples shown in this book, one can develop Java programs using the DMelt IDE. Any other Java IDE, such as Eclipse or NetBeans, will also work. Applications based on the DMelt libraries can be converted into a Java jar file and deployed together with other jar libraries that come with DMelt.

So, how to convert the Jython examples into Java program? To do this you should remember to create the main() method, which must be encapsulated into a class with the same name as the file name, which must have the file extension "java". Then you should add the "package" line to your code. All functions must be inside of this class.

We should remind that Python can use single quotes for strings. But Java interprets a string defined by the single quotes as a character. So, do not use single quotes in your Jython codes, otherwise, you should rewrite such places of the code before converting it to Java. In this book, we made sure that we used double quotes for Python strings, so that all such strings can be easily converted.

Unlike Jython, you should explicitly declare all variables and put the standard new operator when instantiating a Java class. Do not forget changes to be made for loops. Lastly, put semicolons at the end of each Java statement.

Our next example illustrates the difference between the Jython and Java syntax.
Let us write our example code in Python/Jython as

```
>>> from jhplot import *
>>> p0=POD("test")
>>> for i in range(5):
...     p0.add(i)
...     print i
```

The same code snippet in Java is

```
import jhplot.*;
POD p0 = new POD("test");
for ( i=1; i<5; i++ ) {
        p0.add(i);
        System.out.println(i);
        }
```

One should put these lines in a file (with the same name as the class name), compile
it using the `javac` compiler, and run it as you would normally do for any Java
application. One can also include this Java bytecode file into a "jar" file.

By convention, Java documentation comments are set inside the comment delimiters `/** ... */`. The single-line comment starts with two forward slashes and
continues to the end of the line, instead of the Python symbol "#".

If you are using an external Java IDE, do not forget to add the directory "lib" from
the DMelt installation into the `CLASSPATH` variable, so all DMelt libraries should
be visible for the Java virtual machine.

Let us show a more detailed example. Section 7.1 discussed how to plot three
histograms using Jython. We will rewrite this code in Java as

```
import java.awt.Color;
import jhplot.*;

class Example {
  public static void main(String[] args) {

  HPlot c1=new HPlot("Canvas",600,400,2,1);
  c1.visible();
  c1.setAutoRange();

  H1D h1=new H1D("First",20,-2.0,2.0);
  h1.setFill(true);
  h1.setFillColor(Color.green);
  H1D h2=new H1D("Second",10,-2.5,2.5);
  h1.fillGauss(500,0,1);   // mean=0, sd=1
  h2.fillGauss(1000,1,2);  // mean=1, sd=2

  H1D h3=new H1D("Third",20,0.0,10.0);
  h3.setFill(true);
  h3.fillGauss(50000,2,1); // mean=2, sd=1
```

```
    c1.cd(1,1);
    c1.setAutoRange();
    c1.draw(h1);  c1.draw(h2);

    c1.cd(2,1);
    c1.setAutoRange();
    c1.draw(h3);
    }
}
```

Listing 16.2 Showing histograms using "Example.java"

Save these lines into the file "Example.java" and open it inside the DMelt IDE. Now you can: (1) Compile and run the program by pressing the [Run] button. (2) Or, one can just compile the file loaded to the editor into a bytecode file using the menu [Run]→[Javac current file]. Then you can run the compiled code as: [Run]→[Run Java]. As you will see, the output will be identical to that shown in Fig. 7.1 of Sect. 7.1.

We should note that the above code can be compiled manually, without any IDE. One would only need to import all Java jar libraries from the DMelt installation directory. For Unix-like environments (like Linux or Mac OS), execute this command if you want to run the above code without DMelt IDE:

```
bash> dmelt_batch.sh Example.java
```

which will compile and run this code. Or one can make your own compilation program by creating this small bash script:

```
#!/bin/bash
args=$#
if [ $args == 0 ]
then
    echo "did not specify input file!"
    exit 1;
fi

for i in lib/*/*.jar
do
    CLASSPATH=$CLASSPATH:"$i"
done

CP=$CP:$CLASSPATH
javac -classpath "$CP" $1
echo "File $1 compiled!"
```

Listing 16.3 Compilation script "compile.sh"

The script scans all subdirectories inside the directory "lib" of the DMelt installation directory, and appends all jar files to the CLASSPATH variable. Then run this script as

```
bash> compile.sh Example.java
```

This script compiles the code to Java bytecode, creating the file "Example.class". Similarly, one can run the compiled class file "Example.class" using the same script, but replacing the statement javac by java, and passing the name "Example" (not "Example.java").

One can easily convert this program into a Java applet, build your own Java library, or to deploy it as an application. We will discuss how to create a Java applet in the following section.

16.2.1 External Java Libraries

If one needs to include Java external libraries deployed in the form of jar files, one should first create a new directory inside the "lib" directory and then put the jar files inside this directory. Next time when you run the script "dmelt.sh" (for Linux, Unix, Mac) or the script "dmelt.bat" (for Windows), your external classes should be available in the Jython Shell and for the DMelt IDE editor. The start-up script recursively scans all directories inside the "lib" directory.

One can also include additional jar files to the directory "lib/user" (which is empty for the default installation). This directory is slightly special: every jar file located in this directory will be scanned by the Jython engine, so your library will be immediately available for Jython projects that use asterisks for the import statements.

16.2.2 Working Java Projects

In this section, we will discuss how to work with Java projects using the DMelt IDE.

If you are working with a file, "Example.java", which is already opened in the DMelt IDE, press [Run] (or [Run java]) using the tool bar. One can further adjust the class path of external libraries by modifying the BeanShell scripts in run_java* files in the director "macros/user".

One can mix Jython and Java codes for CPU-intensive tasks. Such tasks must be implemented in Java to achieve the best possible performance. Jython can be used as a glue language for various Java libraries with the source files located inside the main project directory.

First, build a jar file that contains the byte-codes of Java classes. For this example, set "proj" as your project directory using the DMelt IDE, and select the menu

[Compile and jar] project files from the [Run] toolbar menu. After executing it, a jar file classes.jar appears in the main project directory, which contains the classes of the source files located in the project directory. Then, one should move the file "classes.jar" to the directory "lib/user", and restart the DMelt IDE. Now your Jython code can import the classes from this jar file.

16.2.3   Embedding DMelt in Applets

DMelt numerical libraries can easily be deployed over the Web in the form of applets. This is one of the most significant advantages of Java compared to other programming languages.

In this section, we will show how to write a small applet which brings up a HPlot canvas with a filled histogram at runtime. If you have already some experience with Java, the code below will be rather trivial:

```java
import java.applet.*;
import java.awt.*;
import java.util.Random;
import jhplot.*;

public class Histogram
    extends Applet implements Runnable {

  private static final long serialVersionUID = 1L;
  private HPlot c1;
  private Thread thread;
  private Random rand;
  private H1D h1;
  private int i;

  public void init() {
    c1 = new HPlot();
    c1.setGTitle("Gaussian numbers");
    c1.visible(true); c1.setAutoRange();
    h1 = new H1D("Random numbers",20, -2.0, 2.0);
    h1.setColor(Color.blue);
    h1.setPenWidthErr(2);
    h1.setFill(true);
    h1.setFillColor(Color.green);
    c1.setNameX("X axis"); c1.setNameY("Y axis");
    c1.setName("100 numbers and statistics");
    rand = new Random();
    i=0;
  }

  public void start() {
    (thread = new Thread(this)).start(); }
```

```
  public void stop() {
     thread = null;
     c1.drawStatBox(h1); }
public void run() {
 try {
  while (thread == Thread.currentThread()) {
     Thread.sleep(50);
     h1.fill(rand.nextGaussian());
     c1.clearData();
     c1.draw(h1);
     i++;
     if (i >100) stop();
     }
   } catch (Exception e) {} }
}
```

Listing 16.4 File "Histogram.java"

Although this file is slightly lengthy (it is not a Python/Jython script!), the code itself
is rather simple: We create the usual HPlot and H1D objects and set some attributes.
Then we fill a histogram with random numbers in a thread. We animate the canvas
by creating random numbers in a loop with some time delay. After the number of
entries reaches 100, we stop filling the canvas and display the statistics.

Compile this code as discussed in the previous section, either inside the DMelt
IDE or with the help of the compilation script discussed in the previous section:

```
bash> compile.sh Histogram.java
```

This creates the file "Histogram.class" with the Java bytecode. The next step is to
embed it in an HTML file, or in a PHP script.

Our preference is to use the PHP language that will allow an automatic scanning
of all jar files inside a given directory. First, copy the library directory "lib" with
the DMelt jar libraries to a location where they can be accessible by the Web server.
Then, write a script:

```
<?php
require("list_files.php");
$list=list_files("lib/system/") . ", " .
      list_files("lib/freehep/");
$html_body  = "";
$html_body .= <<<EOT
<html>
<body>
<h1>Histogram example</h1>
<APPLET
  CODE="Histogram.class"
  WIDTH=0 HEIGHT=0 ARCHIVE="$list">
  Please use a Java-enabled browser.
</APPLET>
```

```
</body>
</html>
EOT;

print $html_body
?>
```

Listing 16.5 PHP script "applet.php"

This PHP script creates the necessary HTML file with our applet "on-the-fly". The only unusual variable is $list which is a comma-separated list of jar files from the directory "lib". This list is created by the function "list_files.php", which returns a list of files in a certain directory. This function is shown here:

```
<?php
function list_files($dir)
{
$list="";
if ($handle = opendir($dir)) {
    while (false !== ($file = readdir($handle))) {
        if ($file != "." && $file != "..") {
            $ff= $dir . $file;
            $list .= $ff . ", ";
        }
    }
    closedir($handle);
}
return $list;
}
?>
```

Listing 16.6 PHP script "list_files.php"

Now you are ready. Copy the created files "applet.php", "list_files.php" and "Histogram.class" to a Web-server accessible directory. Make sure that this directory contains the directory "lib" from the DMelt installation directory (one can also make a link to this directory). One possible option to reduce the start-up time of the applet is to remove unnecessary jar files, if they are not used to run the applet. Then, point the Web browser to the file "applet.php". The PHP module "list_files.php" will scan the jar files inside the directory "lib" and will pass it to the "applet.php" file. This file will build a proper HTML file to be displayed inside the Web browser. You should see a HPlot frame with filled histogram entries. After 100 entries, the animation will stop and you will see the statistics for the filled histogram.

16.3 Using BeanShell Language

The BeanShell programming language [1] simplifies the Java syntax in several ways. BeanShell dynamically compiles and executes Java-like scripts, extending Java with scripting conveniences, such as loose types (similar to Jython). BeanShell does not require strict Java typing, thus there is no need to declare variables with types. It is relatively easy to turn a prototype code written in BeanShell into proper Java programs, or to other scripting languages.

The code examples given in this book can easily be rewritten in BeanShell, calling the same Java classes and methods as for the Jython examples. When using the DMelt IDE for BeanShell files, save BeanShell commands in files with the extension ".bsh". In this case, the DMelt IDE will recognize the BeanShell interpreter. One can edit and run BeanShell macros in the same way as discussed in Sect. 1.4.

Let us show the usual "Hello, World!" example in BeanShell. In addition, the example code will illustrate the variable assignments:

```
print("Hello, World!");
val = 1;
va2= "Hello";
```

Unlike Python, BeanShell always requires a semicolon after each statement. It instructs the compiler that a command has ended. Documentation comments in Bean-Shell are similar to those in Java:

```
// single-line comment
/*
multiline comment
*/
```

The import and loop statements are also similar to the usual Java language specification:

```
// Loop in BeanShell
for (i=0; i<10; i++) {
    print(i);
    }
```

Generally, BeanShell supports all constructs of the Java language, including variable assignments, method calls, mathematical expressions, and for-loops. In addition, it adds several convenient high-level methods, such as the print() statement shown in the above example.

BeanShell imports the Java core classes at start-up, but when importing additional classes, use the standard Java import syntax. For example, import all the classes from the jhplot package of DMelt as

```
import jhplot.*;
c1=new HPlot("Canvas");
```

This code creates an untyped script variable (named "c1") and assigns the HPlot object designed to create a canvas for displaying data. Note that, when creating an object, there is no need to declare its type. But the use of the new statement is mandatory. As for any high-level scripting language, BeanShell variables can change their type at runtime depending on the object.

BeanShell import statements may appear anywhere, even inside a method, and not just at the top of a file. By default, the following packages from the Java platform are imported automatically:

- javax.swing.event
- javax.swing
- java.awt.event
- java.awt
- java.net
- java.util
- java.io
- java.lang

Now let us show a practical example. We will rewrite the Java example with histograms shown in Sect. 16.2 using the BeanShell scripting language:

```
import java.awt.Color;
import jhplot.*;

c1 = new HPlot("Canvas",600,400,2,1);
c1.visible(); c1.setAutoRange();

h1 = new H1D("First",20,-2.0,2.0);
h1.setFill(true);
h1.setFillColor(Color.green);
h2 = new H1D("Second",10,-2.5,2.5);
h1.fillGauss(500,0,1);   // mean=0, sd=1
h2.fillGauss(1000,1,2);  // mean=1, sd=2

h3 = new H1D("Third",20,0.0,10.0);
h3.setFill(true);
h3.fillGauss(50000,2,1); // mean=2, sd=1

c1.cd(1,1);
c1.setAutoRange();
c1.draw(h1); c1.draw(h2);

c1.cd(2,1);
c1.setAutoRange();
c1.draw(h3);
```

Listing 16.7 Showing histograms using "Example.bsh"

Compare this code with the corresponding Jython example in Sect. 7.1. Here we will remind that if you work with a BeanShell script using the DMelt IDE, one can access the following internal variables as

- textArea—the text area with the script
- view—the main GUI window with text in the editor
- SystemDir—the current system directory
- UserMacrosDir—the user directory with BeanShell macros
- DocName—name of the last opened document directory
- DocMasterName—name of the last opened document without the extension
- DocDir—the directory of the last opened document
- DocStyle—the style of the last opened document

Check these variables using the usual print command. For example, this line:

```
print(SystemDir);
```

prints the current system directory.

By reading this section, you may ask yourself the following question. Why is BeanShell needed at all, if one can write data-analysis code using Java, or using Python/Jython with the elegant scripting syntax? The possible answer to this question is this: BeanShell was designed from the ground to prototype Java programs. One can prototype your analysis code with BeanShell scripts quickly, and then transform them to real Java source code. It is more difficult to do this using Jython with the Python syntax, which is quite different from the Java language. For example, the semicolon and braces are parts of the syntax of BeanShell and Java. However, Python uses the indentation instead of braces to structure code blocks. Similarly, the loop command is identical in BeanShell and Java, while range() is a popular method to deal with loops in Python.

As a last comment, one should mention that BeanShell can also interpret ordinary Java source files.

16.4   Using Groovy Language

Groovy [2, 3] is an interpreted, object-oriented, high-level programming language. Similar to BeanShell, it is tightly integrated with the Java platform, and is fully supported by the DMelt software environment.

Groovy is strictly object oriented. This means everything in Groovy, including the scripts themselves, is an object. Every Java construct is a valid Groovy code. Every Java class imported from the DMelt library can be imported in Groovy programs. This makes integration of Groovy with the DMelt numeric libraries relatively smooth.

Groovy source files end with the extension "groovy". The files can contain a Groovy script or a Groovy class. The classical "Hello, world" program can be written as

```
println"Hello World"
```

omitting the usual in Java/BeanShell parentheses. There is another difference: Groovy does not require semicolons at the end of each statement. In some sense, this makes codding similar to Python. But, of course, Groovy is a very different language, with a syntax more similar to Java.

Groovy uses BeanShell (or Java) notion for the import statement to resolve class references. It automatically imports a number of Java packages and classes that can be used in Groovy without specifying the package name. Here is a list with the imported Java packages:

- java.io.*
- java.lang.*
- java.util.*
- java.net.*

Groovy supports the standard Java "for", the "for-each", and the "while" loops, but it does not support the "do while" loop.

Now let us show a practical example. We will rewrite the Java example with histograms shown in Sect. 16.2 using the Groovy scripting language. Let us create the file "Example.groovy" with the following code:

```
import java.awt.Color
import jhplot.*

c1 = new HPlot("Canvas",600,400,2,1)
c1.visible(); c1.setAutoRange()

h1 = new H1D("First",20,-2.0,2.0)
h1.setFill(true)
h1.setFillColor(Color.green)
h2 = new H1D("Second",10,-2.5,2.5)
h1.fillGauss(500,0,1)    // mean=0, sd=1
h2.fillGauss(1000,1,2)   // mean=1, sd=2

h3 = new H1D("Third",20,0.0,10.0)
h3.setFill(true)
h3.fillGauss(50000,2,1)  // mean=2, sd=1

c1.cd(1,1)
c1.setAutoRange()
c1.draw(h1); c1.draw(h2)

c1.cd(2,1)
c1.setAutoRange()
c1.draw(h3)
```

Listing 16.8 Showing histograms using "Example.groovy"

Did you notice changes compared to the BeanShell example in Sect. 16.3? Probably, not. The only change is in removing the semicolons after each statement. Groovy does not need them. But you can also put them back, since you can still use the semicolons to separate statements on one line. They are also useful if one needs to convert Groovy prototype files to Java codes.

Let us illustrate Groovy programming using another example. This time we will rewrite the example shown in Sect. 8.4 which uses the chart classes from the JFreeChart library:

```
import org.jfree.chart.ChartFactory
import org.jfree.data.general.DefaultPieDataset
import jhplot.HPlotChart

def dataset = new DefaultPieDataset()
dataset.with {
    setValue "Apr", 10
    setValue "May", 30
    setValue "June", 40}
def options = [true, true, true]
def chart = ChartFactory.createPieChart("Pie Chart",
    dataset, *options)
def c1 = new HPlotChart( chart )
c1.visible()
```

Listing 16.9   Creating a chart using JFreeChart and Groovy

As you can see, the code remains rather short.

Of course, there are more differences between BeanShell and Groovy. We advise the reader to look at the Groovy manuals [2, 3] to learn more. One important difference we should mention is this: Groovy scripts are better integrated with the JVM, and execution of Groovy scripts can significantly be faster than the equivalent BeanShell programs. Several benchmark tests made with DMelt indicate that the execution of Groovy scripts with long loops is about a factor 2–3 faster than running identical algorithms implemented in BeanShell.

16.5   Using Ruby Language

JRuby [4], an implementation of the Ruby programming language on the Java platform, can also be used with the DMelt programming environment. Similar to Ruby, JRuby supports object-oriented programming and dynamic typing. Similar to BeanShell, Groovy, and Jython, JRuby is tightly integrated with the Java platform, thus allowing embedding the Ruby interpreter into any Java application. JRuby computing engine is a part of the DMelt program. It can invoke the Java classes shipped with the Java platform, as well as the DMelt collection of numerical and statistical libraries.

In order to run a JRuby program, create a file with the extension "rb" containing JRuby code and run this file either in the batch mode or using the DMelt IDE. Let us show how to get started with JRuby using DMelt. Here is our usual "Hello World" example written in JRuby:

```
print"Hello World"   # comment line
puts "Hello World"
```

Note that print does not add a newline at the end, while puts does. A single-line comment begins with the number sign character, and ends at the end of the line.

In order to call Java libraries inside JRuby, one should load JRuby's Java support by adding the statement require 'java'. Let us show how to rewrite our usual example with the histograms discussed in the previous sections in JRuby:

```
require 'java'
java_import Java::jhplot.HPlot
java_import Java::jhplot.H1D
java_import Java::java.awt.Color

c1 = HPlot.new("Canvas",600,400,2,1)
c1.visible(); c1.setAutoRange()

h1 = H1D.new("First",20,-2.0,2.0)
h1.setFill(true)
h1.setFillColor(Color.green)
h2 = H1D.new("Second",10,-2.5,2.5)
h1.fillGauss(500,0,1)   # mean=0, sd=1
h2.fillGauss(1000,1,2)  # mean=1, sd=2

h3 = H1D.new("Third",20,0.0,10.0)
h3.setFill(true)
h3.fillGauss(50000,2,1) # mean=2, sd=1

c1.cd(1,1)
c1.setAutoRange()
c1.draw(h1); c1.draw(h2)

c1.cd(2,1)
c1.setAutoRange()
c1.draw(h3)
```

Listing 16.10 Showing histograms using "Example.rb"

Now you can see several differences between BeanShell and Groovy when initializing Java classes. One important difference is that the new statement is moved behind the class name.

The loop statements in JRuby are also quite different from the Java and BeanShell language specifications. Here is an example showing how to use random numbers from the Java package java.util using the JRuby's loop statement:

```
rand =java.util.Random.new()
print "Use of for loop in JRuby \n"
for i in 1..100
   puts "Test Value is => #{i}"
   h1.fill(rand.nextGaussian())
   end
```

The above script shows that you can use all the goodness of the Ruby programming language on the JVM. One reason to prefer JRuby over other scripting languages is the fact that Ruby is one of the popular languages used for Web applications, such as the Ruby on Rails web framework. But, in order to use the DMelt numerical libraries, you still need to run JRuby on the Java platform.

16.6   Using Octave Language

The support for Octave, a high-level interpreted language similar to MATLAB, is also included in DMelt using the jMathLab [5] package. It should be mentioned that this language is not integrated with the Java platform as tight as other scripting languages, such as Jython, Groovy, BeanShell, and JRuby. This means that you cannot directly access Java classes from Octave programs. We should also mention that most of DMelt Java libraries are not visible in Octave program codes. How to use Octave symbolic calculations on the Java platform using the `Symbolic` class was briefly discussed in Chap. 6.

The current implementation of the Octave support is based on the Jasymca symbolic calculator [6], jMathLib [7] and jMathLab [5] Java projects, so much of the credit should go to these programs and to the descriptions on their original websites. DMelt includes substantially extended libraries based on jMathLab, which gives a larger choice of methods compared to the original Jasymca program.

Let us say a few words about jMathLab, a clone of MATLAB and GNU Octave. Unlike MATLAB from the MathWorks, Inc., jMathLab is free. jMathLab also differs from the GNU Octave: it is available on all computing platforms where Java is installed, and can be easily extended using the standard Java graphical libraries. This program is based on Jasymca code for symbolic calculations and other libraries that substantially extend Jasymca's functionality. For example, jMathLab includes the plotting canvases based on the familiar `HPlot` class from the DMelt core library. It also includes a number of functions from jMathLib [7].

The jMathLab program can be used for:

- Symbolic calculations (simplification, differentials, integration).
- Numeric evaluation of mathematical functions, and special functions.
- Linear algebra with vectors and matrices.
- Displaying data, vectors, matrices, and functions using 2D and 3D interactive plots.

- Saving data (vectors and matrices) in CSV files.
- Random numbers using the major distributions.
- Solving linear and nonlinear equations, as well as systems of equations.
- Basic statistical calculations and histogramming.

This list is rather comprehensive—in fact, it is almost everything we have discussed before in the context of Jython scripting. But jMathLab extends the numeric calculations supported by DMelt with handy symbolic calculations that have not been discussed so far.

The Java scripting using the Octave-style scripting language deserves special attention. A short tutorial on the Octave-style programming is given in Chap. 17.

References

1. BeanShell, Lightweight scripting for Java. http://www.beanshell.org
2. Groovy, A multi-faceted language for the Java platform. http://www.groovy-lang.org/
3. Subramaniam V (2013) Programming Groovy 2: Dynamic Productivity for the Java Developer. Pragmatic Programmers, LLC
4. JRuby, The Ruby programming language on the jvm. http://jruby.org/
5. JMathLab, A multiplaform computational platform for symbolic calculations. http://jwork.org/jmathlab/
6. Dersch H, Jasymca, A Java symbolic calculator.http://webuser.hsurtwangen.de/dersch/jasymca2
7. Sparshatt M, Muller S, Torras A, JMathLib, A Java clone of Octave, SciLab, Freemat and MATLAB. http://www.jmathlib.de/

Chapter 17
Octave-Style Scripting Using Java

This chapter gives a brief introduction to the Octave-style language supported by DMelt. As mentioned previously, much of the credit for this description goes to the jMathLab project [1] that combines many features of the jMathLib [2] Java project and the Jasymca symbolic calculator [3]. We remind that the syntax of GNU Octave is quite similar to MATLAB, therefore, one can reuse some MATLAB code in your programs running on the Java platform.

The high-level interpreted language based on Octave-style syntax is not tightly integrated with the Java platform, unlike other Java scripting languages that can import external Java libraries. But, as discussed in Chap. 6, you can easily integrate Octave-like code with symbolic calculations inside Java programs or Java-supported scripting languages, such as Jython, Groovy, JRuby, and BeanShell. You can find a comprehensive description of GNU Octave in several books [4, 5].

To perform symbolic calculations using the Octave language on the Java platform, we will use the jMathLab console of the DMelt IDE, which is located directly under the main editor. Alternatively, one can create files with the extension ".m" and run them as usual, i.e., either in the DMelt editor, or using the console command to execute macro files with the Octave commands. For example, if your program code is the file "file.m," simply run the command:

```
> dmelt_batch.sh file.m
```

The bash script "dmelt_batch.sh" is included with the DMelt installation. This approach works on all Linux/Unix or Mac OS computers.

17.1 Getting Started

Printing messages in Octave can be done with the command disp() (a shortcut from "display"):

© Springer International Publishing Switzerland 2016

S.V. Chekanov, *Numeric Computation and Statistical Data Analysis on the Java Platform*, Advanced Information and Knowledge Processing, DOI 10.1007/978-3-319-28531-3_17

```
disp("message\n new line")    % a text message
ans=                          % printed answer
   message
   new line
```

The example shows that the Java new line character "\n" is used as a separator in strings, while the symbol "%" is reserved to add comments to your code. In the following discussion, the message ans= will be used to indicate the code output. Note that one can use either double or single quotes for strings. As for many programming languages, semicolons are used to separate statements.

Let us show how to print numbers using a formatted output:

```
printf("%f",100)
```

Comments in the output message are supported by adding a string before the % character:

```
x=log(sqrt(854));    printf("Answer=%f", x)
ans=3.375
```

To run Octave scripts (i.e., files with the extension ".m") inside the jMathLab interpreter, use the function loadfile:

```
loadfile("/path/myfile.m")
```

jMathLab has its own help system. To understand it better, we should note that all programming modules in jMathLab are arranged in groups. One can list all functions which come with the DMelt program by executing this statement:

```
path
```

It prints all modules and functions arranged in external files. The outputs of the above command are the groups of modules:

```
bin/toolbox
toolbox/test
toolbox/statistical_distributions
toolbox/special_functions
toolbox/random_numbers
toolbox/polynomial
toolbox/plot
toolbox/linear_algebra
toolbox/input_output
toolbox/system
toolbox/statistics
```

```
toolbox/trigonometric
toolbox
```

Functions typically end with the extension "m" (as for the usual MATLAB/Octave scripts), or "java" (Java-implemented functions). To add a new directory with the files ("m-files") which implement your code, use the `addpath` statement. For example, this command adds the directory where user-defined modules are searched:

```
addpath("/home/user")
```

There are more than 100 built-in functions included with jMathLab. They are usually implemented as external M-files (with the extension ".m"), or as Java files.

To print the description of a particular module, use the `help` statement:

```
help("acosh")
```

where `acosh` is the name of the inverse hyperbolic cosine. This will create a dialog window showing how to use this function. As an exercise, find how the function `help` is defined by replacing `acosh` with `help` (i.e., with the function which implements the help system).

A few words on string manipulations. One can combine two strings `s1` and `s2` using the method `strcat(s1,s2)`,

```
s="test"
s=strcat("new_",s)
disp(s)
ans="new test"
```

One can search for a particular substring `s2` inside another string `s1` using the method `index(s1,s2)`. The method `substr(s1,i1,i2)` extracts a substring between the positions `i1` and `i2`.

17.2 Variables and Operators

Let us start with the basics. A variable holds a value that may be changed at runtime. As usual, variables are declared using the standard form, `name=value`, with "name" being any character sequence. It may also contain numbers. The value can be any number or expression.

First, we will learn how to print a variable that keeps a simple number. This example shows how to assign 1000 to the variable x and print the result:

```
x=1000; printf("%f",x)
ans=1000
```

Numbers are entered in the usual computer format, with optional decimal point and decimal exponent following the letter e (or E). For example, the number 1.120102×10^{11} should be entered as:

```
x=1.120102e11;  printf("%f",x)
ans=1.1201E11
```

When numbers are printed, only five significant digits are displayed. To preserve the full precision when displaying the numbers, add the statement "format long" before the print statement. One can also use the command "format Base Number", which is used to display numbers in a system with arbitrary Base with any Number of significant digits. To display numbers with 15 digits in the binary system we type:

```
format 2 15
x=1.52102e10;  printf("%f",x)
ans=1.11000101010011E33
```

We can bring back the display mode to its default (short decimal) by entering the statement "format short". It should be noted that the format commands do not influence the internal accuracy of calculations.

The usual symbols "+," "−," "∗," "/" are used to deal with arithmetical operations. Exponentiation is performed with the accent character. Multiplication and division precede addition and subtraction. As for many languages, the order of expression evaluation can be forced by parenthesis. Here are a few examples:

```
a=9.23*(14-2^5)/(15-(3^2-2^3))
a=1.5e-13/0.000001
a=200.29e4/(1.12-17.23e4/(1.1-17.2e8/1.1))
a=100.4^((3-2.10^1.9)^0.12)
```

Run these examples and print the output to see the answers.

Relations can be connected by the logical operators. They are "<" (less), ">" (larger), ">=" (larger or equal), "<=" (less or equal) and "==" (equal). The boolean functions are "&" (and), "|" (or) and "~x" (not). When applied to scalars, a relation is the scalar 1 or 0, depending on whether the relation is true or false. Similar to Jython, logical true is the number 1, false is 0.

There is one useful predefined variable called eps, which is the smallest positive value. Let us take a look at its property:

```
printf("%f\n",eps)
ans=2.2204E-16
a=1+eps>1;              printf("%f",a)
ans=1
a=1+eps/2>1;           printf("%f",a)
ans=0
```

The output will be 2.2204×10^{-16}, 1 and 0, respectively. Another useful predefined constant is π.

Numbers entered without the decimal point and exponent, and which are larger than 10^{15}, are stored as exact rational data type. These numbers are represented as quotient of two variable length integers. This approach allows calculations without any rounding errors. This feature comes from the Jasymca computing engine, so we will quote the example from its documentation:

```
10000000000000001.
ans  =  1.0E16
10000000000000001
ans  =  10000000000000001
```

In the first case of this example, a floating point number is created, while the second line corresponds to an exact rational number. Floating point numbers can be converted to an exact number using the command `rat(x)`. This conversion is accomplished by continued fraction expansion with accuracy determined by the predefined variable `ratepsilon` which has the default value of 10^{-8}.

```
x=rat(0.33333333333333333);  printf("%f",x)
ans=1/3
x=rat(0.33333);  printf("%f",x)
ans=33333/100000
```

As you may notice, `rat(x)` is a useful function for symbolic calculations.

Operations between exact and floating point numbers always lead to the promotion of floating point numbers:

```
x=80/21/125/21/5*17*6*5-1;   printf("%f",x)
ans=  -0.85197
```

Calculations can be performed without rounding errors by rationalizing the first number:

```
x=rat(80)/21/125/21/5*17*6*5-1;   printf("%f",x)
ans=  -3131/3675
```

The command `float(x)` converts numbers into the floating point format. The same command also works for composite data types, such as polynomials and matrices, whose coefficients are transformed in one step. Let us show a few examples:

```
x=sqrt(5)
x=sqrt(rat(5))
x=float(100)
```

One can also work with imaginary numbers. The imaginary part is indicated by either i or j:

```
x=2+3i;  printf("number=%f",  x)
ans=  number=(2+3i)
```

One can extract the real and imaginary parts using the commands `realpart(x)` and `imagpart(x)`. Variables stored in the memory can be displayed by the command `who`. Single variables can be removed from the memory by entering the `clear` statement.

```
x=5+7i;  a=10
who      % show memory variable
clear  % clear the variables
who      % show again variables
```

You can determine the type of the object using the `type` method. Try this example which prints the type of the symbolic expression $x * y$:

```
syms  x,y
a=x*y;  disp(type(a))
ans=jasymca.Polynomial
```

As you can see, the symbolic expression belongs to the Jasymca polynomials. Most operations accept any mixture of numeric and symbolic arguments using the same commands and command syntax.

17.2.1 Symbolic Variables

Symbolic variables are quite different from the numeric variables discussed until now. The symbolic variables are algebraic objects to be used in symbolic expressions. For example, if x is a conventional variable, entering x in the input field makes the program to search in the environment for the corresponding object, which then replaces x. If x is a symbolic variable, the same action leads to a first-degree polynomial with variable x, and coefficients 1 and 0.

A variable (say x) can be declared "symbolic" by entering the command `"syms x"` before using x in an expression. The command "`clear x`" removes the symbolic variable x from the memory. Let us show again the usual "non-symbolic" initialization:

```
x=3                    % non-symbolic  variable
a=x^2-4-3*sin(x);  printf("%f",a)
ans=4.5766
```

Now we create a function with a symbolic value x:

```
syms  x        % x is symbolic variable
y=x^2-4-3*sin(x); printf("%f",y)
ans=(-3*sin(x)+(x^2-4))
```

Here are a few more examples:

```
syms  x,y
a=2*sqrt(x)*exp(-x^2);
ff=subst(3,x,a);   printf("%f",ff)
ans=4.275E-4
ss=trigrat(sqrt(4*y^2+4*x*y-4*y+x^2-2*x+1))
printf("%f",ss)
ans=(y+(1/2*x-1/2))
```

17.2.2   Operators and Commands

Now let us discuss the operators to be used in mathematical expressions. Table 17.1 shows the supported mathematical operators. The operations are executed according to their precedence as discussed in the previous section.

Commands are used for setting modes and various options. In contrast to functions, arguments can be supplied without parenthesis. One should mention that only one command per text input is allowed. A few basic examples are the `short` and `long` commands used for displaying format for float numbers: 5 digits, all digits, or in `base` with count digits. We have already discussed several other useful commands. Let us recall them again:

- `clear`—clear a given variable from the memory
- `who`—shows variables in the memory
- `path`—prints all available functions.

The usage of these commands was discussed previously.

17.3   Functions

jMathLab includes about 200 mathematical functions. Some functions come with the Jasymca engine, other functions are implemented in external modules. You can define your own functions in separate files with the extension ".m," or inline, i.e., in the same script where the main code is.

Table 17.1 Supported mathematical operators

Function	Syntax	Precision	Order
Addition	$x + y$	4	Left–right
Adjunct	x'	1	Right–left
Assignment	$x = y$	10	Right–left
Assignment	$x- = y$	10	Right–left
Assignment	$x/ = y$	10	Right–left
Assignment	$x* = y$	10	Right–left
Assignment	$x+ = y$	10	Right–left
Less or equal	$x <= y$	6	Left–right
Less	$x < y$	6	Left–right
Matrix division (left)	$x \ y$	3	Left–right
Matrix division (right)	x/y	3	Left–right
Post-decrement	$x--$	10	Right–left
Post-increment	$x++$	10	Right–left
Pre-decrement	$--x$	10	Left–right
Pre-increment	$++x$	10	Left–right
Range	$x : y : z$	5	Left–right
Scalar division	$x./y$	3	Left–right
Scalar exponentiation	$x.\widehat{}y$	1	Left–right
Scalar multiplication	$x. * y$	3	Left–right
Subtraction	$x - y$	4	Left–right
Vector/Matrix exponentiation	$x\widehat{}y$	1	Left–right
Vector/Matrix multiplication	$x * y$	3	Left–right

Here is a simple example. Let us compute the inverse cosine of the angle 0.4 rad:

```
printf("%f", acos(0.4))
ans=1.1593
```

The code calls the `acos()` function. We remind that one can get the description of this function as: `help("acos")`. The usual elementary functions, such as $\log(x)$, $\sqrt{x}$, $\sin(x)$, $\cos(x)$, $\tan(x)$ etc. are supported. All such functions can be nested, as shown in the next example:

```
a=log(sqrt(800)); printf("%f",a)  % natural logarithm
ans=  3.3423
```

Let us show a more detailed example using other predefined functions

```
a=float(sin(pi/2))  % argument in radian
a=gammaln(1234)     % logarithm of gamma function
```

```
a=primes(1000000000000000001); printf("%f\n",a)
ans=[ 101   9901   999999000001 ]
a=factorial(12); printf("%f\n",a)  # 12!
ans=4.79E8
```

Table 17.2 lists the available functions for a scalar variable. Table 17.3 shows the built-in functions which can be used for nonscalar variables, i.e., they can accept either a value (number), a vector of numbers or a symbolic value. In the case of vectors, the function can return vectors.

Special mathematical functions are also supported. Table 17.4 shows a set of special functions in jMathLib.

Functions can also be defined inline, i.e., in the same script where they are called from. How to program a mathematical function is demonstrated in our next code snippet which creates a function called `examples(x)`. The goal of this simple function is to multiply its argument by 2. Once the function is defined, we can call it later in exactly the same way as the standard built-in function:

```
function y=example(x) y=2*x; end
printf("%f",example(3.123) )
ans=6.246
```

As you can see, we use the keyword `function` to begin the function definition. The next symbol is the variable y used for the assignment of the name of the function `example`. Then we code the body of the function making sure that it defines the variable y. Note that we do not have the usual statement `return`, as for Java or other programming languages. Instead, the variable y is defined at the very beginning of the function definition. We finish the function code block with the keyword `end`.

Table 17.2 Predefined functions of jMathLab

Name(arguments)	Function
abs(x)	Absolute value of x
angle(x)	Angle of x
cfs(x) [var_T])	Continued fraction expansion of x with accuracy var_T
conj(x)	x conjugate complex
float(x)	x to floating point number
imagpart(x)	Imaginary part of x
primes(VAR)	VAR decomposed into primes
rat(x)	x to exact number
realpart(x)	Real part of x
sign(x)	Sign of x

Table 17.3 Built-in functions in the jMathLab program for numbers, vectors and symbolic variables

Name(arguments)	Function
acosh(x)	Hyperbolic areacosine
acos(x)	Arccosine (radian)
acsch(x)	Hyperbolic areacosecans
acsc(x)	Arccosecans (radian)
asech(x)	Hyperbolic areasecans
asec(x)	Arcsecans (radian)
asinh(x)	Hyperbolic areasine
asin(x)	Arcsine (radian)
atan2(x_1, x_2)	Arc tangent (radian)
atan(x)	Arctangens (radian)
cosh(x)	Hyperbolic cosine
cos(x)	Cosine (radian)
csch(x)	Hyperbolic cosecans
csc(x)	Cosecans (radian)
exp(x)	Exponential
factorial(N)	Factorial $N!$
log(x)	Natural logarithm
nchoosek(N, K)	Binomial coefficient $\binom{N}{K}$
sech(x)	Hyperbolic secans
sec(x)	Secans (radian)
sinh(x)	Hyperbolic sine
sin(x)	Sine (radian)
sqrt(x)	Square root
tan(x)	Tangent (radian)

Here is a more standard way to format the same function:

```
function y=test1(x)
      y=2*x;
      printf("%f",y)
end
test1(10)
```

This code is better structured and easier to read.

If functions are to be reused later, they should be included in a text file, and saved on the file system. The search path should point to the location of this file. The file must have the same name as the function, and it should have the file extension ".m." Functions can be created in the same macro, and can be passed to other functions as arguments. Here is a simple example:

Table 17.4 Built-in special functions supported by the jMathLab program

Name(arguments)	Function
betaln(x)	Logarithm of beta function
beta(x)	Beta function
erf(x)	Error function
gammaln(x)	Logarithm of gamma function
gamma(x)	Gamma function
inverf(x)	Inverse error function
j0(x)	Bessel function of the first kind of order 0

```
function y=fit(a,x) % build function to fit data
   y=a(1)*exp(-(x-a(2)).^2/a(3)^2)
end
d=fit([1 4 5],5)        % testing function
printf('fit=%f\n',d)

% this takes function "ff" as argument
function y=newtest(ff,a,x)
      y=ff(a,x)
end
a=newtest(@fit,[10,20,30],10)
a=newtest(@fit,[20,2,30],10)
printf('test=%f\n',a)
```

Listing 17.1 Creating custom functions in jMathLab scripts

Let us explain this macro: First, we create a function `fit` which takes a list (we will discuss this topic later) and a number. We call this function and print its output. In the next line, we pass the function `fit` to the function `newtest(..)`. Note that we use the symbol "@" to tell the program the type ("user function"). Then we evaluate the function `newtest(..)` using the following three parameters: the function `fit(..)`, a list of values and the value 10.

17.4 Polynomials

A polynomial is a mathematical expression containing symbolic variables and coefficients. The description of polynomials in jMathLab closely follows the Jasymca calculator, therefore, please refer the original source [3] for detailed discussion and examples.

A polynomial of a degree n is codded by a vector with $n + 1$ elements. By convention, the element with index "1" is the coefficient of the highest exponent. A normal polynomial is created using the command `poly(x)`, while `polyval(a,x)`

Table 17.5 Predefined polynomial functions in the jMathLab program

Name(arguments)	Function
poly(vec)	Coefficients of polynomial having roots vector
polyval(vec, var)	Function value of polynomial with coefficient vector in the point var
polyfit(vecX, vecY, N)	Fits Nth-degree polynomial to data points having coordinates vecX and vecY
roots(vec)	Roots of polynomial having coefficients vector
coeff(var, sym, N)	Coefficient of sym in var
divide(var1, var2)	Division var1 by var2 with remainder
gcd(var1, var2)	Greatest common denominator
sqfr(var)	Squarefree decomposition
allroots(var)	Roots

returns function values with coefficients a at the point x. Let us give an example of the third-degree polynomial $(4, -8, 2)$ in the normal form:

$$y = (x - 4) \cdot (x + 8) \cdot (x - 2)$$

We can write it as

```
a=poly([4,  -8,  2]);  printf("%f",a)
ans=  [1  2  -40  64]
```

The output of the `printf` command is the expanded form of the polynomial. Using our convention, this corresponds to $x^3 + 2x^2 - 40x + 64$. You can check this by distributing parenthesis and refining the answer.

Finding roots of a polynomial is done via the command `roots(a)`, while `polyfit(x, y, n)` calculates the coefficients of the polynomial of the degree n, whose graph passes through the points x and y. If their numbers are larger than $n + 1$, then a least-square estimate is performed. Table 17.5 shows the supported polynomial functions.

Polynomials can be represented by the vector of their coefficients. Using a symbolic variable x, now we will create a symbolic polynomial with the name p. We can extract the coefficients from a symbolic polynomial using the function `coeff(p,x,exp)`, and then the command `allroots(p)` finds its roots:

```
a=[3  2  5  7  4]          % coefficients
syms  x
y=polyval(a,x)             % symbolic  polynomial
a=allroots(y);  printf("%f\n",a)    % same  as  roots(a)
a=coeff(y,x,3)             % get  one  coefficient
a=coeff(y,x,4:-1:0);  printf("%f\n",a)    % get  all  at  once
```

Listing 17.2 Finding roots of the polynomial $3x^4 + 2x^3 + 5x^2 + 7x + 4$

The output of this code consists of two lines: one with printed coefficients, and the second with the roots of the polynomial:

```
[ 0.36338-1.3737i      0.36338+1.3737i
 -0.69672-0.41832i   -0.69672+0.41832i]
[3 2 5 7 4]
```

When several variables are used, the symbolic variables are sorted alphabetically, i.e., z is main variable compared to x. The coefficients can be calculated for each variable separately.

```
syms x,z
y=(x-3)*(x-1)*(z-2)*(z+1);  printf("%f\n",y)
ans=((x^2-4*x+3)*z^2+(-x^2+4*x-3)*z+(-2*x^2+8*x-6))
a=coeff(y,x,2);  printf("%f\n",a)
ans=(z^2-z-2)
a=coeff(y,z,2);  printf("%f",a)
ans=(x^2-4*x+3)
```

As before, one can use the command `allroots()` for polynomials with several variables. When searching for the root of an equation, one should use the more general method `solve(p,x)` which solves the expression p=0 assuming x is a symbolic variable. Let us show this:

```
syms x,z
y=x*z^2-3*x*z+(2*x+1)
a=allroots(y);  printf("All  roots=%f\n",a)
a=solve(y,x);   printf("Solves=%f",a)
```

The code prints:

```
All roots=[ sqrt((1/4*x-1)/x)+3/2
                -sqrt((1/4*x-1)/x)+3/2 ]
Solves=(-1/(z^2-3*z+2))
```

The decomposition in linear, quadratic, cubic, etc., factors is accomplished by the command `sqfr(p)`. It returns a vector of factors sorted in ascending order of the exponents:

```
syms x
y=(x-1)^3*(x-2)^2*(x-3)*(x-4)
z=sqfr(y);    printf("sqfr=%f\n",z)
```

which leads to the output:

```
sqfr=[x^2-7*x+12  x-2  x-1]
```

The division of two polynomials p and q is done with the command `divide` `(p,q)`. If polynomials have more than one variable, an optional variable can be specified, which will be used for division. The method `gcd(p,q)` returns the greatest common denominator of two expressions. Both functions also work with scalar values.

Let us show several examples that illustrate how to use the methods discussed above:

```
syms x,z
a=divide(x^3*z-1,x*z-x,x); printf("%f\n",a)
ans=[ x^2*z/(z-1)    -1 ]
a=divide(x^3*z-1,x*z-x,z); printf("%f",a)
ans=[ x^2    x^3-1 ]
```

Here is the output of the `gcd(p,q)` method:

```
syms x,z
a=gcd(32897397,24552502); printf("%f\n",a)
ans=377
a=gcd(z*x^5-z,x^2-2*x+1); printf("%f",a)
ans=(x-1)
```

There is a special method `taylor(f,x,x0,n)` that calculates the nth Taylor polynomial in the symbolic variable x at the point $x0$. For example,

```
syms x
y=taylor(log(x),x,1,1); printf("%f",y)
ans=  (x-1)
```

Here is a more complicated code snippet:

```
syms x
y=rat( taylor(exp(x),x,0,6)); printf("%f",y)
ans=(1/720*x^6+1/120*x^5+1/24*x^4+1/6*x^3+1/2*x^2+x+1)
```

The last line represents the Taylor expansion of the degree $n = 6$ of the function $\exp(x)$. The coefficients are rational numbers since we used the method `rat()`. Removal of this function leads to numeric values for the output.

17.5 Vectors and Matrices

17.5.1 Vectors

To define the vector data types, use square brackets. The elements of vectors are entered as comma-separated list, similar to Python lists. One unusual feature is that the comma can be left if the elements are distinguished in a unique way. The method `length(vec)` returns the size of a vector. Consider this example with the output "5:"

```
x=[1 2 3 4 5]; printf("%f",x)
ans= [1 2 3 4 5]
printf("%f",length(x))  %  print length of the vector
ans= 5
```

The colon ":" punctuation mark and the function line space are used to define the ranges of numbers in a vector.

```
v=1:10; printf("%f",v) % numbers from 1 to 10, with step 1
ans= [1 2 3 4 5 6 7 8 9 10]
```

As you can see, $1:10$ is actually the row vector. Thus, vectors can be used to achieve fairly complex data manipulation effects. The numbers do not need to be integers, nor incremental. For example:

```
v=0.2:0.2:1.2; printf("%f",v)
ans= [0.2 0.4 0.6 0.8 1 1.2]
v=5:-1:1;       printf("%f",v)
ans= [5 4 3 2 1]
```

The mathematical functions discussed in Sect. 17.3 can be used for transformations of vectors:

```
x = [0.0:0.5:2.0];   printf("x=%f",x)
ans=[0 0.5 1 1.5 2]
y = sin(x) ; printf("y=%f",y)  % transformed x
ans= [0 0.47943 0.84147 0.99749 0.9093]
```

Here $\sin(x)$ operates entry-wise, producing a vector y from the vector x. Here is another example:

```
y=linspace(0,2,5)  %   5 values from 0 to 2.5
printf("%f",y)
ans= [0 0.5 1 1.5 2]
```

Individual elements of vectors are extracted by specifying the index i. For example, $x(i)$ corresponds to one element at the position i. The index should be in the range from 1 to `length(x)`. If the colon operator used as an index, all elements of the vector are returned. Additionally, ranges of numbers can be used as indices. Let us give a few examples:

```
y = [1 2 3 4 5 6 7 8 9 10]
a=y(2);      printf("%f",a) % single element
ans=2
a=y(:);      printf("%f",a) % magic colon
```

```
ans= [1 2 3 4 5 6 7 8 9 10]
a=y(2:3);   printf("%f",a) % index between 2 and 3
ans=   [2 3]
a=y(2:length(y));  printf("%f",a) % all from index 2
ans= [2 3 4 5 6 7 8 9 10]
a=y([1,3,4]);  printf("%f",a)     % indices 1,3 and 4
ans= [1 3 4]
a=y([1,3,4]) = [1,2,3]; printf("%f",a)
ans= [1 0 2 3]
```

The method ones (1, N) creates a vector with N values equal 1:

```
y=ones(1,10); printf("%f",y)
ans= [1 1 1 1 1 1 1 1 1 1]
```

If you want to have all elements to be equal to 5, simply multiply this vector by 5. For example, let us create a vector of size 10 with identical values 5:

```
y=ones(1,10) * 5; printf("%f",y);
ans=[5 5 5 5 5 5 5 5 5 5]
```

If we need to make a vector with 0 values, use the zeros () command:

```
y=zeros(1,10)
```

You can perform the usual operations with the vectors, i.e., multiply, divide, add, and subtract. This example shows this:

```
a=zeros(1,10)          % 10 numbers all equal 0
b=a+1                  % add a constant 1
c=b*2                  % multiply by 2
d=c-b                  % subtract vectors
```

The expected output is the vector with all elements equal to one.

17.5.2 Matrices

Matrices are handled as vectors, but one should use two indices: one for the row number and, second, for the column number. Rows are separated by either a semicolon or a line.

Here is a simple matrix:

```
M=[1 2; 3 4; 5 6]; printf("M=%f",M)
ans=
   M=
   1   2
```

```
3    4
5    6
```

One can use colons exactly as for the vectors:

```
M=[9:12  ;  4:6  ;  8:10];  printf("M=%f",M)    % matrix 3x3
ans=
M=
   9     10    11    12
   4      5     6     0
   8      9    10     0
a=M([1  3],:);  printf("%f",a)      % keep 1st  and  2nd  row
ans=
   9     10    11    12
   8      9    10     0
```

The colon notation can be used to access submatrices of a matrix. For example,

```
M=[1:10;  1:10;  1:10;  1:10;  ]
A=M(1:4,3);  printf("%f",A)
ans=
   3
   3
   3
   3
```

is the column vector consisting of the first four entries of the third column. A colon denotes an entire row or column, i.e.,

```
A(:,3)
```

is the third column of A, and $A(1:4,:)$ is the first four rows. Arbitrary integral vectors can be used as subscripts:

```
A(:,[2  4])
```

extracts columns 2 and 4 of this matrix. Such subscripting can be used on both sides of an assignment statement. This example:

```
A(:,[2  4  5])  =  B(:,1:3)
```

replaces columns 2, 4, and 5 with the first three columns of the matrix B.

All the operations discussed for vectors can also be applied for matrices. Table 17.6 lists the available matrix operations:

Table 17.6 Operations on
matrices in jMathLab

Symbol	Operation
+	Addition
−	Subtraction
*	Multiplication
^	Power
'	Transpose
\	Left division
/	Right division

These matrix operations apply to scalars (i.e., 1-by-1 matrices) as well. If the sizes of the matrices are incompatible for the matrix operation, an error message will be thrown. Look at this example:

```
M=[1:3  ;  1:3  ;  1:3]
MM=M*M;  printf("MM=%f",MM)
ans=
  MM=
  6    12   18
  6    12   18
  6    12   18
```

Table 17.7 shows a list of matrix and vector operations. They are almost identical to those from the Jasymca calculator discussed at the beginning of this chapter.

There are several standard methods to create matrices. The usual method `zeros` (m,n) creates a *m*-by-*n* matrix with zero elements, and `zeros(n)` produces an *n*-by-*n* matrix. The method `ones(n,m)` returns a matrix with elements equal to 1, `zrand(n,m)` creates a matrix with random numbers between 0 and 1, and `eye(n,m)` has diagonal elements 1, else 0. Finally, `hilb(n)` creates the *n*th-degree Hilbert matrix. This example illustrates this:

```
A=rand(1,3);  printf("%f",A)    % print random matrix 3x1
ans= [ 0.96401 0.24194 0.50943 ]

B=hilb(4);     printf("%f",B)
ans=
  1      1/2   1/3   1/4
  1/2   1/3   1/4   1/5
  1/3   1/4   1/5   1/6
  1/4   1/5   1/6   1/7
```

Let us write a few examples using the methods described in Table 17.7:

```
A=det(hilb(4));  printf("%f",A)  % determinant
ans=1/6048000
```

```
M=[2 3 1; 4 4 5; 2 9 3]
A=eig(M); printf("eig=%f",A)    % eigenvalues
ans=
    eig=[ 11.531 -3.593 1.062 ]

A=inv(M); printf("Inv=%f",A)    % inverse M
ans=
  Inv=
  0.75          0             -0.25
  4.5455E-2    -9.0909E-2     0.13636
  -0.63636      0.27273       9.0909E-2
```

The nontrivial functions are all based on the LU-decomposition, which is also accessible via the function call `lu(x)`. It has 2 or 3 return values, therefore, the left side of the equation must provide multiple variables, see the example below:

Table 17.7 Operations with vectors and matrices

Name(arguments)	Function
det(*matrix*)	Determinant
diag(*x*, [offset])	If *x* is a vector: matrix with *x* as diagonal, if *x* is matrix: diagonal as vector
eigen(*matrix*)	Eigenvalues (using Lapack)
eig(*matrix*)	Eigenvalues
eye(rows[,columns])	Matrix with diagonal one
find(*x*)	Indices of nonvanishing elements
hilb(rank)	Hilbert matrix
invhilb(rank)	Inverse Hilbertmatrix
inv(*matrix*)	Inverse matrix
length(*vector*)	Number of elements in *vector*
linspace(var_1, var_2, count)	Vector with count numbers ranging from var_1 to var_2
lu(*matrix*)	LU-decomposition
max(*x*)	Largest element in *x*
min(*x*)	Smallest element in *x*
ones(rows[,columns])	Matrix of ones
pinv(*matrix*)	Pseudoinverse
qr(*matrix*)	QR-decomposition (using Lapack)
rand(rows[,columns])	Matrix of random numbers
size(*matrix*)	Number of rows and columns
sum(*x*)	If *x* is a vector: sum of elements, if *x* is a matrix: sum of columns
svd(*matrix*)	Singular value decomposition (using Lapack)
zeros(rows[,columns])	Matrix of zeros

Here "rows" and "columns" are integer values defining either the size of a vector (when only "rows" is given), or the size of a matrix (rows x columns)

```
M=[2 3 1;  4 4 5;  2 9 3]
[l,u,p]=lu(M)                % 2 or 3 return values
```

The operator "*" (dot and multiplication) corresponds to element-by-element multiplication for matrices and vectors:

```
x=[2 1 4];  y=[3 5 6]
c=x.*y;  printf("x.*y=%f",c)   %  with point
ans=
   x.*y=[6 5 24]

c=x*y;  printf("x*y=%f",c)    %  without point
ans=  x*y=35
```

If one of the arguments is a scalar data type, the operation is repeated for each element of the other argument:

```
x=[1 2 3 4]
c=x+3;  printf("%f",c)
ans=  [4 5 6 7]
```

Matrix division corresponds to multiplication by the pseudo-inverse. The back-slash operator leads to left-division, which can be used to solve systems of linear equations:

```
M=[2 3 1;  4 4 5;  2 9 3]
b=[0;3;1]
x=M\b;  printf("M\b=%f",x)  % solution of M*x = b
ans=
  M\b=
  -0.25
  -0.13636
  0.90909
```

Now run the inverse operation to recreate the vector b:

```
c=M*x;  printf("M*x=%f",c)      % control
ans=
  M*x=
  0
  3
  1
```

The application contains a Java-port of the LAPACK library with routines for matrix calculations. However, these are limited to matrices with real coefficients in the floating point format. The LAPACK matrix routines are accessed by the following functions:

- `svd(A)`—Singular value decomposition of a matrix A. Here is a simple example:

```
A=[2 3 1; 4 4 5; 2 9 3]
a=svd(A); printf("svd=%f",a)
ans=
    svd=[ 12.263 3.697 0.9705 ]
```

- `eigen(A)`—eigenvalues of the matrix A.

```
A=[2 3 1; 4 4 5; 2 9 3]
a=eigen(A); printf("eig=%f",a)
ans=
 eig=[11.531   1.062   -3.593]
```

- `qr(A)`—Compute the QR factorization of a matrix A. The example below shows how to run this decomposition.

```
A=[2 3 1; 4 3 6; 2 4 3]
[Q,R]=qr(A); printf("Q=%f",Q)
ans=
  Q=
  -0.40825   0.34503   -0.84515
  -0.8165   -0.55205   0.16903
  -0.40825   0.75907   0.50709
```

- `linsolve(A, b)` and `linsolve2(A, b)`—Solves a linear equation of the form $A \cdot x = b$. Here A is a quadratic matrix of the system of equations, and the vector b represents the right side of the equation.

How to solve systems of linear equations will be discussed later in Sect. 17.9.1.

17.6   Flow Control

Control flow in jMathLab programs is similar to the Octave/MATLAB language. As for any language, the simplest form of flow control is the branch statement:

```
if test1
     code1;
else
     code2;
end
```

One can use the usual "if x A else B end" statement. Here is an example with the "if" statement

```
function y = H2(f, e)
   if (f==1)
      y = e+f;
   else
      y = e+2*f;
   end
end
y=H2(1,2) % evaluate at 1,2
```

Loops with a condition "x" and statement(s) "A" are programmed using the "while x A end" blocks. The while-loop is repeated until x becomes false (0):

```
x=1; y=1
while(x<10) y=x+y; x++; end
printf("%f",y)
ans= 46
```

Loops can contain a counter "z" and statement(s) "A", i.e., they have the form "for z = vector A end". In the "for-loop", the counter is initialized by a vector. In each execution of a statement inside the loop, the counter takes the value of the next element of the vector.

```
x=1; y=1
for(x=1:0.1:100) y=x^2+y; end
printf("%f",y)
```

Jumps in an Octave program can be done with the `return`, `continue` and `break` statements. Mathematical functions can be a part of the loops. The keyword `break` permanently leaves the loop:

```
x=1;
while( 1 )
      if(x>1000)
          break;
      end
      x++;
end
```

Let us summarize what we learned so far in Table 17.8.

17.6.1 Benchmarks

You can benchmark programs to ensure that your codding is well optimized. To do this, use the keyword `time(0)` to return the number of milliseconds since 1970. Here is the small example written in Octave to calculate the following expression $\sum_{i=0}^{N} \cos(i) \sin(i^2)$ for $N = 1,000,000$:

Table 17.8 Program statements in jMathLab

Definition	Example
Branch	`if x==0 xp=1; end; if x==0` `xp=1; else xp=2; end`
Loop	`while x<17 x=x+1; end; for` `i=0:100 x=x+i; end`
Jump	`return, continue, break`
Function	`Function y=ttwo(x) y=2*x; end`
Evaluate	`eval("x=24")`
Text output	`printf("Number=%f", x)`
Error message	`error("Test")`

```
d1=time(0)
s=0; n=0
for(x=0:1:1000000)
    s=s+cos(x)*sin(x^2);
    n=n+1;
end
d2=time(0)
printf("\nNr of iterations=%f\n",n)
printf("Time=%f",(d2-d1)/1000)
```

Listing 17.3 Benchmarking Octave calculations in jMathLab

Now let us compare the time needed for the same calculation using the equivalent Python/Jython program:

```
import math,time

start=time.clock()
s=0; n=0
for x in range(1000000):
  s=s+math.cos(x)*math.sin(x*x)
  n=n+1
print "CPU time (s)=",time.clock()-start
```

Listing 17.4 Benchmarking Jython calculations

The output of each of these scripts will give you an idea about the comparative performance of jMathlab and Jython codes. On the author's computer, the Octave program is about a factor 3 slower than the Jython code, since parsing jMathLab mathematical statements is relatively CPU intensive compared to the Jython computing engine. This gives you a warning about using jMathLab scripting for long loops. Of course, when it comes to usage of numerical libraries, jMathlab is as fast as any other program since it calls Java-implemented libraries.

17.6.2   File Input and Output

Matrices and vectors can be saved into comma-separated values (CSV) files. One can read these objects back using several convenient jMathLab methods. Note that we cannot use the class `jhtplot.IO` discussed before in the context of other Java-based languages. The jMathLab program runs on the Java platform, but jMathLab is not integrated into the Java API, and thus it cannot import Java classes.

Let us write a matrix into a file called "test1.csv:"

```
M=[1:3 ; 4:6 ; 7:9]
printf("%f\n",M)
csvwrite("test1.csv",M)
```

You can read the matrix from the file as:

```
M=csvread("test1.csv")
printf("Read matrix=%f\n",M)
```

Similarly, you can write and read vectors:

```
A=[1:10]
printf("%f\n",A)
csvwrite("test2.csv",A)
A=csvread("test2.csv") % let us read it back
printf("Read vector=%f\n",A)
```

If you need to download a text file from the Web using the URL or FTP protocols, use the function `urlread(url)` which reads a text file from the Web using a given `url` address.

Let us also remind that one can put your program into the M-files (MATLAB-like files with the extension ".m") and run such files using the method:

```
loadfile("file.m")
```

where "file.m" is an Octave file.

17.7   Calculus

17.7.1   Differentiation

Differentiation of a function "f" with respect to the symbolic variable x can be done with the method `diff(f,x)`. The input can be constructed from predefined functions, or by a user code. Let us start from a simple example that shows how to differentiate the function ax^3, where a is a symbolic constant.

```
syms a,x      % symbolic x and a
y=diff(a*x^3,x); printf("%f",y)
```

The code prints $3 \cdot a \cdot x^2$. Here are a few more examples:

```
syms x
y=diff(3*sqrt(exp(x)+2),x)
```

and, for a user-defined function ftwo, the code will look as

```
syms x
function y=ftwo(x) y=2*x; end
y=diff(ftwo(sin(x)),x)
```

One can easily plot the result of differentiation as shown in the next example:

```
syms x                % symbolic x
y=diff(x*cos(x),x) %   differentiate x*cos(x)
plot2d("minx=0;maxx=10;miny=-1;maxy=10")
draw2d(y)              % plot the result
```

We will come back to the subject of plotting in the following sections.

17.7.2 Integration

For numerical integration, use the method quad(exp,low,up), where an expression exp, a lower (low) and upper (up) value for the integration region must be defined. The algorithm uses the Simpson's integration method. This example shows how to use this function:

```
y=quad("exp(-x.^2)",0,1);   printf("%f",y)
ans=  0.74682
```

Note that there is no need to define x as a symbolic variable.

Alternatively, the method romberg uses the well-known Romberg's method of estimation of the definite integrals. Unlike the previous method, it requires the use of symbolic variables:

```
syms x            % define symbolic x
y=romberg(exp(-x^2),x,0,1);  printf("%f",y)
ans=  0.74682
```

The maximum number of iterations for the `romberg` method is set by the variable `rombergit` (default 11), and the integration accuracy by `rombergtol` (default: 10^{-4}).

Let us show an example where we will increase the precision for numeric integration. First, we calculate the integral $\exp((-x*\cos(x))^2)$ using the default setting, and then we will increase the numeric precision. We will also apply the `long` formatting to see the changes:

```
syms x
format long
% default settings
y=romberg(exp((-x*cos(x))^2),x,0,1);  printf("%f",y)
rombergtol=0.0000001;   rombergit=1000
y=romberg(exp((-x*cos(x))^2),x,0,1);  printf("%f",y)
```

Listing 17.5 Numeric integration in jMathLab

The result of this code is shown below:

```
1.2007246534547384   1.2007246147959552
```

17.7.3 Indefinite Integral

The method `integrate(f,x)` is designed for indefinite integrals. It integrates a function "f" with respect to the symbolic variable x. The program recognizes all built-in functions and polynomials:

```
syms x
a=integrate(x^2+x-3,x)
printf("%f",a)
```

The code prints the expected answer $0.33333x^3 + 0.5x^2 - 3x$. All trigonometric functions are supported as shown in the next example:

```
syms x
y=integrate(cos(x)*sin(x),x)
```

The integration can be verified by applying the method `diff()` on the result.

```
syms x
y=(x^3+2*x^2-x+1)/(x^3+3*x^2+x+3)
a=integrate(y,x)
b=diff(a,x)       % cross check
printf("%f",b)
ans=((x^3+2*x^2-x+1)/(x^3+3*x^2+x+3))
```

The last line prints the original function. Some substitutions may be used through the command `subst()`. If they all fail, the expression can be integrated numerically using the `quad` or `romberg` methods.

If the function is a rational (i.e., quotient of two polynomials whose coefficients do not depend on x), one can use the standard approach of separating a polynomial part. Then one can integrate the rest using partial fractions. The final terms are collected to avoid complex expressions.

Finally, let us consider an example that shows the original function and its integral plotted on a single 2D plot:

```
syms x
w1 = x^2 + x
w2 = integrate (w1, x)
plot2d ()
draw2d (w1, "name = x^{2}+x; linestyle = 1")
draw2d (w2, "name = integral; linestyle = 2; color = blue; ")
```

Listing 17.6 Plotting a function and its indefinite integral

The result is shown in Fig. 17.1. We will discuss how to use the canvas `plot2d()` in the next sections.

17.7.4 Transformations

As we have discussed before, symbolic transformations can be done by specifying the statement "`syms x`" before using the symbolic variable x.

It is possible to define a variable whose value is a function. In this case, the name of the function must be preceded by the character $ to avoid evaluation. These variables can be used like the function they stand for. For example, `realpart(x)` name can be shortened as

Fig. 17.1 Showing the function $x^2 + x$ and its indefinite integral

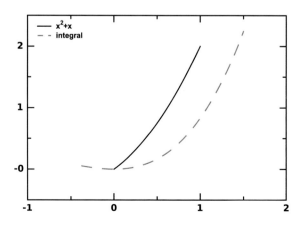

```
real=$realpart      % redefine function name
printf("%f",real)
a=real(50+3i)       % same as realpart(x)
printf("%f",a)
```

The last line prints "50."

Parts of an expression may be replaced by other expressions using the method subst(a,b,c), where a is a substitute for b in c. This approach has many uses. Let us consider an example where we substitute x into the formula $2 * \sqrt{x} * \exp(-x^2)$:

```
syms x
a=2*sqrt(x)*exp(-x^2)
a=subst(3,x,a)
printf("%f",a)
```

This code prints 4.275×10^{-4}.

One can replace a symbolic variable by a complex term. The expression is automatically updated to the standard format. In the following example, the expression $z^3 + 2$ replaces x in the expression $x^3 + \cos(x)^2 + x + 1$:

```
syms x,z
p=x^3+cos(x)+x+1
a=subst(z^3+2,x,p)
printf("%f",a)
ans=  cos(z^3+2)+(z^9+6*z^6+13*z^3+11))
```

Here is a more complicated example:

```
syms x,y,z
c=(z*x^3)/sqrt(z^2+1)
d=subst(y,z^2+1,c)
d=trigrat(d)
printf("%f",d)
ans=   [ x^3*sqrt(y-1)/sqrt(y)    -x^3*sqrt(y-1)/sqrt(y)
       ]
```

We will discuss the function trigrat(exp) in the following section.

17.7.5 Simplifying Expressions

Let us take a closer look at the method trigrat(exp) mentioned in the last code snippet. The function trigrat(exp) applied to a symbolic expression exp converts trigonometric functions into exponential functions, normalizes, and collects roots. It runs a number of algorithms from the Jasymca engine following these procedures:

- All numbers are transformed to exact format;
- Trigonometric functions are expanded to complex exponential functions;
- Basic trigonometric identities are applied;
- Square roots are calculated and collected (if possible).

In addition, numbers are rationalized using the `rat` method.

Here are examples with the function `trigrat(exp)`.

```
syms x
a=trigrat(sin(x)^2+cos(x)^2)
printf("1)= %f\n",a)
b=sin(x)^2+sin(x+2*pi/3)^2+sin(x+4*pi/3)^2;
a=trigrat(b)
printf("2)= %f\n",a)
a=trigrat(i/2*log(x+i*pi))
printf("3) %f\n",a)
```

Listing 17.7 Example of simplifying an expression

The code prints the answer with the simplified expressions:

```
1)= 1
2)= 3/2
3) (1/4*i*log(x^2+pi^2)+(1/2*atan(x/pi)-1/4*pi))
```

Here is another example with two variables, x and y:

```
syms x,y
a=trigrat(sin((x+y)/2)*cos((x-y)/2))
printf("%f\n",a)
a=trigrat(sqrt(4*y^2+4*x*y-4*y+x^2-2*x+1))
printf("%f\n",a)
```

Running this code gives the following answer:

```
(1/2*sin(y)+1/2*sin(x))
(y+(1/2*x-1/2))
```

The method `trigexp(exp)` expands trigonometric expressions to complex exponential functions. In many cases, you will need this function first before calling the function `trigrat(exp)`. As an illustration, consider this small example:

```
syms x
a=trigexp(i*tan(i*x))
printf("%f\n",a)
a=trigexp(atan(1-x^2))
printf("%f",a)
```

17.8   Data Visualization

Scientific data visualization in the Octave mode is based on several Java packages. The original method of the Octave language to display $X - Y$ data is `plot()`. Its implementation stems from the Jasymca program. In addition, jMathLab adds other canvases discussed in the context of the DMelt Java project. This chapter gives an introduction to plotting methods in the jMathLab program.

17.8.1   Plotting Data

Data can be displayed using `plot(x,y)`, where x and y are vectors with the coordinates of data points. The sizes of the vectors should be the same. A third optional argument `opt` in the method `plot(x,y,opt)` specifies plotting options. For example, the option can define the symbol color, such as `r` (red), `g` (green), `b` (blue), `y` (yellow), etc. The default setting shows data using the blue color. The symbol types can be defined using the characters (+, *, o, x).

The graphic canvas can be decorated with an axis title and global title. The output images are exported to the EPS (encapsulated postscript) format. When `hold` method is added, the graphics gets locked and subsequent plot commands will use the same canvas. Repeating the method `hold` deletes the graphic. Logarithmic and semi-logarithmic plots are provided with the functions `loglog`, `linlog` and `loglin`.

The major methods for the command `plot(x,y,opt)` are listed in Table 17.9.

Now we will show a simple example of plotting the function $\sin(x)$ between 0 and 10 with Step 0.1:

```
x=0:0.1:10;  y=sin(x)
plot(x,y)
```

One can overlie these plots using the `hold` method:

Table 17.9 Plotting commands used together with the `plot(x,y,opt)` command

Name(arguments)	Function
linlog(x, y, opt)	Semi-logarithmic plot
loglin(x, y, opt)	Semi-logarithmic plot
loglog(x, y, opt)	Logarithmic plot
print("file.eps")	Saves image in the EPS file
title('string')	Set a global title
xlabel('string')	Title for the X-axis
ylabel('string')	Title for the Y-axis

```
x=0:0.1:10
plot(x,sin(x))
hold
plot(x,x*x)
```

You can plot symbols as well:

```
x=0:0.1:10;  y=sin(x);  z=cos(x)
plot(x,y,  "or")
hold % hold this plot for overlay
plot(x,y,  "r")
plot(x,z,  "+b")
```

Run this plot to see its output.

Let us give a complete example that shows points with uncertainties. Then, we overlie them with the Gaussian function $y = A * \exp(-((x - x0)/\sigma)^2)$. First, we will initiate a function fit, draw data points, and overlie the Gaussian function at fixed values [1.4912 3.9911 1.4481]. The data points are generated in the range [0–10] with the step size 0.2.

```
function y=fit(a,x) % function to fit data
y=a(1)*exp(-(x-a(2)).^2/a(3)^2); end;

x  =  0:0.5:10   %   vector  of  with   data
for  i=1:length(x),
    y(i)=1.5*exp(-(x(i)-4)^2/2)+rand(1)/5-0.1;  end;
e  =  0.03  *  ones(1,  length(x)  )   % uncertainties

errorbar(x,y,e,"+") % draw error bars
hold
xi=0:0.2:10
yi=fit([1.4912    3.9911    1.4481],xi)
plot(xi,yi,"g")
plot(x,y,"r*")
xlabel("X-values")
title("Data vs Gaussian")
print("plot1.eps")
```

Listing 17.8 Showing data and a function on the same plot

The output of this code is shown in Fig. 17.2.

Let us show another example. Now we will plot the function $y = 1/(1 + 0.5x^2)$ in the range $x = 0.01 \ldots 100$. We will use the linear and logarithmic scales. The code is given below:

```
x=0.01:0.01:100;  y=1./(1+0.5*x.*x);  plot(x,y)
x=0.01:0.01:100;  y=1./(1+0.5*x.*x);  loglog(x,y)
```

You may notice that the plotting options of the method plot() are quite limited.

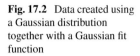
Fig. 17.2 Data created using a Gaussian distribution together with a Gaussian fit function

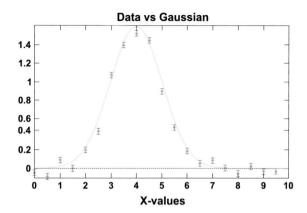

17.8.2   Plot2D

The method `plot2d()` is used to plot functions and arrays in 2D. It has the same look and feel as the `HPlot` canvas discussed earlier in this book in Chap. 8. This graphical canvas is significantly more versable compared to the `plot()` method originating from the Jasymca calculator. With the canvas `plot2d()` one can add a title, labels, change the minimum, and maximum ranges for the X-axis or Y-axis. You can also apply different plotting styles.

To initialize this plotting canvas, run the command:

```
plot2d()
```

which will bring up a plotting window. There are several methods associated with this canvas:

- `clear2d`—clear the canvas from data.
- `export2d(file)`—export the plot to the vector format images.

Depending on the file extension, the last method recognizes all major image formats, including the SVG and EPS vector-graphics formats. When exporting images to a file, use the file extension that defines the format. For example, `export2d("image.pdf")` indicates that the plot will be saved into the PDF format, while `export2d("image.eps")` shows that the plot will be saved into the EPS/PS format. One can also use the GUI to edit and export plots, exactly in the same way as for the `HPlot` Java canvas.

To change the plot default settings, pass a string with key-values separated by the semicolons. For example:

```
s="logx=true;logy=true;"
plot2d(s) % plot logx and logy
s="logx=true;logy=true;title=Name;titlex=X;title=Y"
plot2d(s) % plot logx and logy and title, X and Y labels
s="minx=-1;maxx=1;miny=-1;maxy=1"
plot2d(s) % this string sets min and max value for X and Y
```

The general constructor is

```
plot2d("title=Name;titlex=X;titley=Y;
        minx=-10;maxx=10;miny=-1;maxy=1")
```

The style for presenting data is the same as for the P1D class discussed in the previous sections. There is a direct mapping between the style options of the P1D Java class and the option string passed to the draw2d method:

- name—calls setTitle(..)
- style—calls setStyle(..)
- symboltype—calls setSymbol(..)
- linewidth—calls setPenWidth(..)
- linestyle—calls setLineStyle(..)
- symbolsize—calls setSymbolSize(..)
- color—calls setColor(..)

In addition to scattered data, one can use the draw2d method to display histograms.

To draw a function, first create a mathematical expression using a symbolic variable x, initialize the canvas, and then use the method draw2d(o) to plot the function:

```
syms x
plot2d()
draw2d(x*cos(x))
```

The functions can accept not only a single value but also vectors.

One can also set the ranges for the X or Y axis. In the next example, we differentiate an analytic function and then plot it in the range [0–10] (in X) and [−1–10] (in Y):

```
syms x
y=diff(x*cos(x),x)      % differentiate x*cos(x)
plot2d("minx=0;maxx=10;miny=-1;maxy=10")   % make canvas
draw2d(y)                                  % plot the result
```

Plotting special functions is also possible. Let us display the error function:

```
x=0.01:0.05:1 % a vector  from  0.01  to  1 with  step  0.005
y=erf(x);  printf("%f",y)
ans= [1.1283E-2   6.7622E-2,  ...  0.80188   0.82542]
plot2d()        % make  canvas
draw2d(x,y)     % draw  data
```

Functions can also be plotted by parsing a string with a function definition:

```
plot2d()
draw2d("x*cos(x)")
```

Let us show a simple example of plotting $\sin(x)$ between 0 and 10 with step 0.1

```
x=0:0.2:10;  y=sin(x)
plot2d("titlex=X;titley=Y;")
draw2d(x,y,"name=Test;style=1;linestyle=0;color=black;")
draw2d(x,y*2,"style=1;linestyle=0;color=blue;linewidth=4")
export2d("plot2d.eps") % export  to  EPS  image
```

Listing 17.9 Drawing two functions using the plot2d canvas

Note that the second function is shown with a different style: it uses the solid line style (linestyle=0) in blue and wider line compared to the first plot. Figure 17.3 shows the resulting image.

You can plot data with uncertainties (errors) too. Error bars are shown as lines, and can be attributed to y-values (upper and lower errors) and to x-values (left and right error). Errors can be asymmetric.

Our next code plots data with symmetric errors for the y values. For the sake of illustration, we assume that the sizes of errors correspond to 30 % of the y-values. The array "z" represents symmetric errors on the y-values.

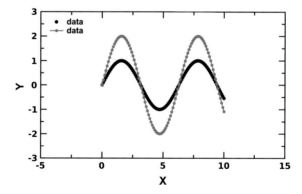

Fig. 17.3 Two overplayed functions with different styles using the jMathLab Java package

```
x=0:1:10;  y=sin(x);
z=ones(1,length(x))*0.3 % vector with errors (=0.3)
plot2d()
draw2d(x,y,z)
```

In the next example, we plot data with asymmetric errors one the y values. The upper errors are 40 % of the value y, while the lower error is 80 % of the value y:

```
x=0:1:10;  y=sin(x);
z=ones(1,length(x))            %  all values= 1
plot2d()
draw2d(x,y,  0.4*z,  0.8*z) % errors  are  asymmetric
```

Now let us plot errors in both directions, X and Y. We assume that the left error is 30 % of the y value, and the right value is 60 % of the x value.

```
x=0:1:10;  y=10*sin(x);
z=ones(1,length(x))   % all  values  =1
plot2d()
draw2d(x,  y,  0.3*z,  0.6*z,  3*z,  5*z)
```

Finally, here is a more custom example showing data with uncertainties:

```
x=[1,  2];  y=[10,  20]
exLeft=[0.5,  0.5]
exRight=[0.3,  0.3]
eyUpper=[5,  5]
eyLower=[2,  2]
plot2d()
draw2d(x,  y,  exLeft,  exRight,  eyUpper,  eyLower)
```

17.8.3 Plot3D

You can plot 3D surface plots (in the case of functions) or $X - Y - Z$ data points using the method plot3D(). The plots are fully interactive, i.e., one can rotate and zoom-in particular regions. This canvas is based on the HPlot3D Java canvas discussed in Chap. 8.

First, let us initialize a canvas and prepare it for plotting 3D data:

```
plot3d(opt)
```

where opt is a string representing the canvas styling. In order to remove all data from the canvas, use the clear3d method. One can export the plotted data into the PDF, EPS, PS, PNG, JPG image formats using the method export3d("file").

The file extension is important since it defines the actual format. For example, export3d("image.pdf") exports the graph to a PDF file. You can also add a title and label, or change the minimum and maximum ranges for the X and Y axes. Let us show examples of the string that defines the plotting options:

```
s="title=Name;titlex=X;titley=Y;titlez=Z;"
plot3d(s) % show title, X,Y,Z labels
s="minx=-1;maxx=1;miny=-1;maxy=1;minz=-1;maxz=1;"
plot3d(s) % sets min and max value for X,Y,Z
```

Once the canvas is created, apply the method draw3d(obj,opt) to show an object obj using the options given by a string opt.

Functions can be shown in 3D as surface plots. You will need to define symbolic variables x and y, and then construct a polynomial using predefined (or user) functions. Then pass it to the method draw3d(). Consider this small example with a surface function shown in 3D:

```
syms x,y
plot3d("minx=-2;maxx=2;miny=-2;maxy=2;minz=0;maxz=1;")
draw3d(x*x+y*y)
export3d("image.pdf")
```

Listing 17.10 Showing a surface plot of the function $x^2 + y^2$ in 3D

The ranges for the axes can be specified using the option string during the initialization. Note that the function is not a string, but a polynomial, so you can do the usual

Fig. 17.4 The function $x^2 + y^2$ shown in 3D using the plot3d method

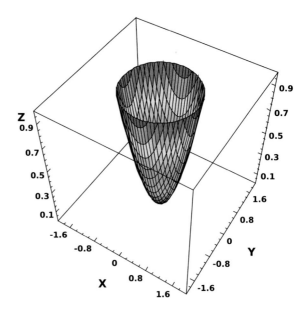

symbolic manipulations, such as simplification, substitution, and so on. Figure 17.4 shows the function $x^2 + y^2$ in 3D.

Data arrays can also be graphed using the plot3d() function. In this case, you need to pass 3 vectors. The drawing options are derived from the methods of the P2D class discussed in the previous chapters. The values used by the option string are mapped to the P2D methods of the DMelt package as

- symbolsize—calls setSymbolSize(..) of P2D Java class
- symbolcolor—calls isetSymbolColor(..) of P2D Java class

Let us show another example of plotting the equation $y = \sin(x)$, $z = \cos(x)$, where x runs from 0 to 10 with step 0.2. We overlie this dataset with a second dataset which has y and z scaled by a factor 0.5:

```
x=0:0.2:10;  y=sin(x);  z=cos(x)
plot3d("maxx=10;miny=-1;maxy=1;minz=-1;maxz=1;")
draw3d(x,y,z)
x=0:0.2:10;  y=sin(x)/2;  z=cos(x)/2
draw3d(x,y,z,"symbolcolor=red;symbolsize=6")
```

Listing 17.11 Showing data in 3D suing the same canvas

Note that the second dataset uses the option string "symbolcolor" and "symbolsize". These are the only options that can be used to show data arrays in 3D. The output of this example is shown in Fig. 17.5.

Fig. 17.5 Data sets shown on the 3D canvas. Equation $(y = \sin(x), z = \cos(x))$ was used to create the data, where x runs from 0 to 10 with a step 0.2. The second data set has y and z scaled by a factor 0.5

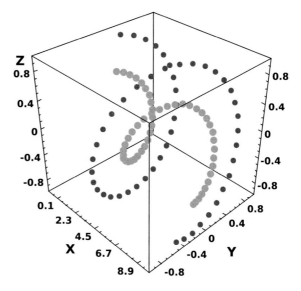

17.9   Equations

The packages dealing with linear and nonlinear equations are included from several Java libraries. The core computational capabilities are provided by the Jasymca engine, while other methods are implemented in external Java packages or as M-files. This section covers topics related to solving systems of linear, differential, and nonlinear equations.

17.9.1   Systems of Linear Equations

The topic of linear equations was earlier discussed in Sect. 17.5.2 dedicated to matrices. Mathematically, a system of equations is represented as $A \cdot z = b$, where z is the solution to be found.

We will recall that systems of linear equations can be solved using the function `linsolve(A,b)`, where A is the quadratic matrix of the system of equations, and b is a (row or column) vector representing the right-hand side of the equations.

Similarly, `linsolve2(A,b)` can also be used for solving systems of linear equations. This method uses the LAPACK library [6], which is well suited for large matrices.

Let us show an example of solving the following system of equations:

$$x - 2y + \ 1 = 7$$
$$2x - 2y + \ z = 4$$
$$-3x + 2y - 2z = -12$$

This system of equations can be coded as

```
A=[1  -2  1;  2  -2  1;  -3  2  -2;]
b=[7;4;-12]
a=linsolve(A,b)
printf('%f',a)
```

Listing 17.12   Solving a system of linear equations

The expected solution printed by this code is $x = -3$, $y = -0.5$ and $z = 11$. To use the LAPACK library, replace the method `linsolve(A,b)` with `linsolve2(A,b)`.

The symbolic engine also handles matrices containing exact or symbolic elements. To avoid rounding errors in these cases, it is advisable to work with exact numbers when possible. The next example includes two symbolic variables, x and y, and finds the solution of a system of four equations:

```
syms x,y
A=[x,1,-3,-2;1  2  3*y  4;1  5  2  0;9  1  6  0;]
```

```
printf('%f\n',A)
b = [1  -2   1    2]
a=trigrat( linsolve( rat(A), b) )
printf('%f\n',a)
```

Listing 17.13 Solving a system of equations symbolically

The code prints the found solution expressed in terms of the symbolic variables x and y:

```
(9/44*y-23/66)/(y+(-14/33*x-85/33))
(7/44*y+(-1/33*x-1/3))/(y+(-14/33*x-85/33))
(-3/22*x-37/132)/(y+(-14/33*x-85/33))
((9/88*x-37/88)*y+(5/22*x+37/24))/(y+(-14/33*x-85/33))
```

17.9.2 Nonlinear Equations

This is a fairly small section as we are going to show only a couple of examples to illustrate how to solve nonlinear equations.

Nonlinear equations can be solved using the function `solve(exp,x)`, where exp is a symbolic expression and x is a symbolic variable. The expression can be constructed from trigonometric functions.

For completeness of this section, we show a few examples that come with the Jasymca symbolic engine:

```
syms x,b
a=solve(x^2-1,x)
printf("%f",a)
a=solve(x^2-2*x*b+b^2,x)
printf("%f",a)
```

This prints the solution `[1,1]` of the first equation, and b of the second equation. Let us take a look at the example constructed from several trigonometric functions:

```
syms x
a=float( solve(sin(x)^2+2*cos(x)-0.5,x) )
printf("%f",a)
```

The found solution is:

```
[ 1.438i  -1.438i  -1.7975  1.7975 ]
```

Table 17.10 Methods for solving equations using the Jasymca computing engine

Name(arguments)	Function
algsys([$var_1, var_2, \ldots$], [$sym_1, sym_2, \ldots$])	Solves the system of equations $var_1 = 0, var_2 = 0, \ldots$ for $sym_1, sym_2, \ldots$
linlstsq($matrix, vector$)	Solves $matrix \cdot x = vector$ for x, overdetermined (using LAPACK)
linsolve2($matrix, vector$)	Solves $matrix \cdot x = vector$ for x (using LAPACK)
linsolve($matrix, vector$)	Solves $matrix \cdot x = vector$ for x
solve(x, Sym)	Solves $var = 0$ for Sym
subst(var_x, var_y, var_z)	Substitute var_x for var_y in var_z
trigexp(x)	Trigonometric expansion
trigrat(x)	Trigonometric and other simplifications

17.9.3 Systems of Equations

A coupled system of equations can be solved using the function `algsys(exp, vars)`, where `exp` is the actual expression, followed by a list of symbolic variables denoted by `vars`. Table 17.10 summarizes the available methods for solving linear and nonlinear equations.

All linear equations are solved using the Gauss method. Then each equation is fed through `solve()` to eliminate one variable in all other expressions. The equations are treated in the order they are supplied for simple systems. The solution is provided as a vector of vectors.

It is time for a small example. Let us solve this equation:

$$2 - a0 = a2$$
$$a1 - 10 = a1$$
$$a2 * xs^2 + a1 * xs + a0 - 3 - xs = a0$$
$$2 * a2 * xs + a1 + 1 = xs$$

The code to solve this system can look as

```
syms  xs,a0,a1,a2
exp=[2-a0,a1-10,a2*xs^2+a1*xs+a0-3-xs,2*a2*xs+a1+1]
a=algsys(exp,[a2,a1,a0,xs])
printf("%f\n",a)
```

The printed solution is shown in the form of the vector:

```
[[  xs-2/7   a2+77/4   a0-2   a1-10  ]]
```

i.e., $xs = 2/7$, $a2 = -77/4$, $a0 = 2$, $a1 = 10$.

The second example has two solutions:

```
syms a,xs
a=algsys([a*xs+3*a-(3-xs^2),a+2*xs],[a,xs])
printf("%f\n",float(a))
```

The answer is shown as the double vector:

```
[[ xs+0.55051    a-1.101 ]    [ xs+5.4495    a-10.899 ]]
```

17.9.4 Differential Equations

Linear first-order differential equations can be written as $y' = f(x) \cdot y + g(x)$. They can be solved using the method ode (exp, y, x), where exp is the actual expression on the right-hand side describing the equation, while x and y are symbolic variables. All functions, including trigonometric functions, are supported in definition of the expression. By convention, C is reserved for a constant.

Let us solve the ordinary differential equation of the form $y' = 5y - 3$:

```
syms x,y,C
a=ode(5*y-3,y,x)
printf("%f",a)
```

The code prints the expected answer $C * \exp(5 * x) + 0.6$.

Here is the example of the equation $y' = -ky$, where k is a constant:

```
syms x,y,k,C
a=ode(-k*y,y,x)
printf("%f",a)
```

The answer is $C * \exp(-k * x)$.

A more sophisticated equation $y' = y \tan(x) + k$ can be solved as

```
syms x,y,k
a=ode(y*tan(x)+k,y,x)
printf("%f",a)
```

The solution is $(k * \sin(x) + C)/\cos(x)$.

17.10 Statistics

The Octave interpreted programming language can be used for statistical calculations with graphical outputs. The jMathLab program supports all major statistical packages for matrices and vectors, nonlinear regressions, and histograms.

17.10.1 Descriptive Statistics

We will start from the descriptive statistics which provides simple summaries of data in the form of vectors and matrices.

Let us calculate the mean, variance, and standard deviations of a vector:

```
v=[1,10,20,2,4,7,4,3]
a=mean(v); printf("%f",a)  % mean (average)
ans=6.375
a=var(v); printf("%f",a)   % variance
ans=38.554
a=std(v); printf("%f",a)   % standard deviation
6.2092
```

Similarly, you can perform calculations for matrices:

```
v=[1,10,20,2,4,7,4,3;  6,1,2,20,42,7,41,3;]
printf("%f",v)
   1    10   20   2    4    7    4    3
   6    1    2    20   42   7    41   3
a=mean(v); printf("%f",a)
ans= [3.5 5.5 11 11 23 7 22.5 3]
a=var(v); printf("%f",a)
ans= [12.5 40.5 162 162 722 0 684.5 0 ]
a=std(v); printf("%f",a)
ans= [3.5355 6.364 12.728 12.728 26.87 0 26.163 0]
```

Note that the calculations are applied to columns of matrices.

You can also calculate correlations between two vectors or two matrices. For example, let us calculate the covariance and the correlation coefficient for two vectors:

```
v1=[1,10,20,2,4,7,4,3]
v2=[3,11,10,3,5,5,7,3]
a=cov(v1,v2);    printf("Covariance=%f\n",a)
ans= Covariance=15.911
a=correlation(v1,v2);    printf("rho=%f\n",a)
ans= rho=0.8053
```

Table 17.11 Probability distributions supported by jMathLab

Function name	Description
erf(x)	The error function of the normal distribution
chi2_prob(n, x)	Area under the left hand tail (from 0 to x) of the χ^2 probability density function with n degrees of freedom
normal_prob(x)	Area under the normal (Gaussian) probability density function, integrated from $-\infty$ to x. It assumes mean $= 0$ and $\sigma = 1$
normalinv_prob(x)	Value, x, for which the area under the normal (Gaussian) probability density function (integrated from $-\infty$ to x) is equal to y. This assumes mean $= 0$ and $\sigma = 1$
student_prob(x, t)	Integral from $-\infty$ to t of the student-t distribution with $k > 0$ degrees of freedom
poisson_prob(x, mean)	Sum of the first x terms of the Poisson distribution
student_prob(α, size)	Value, t, for which the area under the student-t probability density function (integrated from $-\infty$ to t) is equal to $1 - \alpha/2$

The argument x can be vectors and matrices

The obtained value of the coefficient of correlation is close to one, supporting the statement that the values in these two vectors are correlated. Analogously, you can apply the same methods to matrices.

JMathLab supports custom tailored numerical integration for a number of popular probability distributions. Table 17.11 lists the supported probability distributions. For example, the method `normal_prob` returns the area under the normal (Gaussian) probability density function, integrated from minus infinity to x. The argument can be a vector. In the latter case, one can plot such integrals as for any sequence of numbers.

Here is an example that illustrates how to apply the normal probability distribution to a vector, and how to draw an area under this distribution:

```
a = -0.001:0.0001:0
y = normal_prob(a)
plot2d()
draw2d(a,y)
```

Table 17.12 The random number generators supported by jMathLab

Function name	Description
beta_rnd(α, β, nrow, ncol)	Beta distribution with α and β
cauchy_rnd(m, s, nrow, ncol)	Cauchy distribution with the location parameter m and the scale parameter s
chisqr_rnd(ndf)	χ-square distribution with ndf degrees of freedom
exponential_rnd(m, nrow, ncol)	Exponential distribution with the mean m
gamma_rnd(α, β, nrow, ncol)	Gamma distribution with α and β
lognormal_rnd(m, std, nrow, ncol)	Log-normal distribution with the mean m and standard deviation std
nbinomial_rnd(n, p, nrow, ncol)	Negative binomial distribution with n (the number of trials) and p
normal_rnd(m, std, nrow, ncol)	Normal distribution with the mean m and standard deviation std
poisson_rnd(m, nrow, ncol)	Poisson distribution with the mean m
rand(nrow, ncol)	Uniform random numbers in the interval [0, 1]

Here, nrow rows and ncol columns are used to define sizes of vectors and matrices

17.10.2   Random Numbers

The jMathLab program can be used to create vectors and matrices with random numbers using the standard distributions. For example, let us create a vector and a matrix with the random numbers distributed in accordance with the Poisson distribution assuming the mean 2:

```
v=poisson_rnd(2,10)     % create vector with 10 elements
m=poisson_rnd(2,10,3)   % create matrix 10x3
```

The standard set of random numbers widely used in statistics is supported. For example, normal_prob(a) calculates the probability from a normal (Gaussian) distribution with the mean a. Table 17.12 summarizes the most popular random distributions used with the Octave program language.

17.10.3   Data Fitting

Data fitting is also available for the Octave language. In this section we will show a simple nonlinear regression example. First, we will create two arrays, x and y. Then we will fit these arrays with a second-order polynomial:

```
x = 0 : 5
y = [1 , 10 , 20 , 38 , 69 , 81]
plot ( x , y , " +r " )      % Symbol : o , x , + ,
hold                        % hold the plot
plot ( x , polyval ( polyfit ( x , y , 2 ) , x ) )
xlabel ( " X " )
ylabel ( " Y " )
print ( " plot . eps " )
```

Listing 17.14 Nonlinear regression of $x - y$ data array

The key part is the function polyfit(x,y,2), where "2" indicates a second-order polynomial. Replace 2 by 1 to obtain a linear function. Obviously, the latter will fail for this example since the input data have a clear nonlinear trend.

It should, however, be said that the Octave-like language is less versable, in its current Java implementation, compared to the Java and Jython scripts discussed earlier in this book.

17.10.4 Histograms

A limited support for histograms in Octave scripts is also provided. The main method to created a 1D histogram is called h1d(..) (similar to the name of the H1D Java class discussed earlier in this book). As usual, a histogram initialization requires two arguments: the number of bins, and the minimum and maximum value of the considered range. One can fill histograms with data from vectors and matrices, and plot the histograms using the usual plot2d() method. This short example illustrates how to fill and plot a histogram:

```
y = poisson_rnd ( 2 , 100 )
h = h1d ( " Poisson " , 20 , 1 , 10 , y )
plot2d ()
draw2d ( h )
```

Listing 17.15 Plotting a histogram in Octave

The code creates a vector with random numbers distributed using the Poisson statistics. Then we initialize a histogram with the title "Poisson", 20 bins, and assuming the range between 1 and 10. Finally, we use the method draw2d() to display the histogram. You can access the descriptive statistics as for any vector or matrix.

17.11 Again About Integration with Java

Programming in Octave can be done using the interactive jMathLab Shell from the DMelt IDE, or executing the files with the extension ".m" that contain your code. But, as we have discussed in Sect. 6.1 of Chap. 6, one can relatively easy integrate the Octave scripting with the Java program, or with any scripting language running on the Java platform, since the programming environment behind all these programs is 100 % Java.

Let us simplify the expression $(x^3 + x^2 - x - 1)/(x^2 + 2 * x + 1)$ using x as a symbolic variable. Here is a Java code snippet that calls the Octave calculation:

```
import jhplot.math.*;

class JavaOctave {
  public static void main(String[] args)  {

  Symbolic j=new Symbolic("jasymca");
  String exp="syms x; trigrat((x^3+x^2-x-1)/(x^2+2*x+1));";
   try {
      String ans=j.eval(exp);
      System.out.println(ans);
   } catch (Exception e) {
       e.printStackTrace();
   }
  }
}
```

Listing 17.16 Integrating an Octave calculation with Java

Save this code in the file "JavaOctave.java," compile it to the Java bytecode and run it. The output of this source code is $x - 1$. You can intercept the string with the answer to make an appropriate output for your Java application.

To conclude this chapter, we will show how the above example will look using the Python/Jython language:

```
from jhplot.math import *

j=Symbolic("jasymca")
exp="syms x; trigrat((x^3+x^2-x-1) / (x^2+2*x+1));"
ans=j.eval(exp)
print ans
```

Listing 17.17 Integrating an Octave calculation with Python/Jython

The code looks rather simple. As a last exercise, try to rewrite this example in Groovy, JRuby or BeanShell.

References

1. JMathLab, a multiplatform computational platform for symbolic calculations. http://jwork.org/jmathlab/
2. Sparshatt M, Muller S, Torras A, JMathLib, a Java clone of Octave, SciLab, Freemat and MATLAB. http://www.jmathlib.de/
3. Dersch H, JASYMCA. A Java symbolic calculator. http://webuser.hs-furtwangen.de/~dersch/jasymca2

4. Eaton JW, Bateman D, Hauberg S (2008) GNU Octave manual version 3. Network Theory Ltd., London
5. Quarteroni F, Saleri F, Gervasio P (2014) Scientific computing with MATLAB and Octave. Springer, Berlin
6. Dongarra J, Downey A, Seymour K, JLAPACK and the F2J project. http://icl.cs.utk.edu/f2j/

Index

© Springer International Publishing Switzerland 2016
S.V. Chekanov, *Numeric Computation and Statistical Data Analysis
on the Java Platform*, Advanced Information and Knowledge Processing,
DOI 10.1007/978-3-319-28531-3